# International Conference on Advanced Environmental Engineering (ICAEE2017)

AA006059

IOP Conference Series: Earth and Environmental Science
Volume 111

Shenzen, China
15 - 17 December 2017

**Editors:**

**Shaocai Yu**
**Gordon Huang**

ISBN: 978-1-5108-5947-0
ISSN: 1755-1307

**Printed from e-media with permission by:**

Curran Associates, Inc.
57 Morehouse Lane
Red Hook, NY 12571

**Some format issues inherent in the e-media version may also appear in this print version.**



Printed by Curran Associates, Inc. (2018)

For permission requests, please contact the Institute of Physics
at the address below.

Institute of Physics
Dirac House, Temple Back
Bristol BS1 6BE UK

Phone: 44 1 17 929 7481
Fax: 44 1 17 920 0979

techtracking@iop.org

**Additional copies of this publication are available from:**

Curran Associates, Inc.
57 Morehouse Lane
Red Hook, NY 12571 USA
Phone: 845-758-0400
Fax: 845-758-2633
Email: curran@proceedings.com
Web: www.proceedings.com

# Table of contents

## Volume 111

**2017 International Conference on Advanced Environmental Engineering (ICAEE2017)**

**15–17 December 2017, Shenzen, China**

**Accepted papers received: 11 January 2018**
**Published online: 1 March 2018**

**Global Environmental Change and Ecosystems Management**

**Environmental Management and Policy**

IOP Conf. Series: Earth and Environmental Science **111** (2018) 011001 doi:10.1088/1755-1315/111/1/011001

The 2017 International Conference on Advanced Environmental Engineering is an academic conference on advanced environmental engineering. The conference includes the fields of environmental engineering, and is of great significance to the development of environmental engineering

The primary goal of the conference is to promote research and developmental activities in environmental engineering and another goal is to promote scientific information interchange between researchers, developers, engineers, students, and practitioners working all around the world. The conference will be held every year to make it an ideal platform for people to share views and experiences in environmental engineering and related areas and to further promote the development of environmental engineering and strengthen domestic and international academic exchanges and cooperation.

The 2017 International Conference on Advanced Environmental Engineering (ICAEE2017) was held in Shenzhen from December 15 to 17, 2017. I would like to thank the committee members of ICAEE2017 and the reviewers for their careful reviews of the papers and the suggestions for improvements.

The conference received 39 full-papers and abstract submissions from China and many other countries or regions over the world. Through a rigorous peer-review process, all submissions were refereed based on their quality of content, level of innovation, significance, originality and legibility. 39 papers were accepted and their authors were invited to present their papers at the conference. We invited two keynote speakers, who are internationally recognized leading experts in their own research areas, to deliver their recent research progresses. The proceedings are published by IOP Conference Series: Earth and Environmental Science (EES)

Let us hope that the participants of ICAEE2017 will have fruitful interactions among each other and I look forward to meeting you all in the forthcoming years.

**Prof. Shaocai Yu**
**Zhejiang University**

IOP Conf. Series: Earth and Environmental Science **111** (2018) 011002 doi:10.1088/1755-1315/111/1/011002

**Keynote Speaker 1: Prof. Zhi-Min Ao**

Distinguished Professor, "1000 Plan" for Young Professionals of Chinese Government Institute of Environmental Heath and Pollution Control School of Environmental Science and Engineering Guangdong University of Technology

**Research Interest:**

• Bioinformatics

• Biomedical Engineering

**Biographical sketch:**

Zhi-Min Ao is working with "1000 Plan" for Young Professionals of Chinese Government Institute of Environmental Heath and Pollution Control School of Environmental Science and Engineering Guangdong University of Technology. He has won the 2009 JMBE Best Annual Excellent Paper Award and the dragon thesis award that sponsor is Acer Foundation. He has published a number of international journal and conferences article related to these areas.

IOP Conf. Series: Earth and Environmental Science **111** (2018) 011002 doi:10.1088/1755-1315/111/1/011002

**Keynote Speaker 2: Dr. Elammaran Jayamani**

Program Coordinator Mechanical Engineering, Faculty of Engineering, Computing and Science, Swinburne University of Technology Sarawak campus, Malaysia

**Research Interest:**

- Materials Science and Engineering
- Sustainable Energy and Environmental Materials

**Biographical sketch:**

Elammaran Jayamani is working with Program Coordinator Mechanical Engineering, Faculty of Engineering, Computing and Science, Swinburne University of Technology Sarawak campus, Malaysia. He has won Best Annual Excellent Paper Award in international conference. He has published a number of international journal and conferences article related to these areas.

IOP Conf. Series: Earth and Environmental Science **111** (2018) 011003 doi:10.1088/1755-1315/111/1/011003

General Chair

Prof. Shaocai Yu, Lab for Sources and Causes of Atmospheric Pollution, College of Environmental and Resource Sciences, Zhejiang University, China; Department of Marine, Earth and Atmospheric Sciences, North Carolina State University, USA

Co-Editor

Prof. Gordon Huang, Executive Director, Institute for Energy, Environment and Sustainable Communities, University of Regina, Regina, Canada

Technical Program Committee

Prof. Gordon Huang, Executive Director, Institute for Energy, Environment and Sustainable Communities, University of Regina, Regina, Canada

Prof. Yaxin SU, School of Environmental Science and Engineering, Donghua University,China

Prof. Mohamed Benbouzid, University of Brest, France

Prof. Hem Chander Garg, Guru Jambheshwar University of Science &Technology, India

Prof. Philippe Dubois, University of Mons UMONS, Belgium

Prof. Eurgen Axinte, Gheorghe Aasachi Technical University of Iasi, Roman

Prof. Oronzio Manca, Seconda Universita' degli Studi di Napoli, Italy

Prof. AbdelGhani AISSAOUI, Professor, University of Bechar, Algeria

Prof. Ching-Ting Lee, National Cheng Kung University Taiwan, China

Prof. Adonias Magdiel Silva Ferreira, Federal University of Bahia, Brazil

Prof. Aziz Naamane, Aix MArseille universite, France

Prof. Hui Tong Chua, Taiyuan University of Technology, The University of Western Australia, Singapore

Assoc. Prof. Tanakorn Wongwuttanasatian, Khon Kaen University, Thailand

Assoc. Prof. Fa-bin Cao, the Key lab of Metallurgical Separation & Resources Recycling, Ministry of Education (Anhui University of Technology）, China

Assoc. Prof. Yujia Song, Department of Environmental Engineering, Changchun University of Science and Technology, Changchun, P.R. China

Assoc. Prof. Ľuboš Buzna，University of Žilina, ERAchair for Intelligent Transport Systems, Senior Researcher，Slovak

Dr. Emanuele Quaranta, Hydraulic Engineer, Scientific Research and Professional Scientific Consultant, Politecnico di Torino, Department of Environment, Land and Infrastructure Engineering, Hydraulics, Turin (Italy)

Dr. Li Sun, University of Manchester, China

Dr. M. Satyanarayana, JIMMA UNIVERSITY, India

Dr. Mohd Elias Bin Daud, Politeknik Sultan Salahuddin Abdul Aziz Shah, Malaysia

Dr. Elammaran Jayamani,Program Coordinator Mechanical Engineering, Swinburne University of Technology Sarawak campus, Malaysia

Dr. Yu-Lin (Eugene) Song, Department of Bioinformatics and Medical Engineering, Asia University, Taiwan

Dr Malka N. Halgamuge, The University of Melbourne, Australian

Dr. Hassan Koohestani, Faculty of Materials and Metallurgical Engineering, Semnan University, Iran

IOP Conf. Series: Earth and Environmental Science **111** (2018) 011003 doi:10.1088/1755-1315/111/1/011003

IOP Conf. Series: Earth and Environmental Science **111** (2018) 011004 doi:10.1088/1755-1315/111/1/011004

# Peer review statement

All papers published in this volume of *IOP Conference Series: Earth and Environmental Science* have been peer reviewed through processes administered by the proceedings Editors. Reviews were conducted by expert referees to the professional and scientific standards expected of a proceedings journal published by IOP Publishing.

IOP Conf. Series: Earth and Environmental Science **111** (2018) 012001 doi:10.1088/1755-1315/111/1/012001

# Simulating Water Resource Disputes of Transboundary River: A Case Study of the Zhanghe River Basin, China

**Liang Yuan [1,2], Weijun He [1,2], Zaiyi Liao [3,4], Dagmawi Mulugeta Degefu [3,4], Min An [1] and Zhaofang Zhang [1]**

[1] School of Business, Hohai University, Nanjing, Jiangsu, 210098, China
[2] College of Economics and Management, China Three Gorges University, Yichang, Hubei, 443002, China
[3] Department of Architecture Science, Ryerson University, Toronto, ON, M5B 2K3, Canada
[4] College of Hydraulic and Environmental Engineering, China Three Gorges University, Yichang, Hubei, 443002, China
E-mail: heweijun1519@ctgu.edu.cn

**Abstract.** Water resource disputes within transboundary river basin has been hindering the sustainable use of water resources and efficient management of environment. The problem is characterized by a complex information feedback loop that involves socio-economic and environmental systems. This paper presents a system dynamics based model that can simulate the dynamics of water demand, water supply, water adequacy and water allocation instability within a river basin. It was used for a case study in the Zhanghe River basin of China. The base scenario has been investigated for the time period between 2000 and 2050. The result shows that the Chinese national government should change the water allocation scheme of downstream Zhanghe River established in 1989, more water need to be allocated to the downstream cities and the actual allocation should be adjusted to reflect the need associated with the socio-economic and environmental changes within the region, and system dynamics improves the understanding of concepts and system interactions by offering a comprehensive and integrated view of the physical, social, economic, environmental, and political systems.

## 1. Introduction

A river basin shared by multiple administrative regions is called transboundary river. Like other fresh water resources, these rivers can also face water scarcity [1]. When water scarcity prevails in transboundary river, water resources disputes occur because the river water is not adequate to fully satisfy the total demand of all claimants.

The water dispute on a transboundary river can be regarded as a three-dimensional systemic issue. Horizontally, the level of water dispute is related to the recreation strategy of each sharing region and the relationship among different regions [2]. Vertically, the degree of water dispute mainly depends on the water demand, water shortage, risks of the riparian regions facing water stress and the strategy as well as adaption capacity of the regions to the changing socio-economic environmental factors [3]. Furthermore, all these horizontal and vertical factors change over time, making water dispute a dynamic process that varies with the internal and external factors of each sharing regions. As a result, a better understanding of water dispute among different administrative regions should be a complex system which consists of the three-dimensional elements linked through positive and negative feedback loops [4]. Thus, finding the causes, processes and characters of the water dispute is vital to solving water sharing problems.

Forrester developed System Dynamics (SD) for the first time by combining applied control theory, information theory, decision theory, and other relevant theories [5]. System dynamics method can not only properly describe the complicated issue of water dispute but also can provide a unique and comprehensive structure to simulate the degree of dispute among different stakeholders [6]. In this paper, the authors present a strategic analysis of a water allocation and dispute within transboundary river basins and an SD based model for simulating the regional water resources allocation and dispute.

Zhanghe River basin is a tributary of the southern Haihe River basin in China. The upstream consists of two rivers, the Qing Zhanghe and Zhuo Zhanghe rivers which originate from Shanxi Province and merge at the provincial border between Hebei and Henan Province. It is joined by a third branch in the downstream, namely Weihe river at Handan City of Hebei Province, forming a river basin called Zhangwei River basin (figure 1).

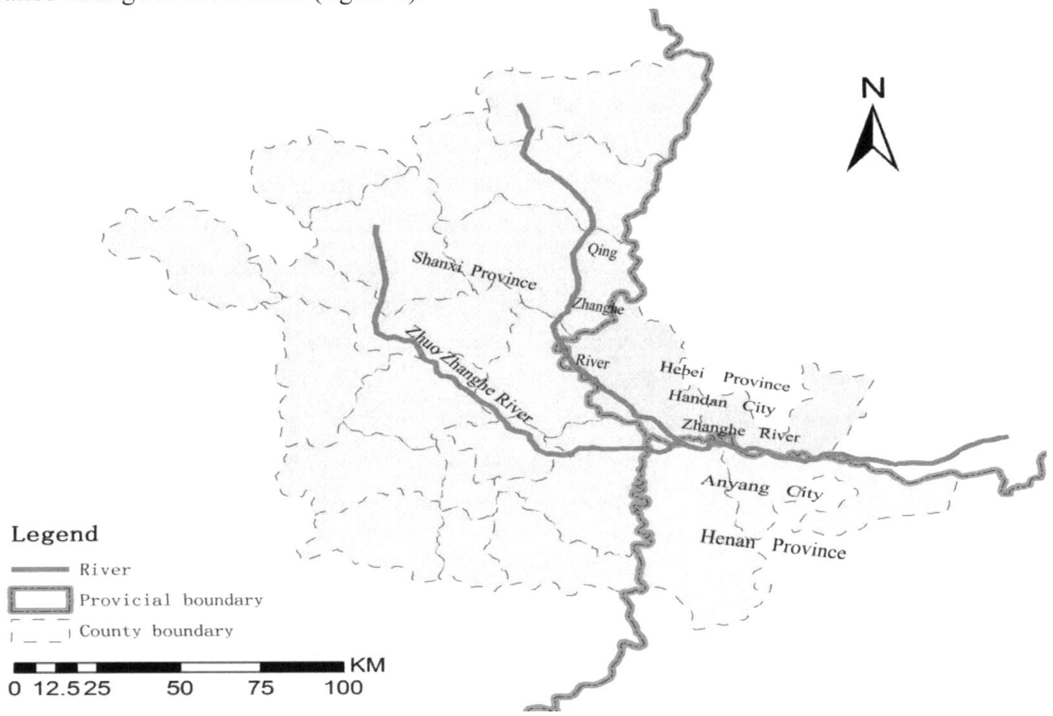


**Figure 1.** Zhanghe River Basin

Zhanghe River basin covers an area of 18284 km$^2$, travels a total distance of 460 km. 86.7%, 9.9% and 3.4% of the basin's area lies in Shanxi, Hebei and Henan provinces respectively. Zhanghe river basin includes 18 counties located in 4 administrative regions of three provinces. 14 counties located in Changzhi and Jinzhong cities of Shanxi Province constitutes the upstream of the basin. In the downstream, 2 counties are located in Handan City of Hebei Province and the other 2 counties are located in Anyang City of Henan Province.

The contribution of Zhanghe River to the surface water supply for the upstream region, Anyang City and Handan City is 30%, 20% and 15% respectively. However, their demands are largely influenced by the changing socio-economic, environmental, and demographic conditions. Due to increased water demands caused by mushrooming population and socio-economy, more than five major incidents related to water resource allocation have occurred since 2008. The national government has involved in seeking solutions for all these cases [7]. Handan and Anyang cities always complain that the upstream region withdraw excessive amounts of water resources, leaving insufficient water for them [8].

At present, the national government is yet to rule on the allocation of the river's water among the three provinces in the basin. The State Council has only ruled that 48% of the available water resources of downstream Zhanghe River allocated to the Anyang City of Henan Province and 52% allocate to the Handan City of Hebei Province in 1989. But the allocation of water resources in the other areas of Zhanghe River basin was not clearly stipulated. The change in the socio-economic and environmental status of the riparian regions makes the water allocation scheme that was put into effect in 1989 inapplicable and outdated. It's important to forecast the degree of water resources dispute in the Zhanghe river basin in the future.

## 2. System Dynamics Modeling

The water dispute system can be represented by a complex information cycle with feedback loops mainly consisting of socio-economic-environmental drivers. Hence, the water dispute system in the Zhanghe River basin is driven by these factors that are interrelated through a complex feedback system. Therefore, SD was employed to develop the model that can not only effectively organize and reveal the internal interaction and feedback mechanism among the various factors within the nonlinear composite system, but also reflect the essential relationship among the factors by causal feedback loops. The steps taken during system dynamics modeling of the water dispute within Zhanghe River basin are following [9]:

### *2.1. Defining Simulation Objectives*

Applying system dynamics model to analyze the Zhanghe river basin water resources dispute through the simulation of riparian provinces water demand, supply and allocation. This might provide the basis for water resources planning and management.

### *2.2. Determining System Boundary*

The riparian provinces and their respective cities are categorized as upstream and downstream. The upstream region comprises 14 of counties located in Changzhi and Jinzhong cities of Shanxi Province, while downstream region consisting of the Handan City of Hebei Province and Anyang City of Henan Province. The simulation period for this study is from 2000 to 2050.

### 2.3. *Designing an Interface Graphical Structure of the System*

According to the water demand of the three riparian provinces' cities located within the basins, divided the complex water dispute system into six sub-systems [11]: domestic water demand subsystem, industrial water demand subsystem, irrigation water demand subsystem, ecological water demand subsystem, water resources allocation and dispute subsystem. The domestic water demand subsystem simulates the riparian cities water demand changes which is the result of both the rural and urban population change. The industrial water demand subsystem simulates the water demand changes due to the influence of industrial scale water consumption. The irrigation water demand system simulates the impacts of the irrigated agricultural area on water demand. The ecological water subsystem simulates the impact of changes in forest cover on water demand. The water allocation subsystem stimulates the available water resources supply under allocation policies. The causal loop diagram of the water dispute in the Zhanghe River basin is shown in Figure 2.

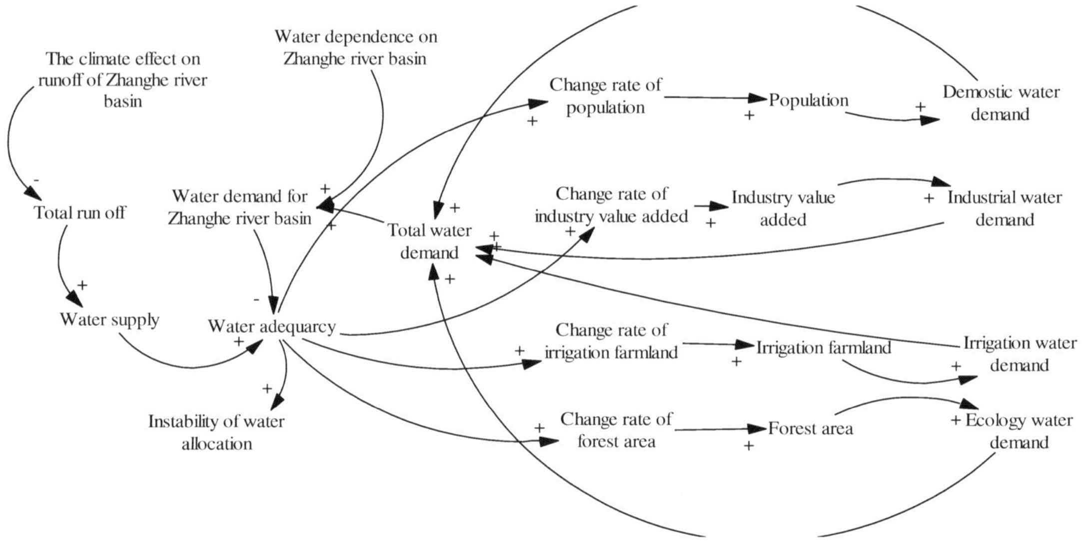


**Figure 2.** Causal loop diagram of the basic model

In Figure 2, each arrow indicates the influence of one element on another. This influence is considered as positive (+) if an initial change is amplified in the same direction as all components are traced through the loop, or negative (-) if an increase in one element causes a decrease in another [12]. The decrease in water dependence rate of the Zhanghe River basin could decrease the regional water demand on the Zhanghe River basin and increase the regional water adequacy of the basin's riparian cities. On the other hand, the climate change due to global warming could decrease the river's runoff as well as the water supply of riparian regions, thereby decreasing the water adequacy of basin's riparian cities. Water adequacy, population, industrial production, irrigated farmland area, forest area, and water demand are the parts of the balanced causal loop.

The water dispute was represented by the instability of water allocation. Higher instability of water allocation is translated as more intense water dispute. The regional water adequacy $A_i$ can be represented as equation (1), $i=1,2,...,n$ refers to the regions.

$$A_i=(WA_i-WD_i)/WD_i \tag{1}$$

the water resources allocated to each region is $WA_i$, $WA_i$ determined by the allocation policy, the total water demand of each region is $WD_i$, which calculated by the water demand.

Assuming that the average value of water adequacy of different regions is $V_i$, The instability of water allocation $S_A$ described as equation (2), higher value of $S_A$ means that water allocation is highly unstable and the water dispute is grave.

$$S_A = \begin{cases} 0, \ A_i > 0 \\ \sqrt{(1/n)\sum_{i=1}^{n}(A_i - V_i)^2}, else \end{cases} \tag{2}$$

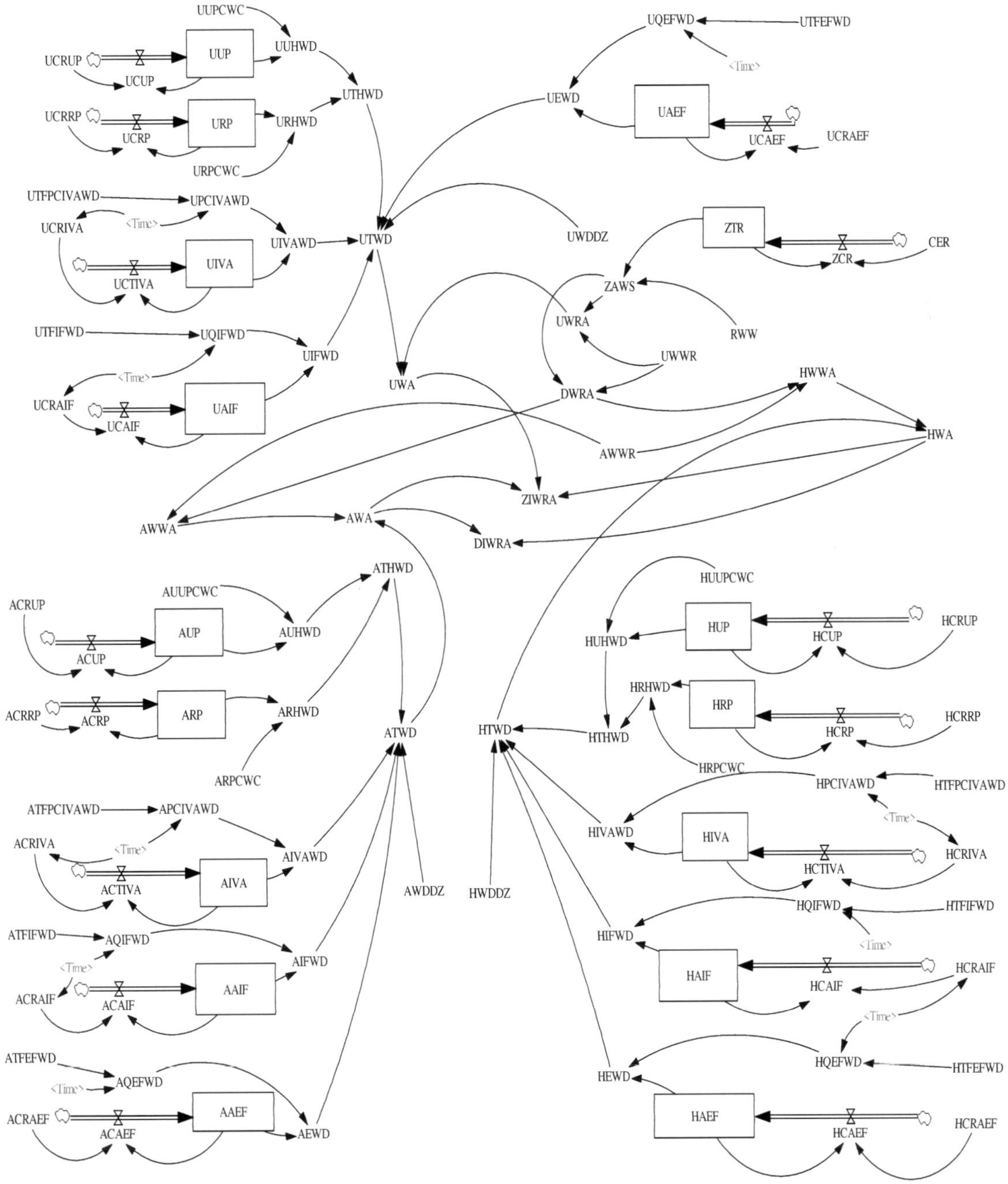


**Figure 3.** Water dispute model of Zhanghe River basin

### *2.4. Developing Stock-Flow Diagram*

The simulation model used in this article was developed using Vensim PLE software. This software is capable of developing a complex water resources system dynamics model with less programming efforts, in addition, the software makes it easy for model expansion. The model comprises of three main sectors capable of simulating the riparian cities water demand, water allocation and water dispute. The model includes 94 parameters and 15 state variables. The flow diagram for water dispute of Zhanghe River basin is shown in Figure 3. It is originated from a basic demand and supply framework based on cause loop. On the demand side, population level, agriculture and industry development, economic growth and ecological requirements determines the water demand of six sub-systems in three different regions. On the supply side, water supply determines by simulated the current water allocation policies. On the dispute side, the water dispute was represented by the instability of water allocation $S_A$.

### *2.5. Formulating a Mathematical Model*

All information used in the model was collected from peer reviewed journals, government records, newspapers, and publicly shared interviews. All elements contained in the model and their relationships are presented in Appendices.

## **3. Simulated Results**

Until now, the national government has only ruled on the water sharing ratio for the sub-basin of Zhanghe River basin located in the downstream section of the basin. The water sharing ratio which considers every riparian city from the upstream to the downstream of the river is yet to be established. Defining the base scenario according to the current water allocation policy of the upstream and downstream riparian regions: 60% of the basin's water was allocated to upstream and 40% allocated to the downstream for the time period from 2000 to 2050 [13]. Through the simulation of the base scenario, it could predict the water resource adequacy of the riparian cities and the instability of the existing water allocation scheme under the current policy for the time period from 2018 to 2050.

With the increase in water demand in the riparian regions, the water adequacy began to decrease since 2000. From 2000 to 2012, upstream regions obtained enough water by constructing water reservoirs and many other water conservative hydraulic structures, but leaving the downstream cities with the problem of water scarcity. The difference among the three riparian regions in terms of water adequacy is expected to result in the instability of water allocation. Figures 4, 5, and 6 also showed the significant gap among the riparian region's water adequacy and the instability water allocation. Such high water allocation instability raises water sharing problems and possible water conflict within the river basin.

From 2012 to 2021, the water adequacy of all the regions continues to decrease and will reach its lowest point in 2021; the gap among the water adequacy of the riparian areas shows a clear decreasing trend (figure 4). As the result, the instability of water allocation is also expected to decrease from 2012 to 2021(figures 5 and 6). During the time period from 2021 to 2040, the water adequacy will increase because the water utilization efficiency of the three riparian regions is expected to increase. But it is not enough to fully satisfy their water demand. The instability of water allocation will increase after 2021, but the extent of the instability of water allocation among three riparian regions is lower than the one during the time period from 2000 to 2013, and the instability between two downstream regions

IOP Conf. Series: Earth and Environmental Science **111** (2018) 012001 doi:10.1088/1755-1315/111/1/012001

will increase. Figures 4 and 5 shows that upstream regions have adequate available water to satisfy their demand but downstream riparian cities will be water scarce under base scenario after 2014. The instability of water allocation among three riparian regions will keep growing. On the contrary, the instability of water allocation between two downstream regions will decrease.

There are two water dispute systems within Zhanghe River basin. The first one involves the upstream and downstream sections of the river. The second one is among the cities located in the downstream reach. Hence the decision-makers should consider both water dispute systems when they try to design methods to solve the water sharing problem within the river basin. Under the base scenario, the degree of water dispute among three riparian regions will decrease from 2018 to 2022, and then increase between 2022 and 2050.

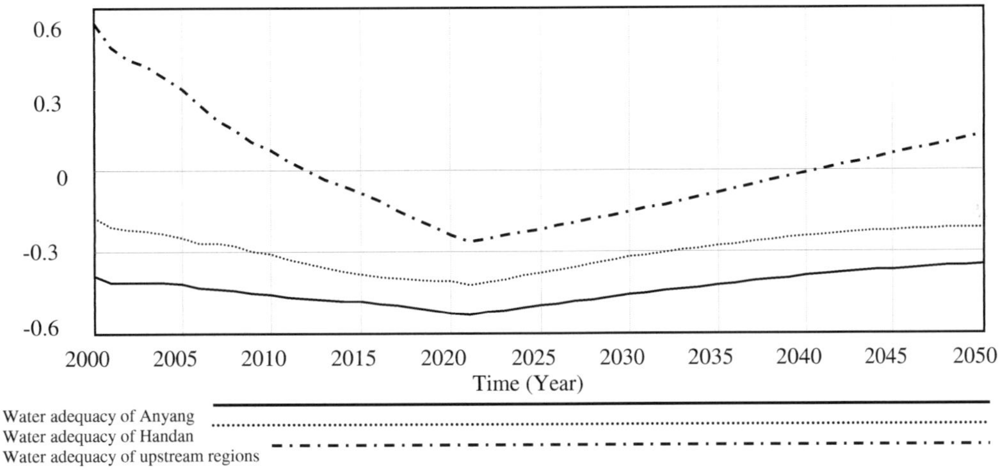


**Figure 4.** Water adequacy of three regions under base scenario

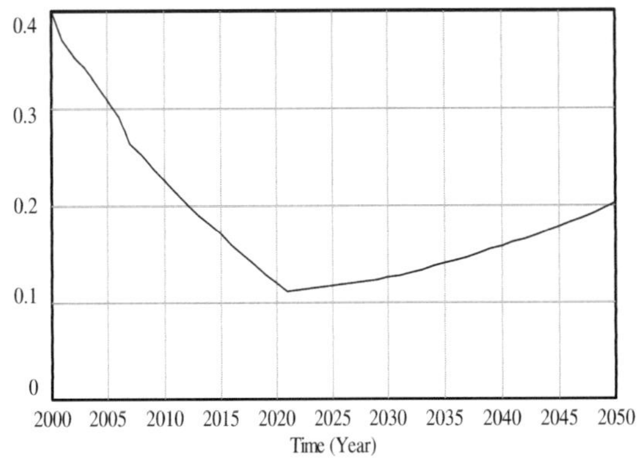


**Figure 5.** The instability of water allocation in Zhanghe River basin under base scenario

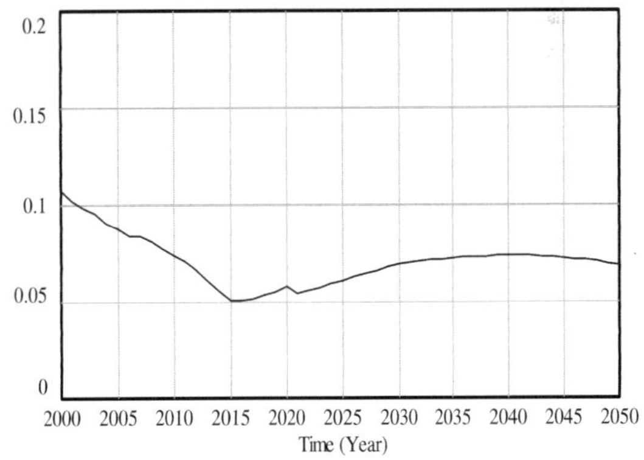


**Figure 6.** The instability of water allocation in downstream under base scenario

## 4. Conclusion

In this paper, system dynamics method not only provides the ability to implement system dynamics thinking, but also apply nonlinear causal thinking to water resources allocation and disputes problems. During the model development process, system dynamics help to comprehend the interactions among the different drivers of the issue via the experiences conveyed, lessons learned and perceptions gained.

It also improves the understanding of concepts and system interactions by offering a comprehensive and integrated view of the physical, social, economic, environmental and political systems, stimulates the issue of water resource system for better understanding of the system and allows managers and stakeholders to better visualize the key hydrologic elements and management constraints within river basins.

The level of water dispute among basin sharing parties depends on their socio-economic and environmental conditions [14]. The situation in the Zhanghe River basin is not different. The study presented in this paper indicates that in order to maintain the regional sustainable development and avoid water conflicts in Zhanghe River basin the national government should introduce a new water allocation management scheme that takes into account the two water disputes simultaneously. In the long term, the national government should design system dynamic water management and planning scheme to manage the water issues of Zhanghe River basin and consider the overall conditions crossing the entire river basin while making a new water allocation management scheme. In the short term, the national government should establish a new water allocation scheme to replace the 1989 non-basin wide water allocation scheme, more water needs to be allocated to the downstream cities and the actual allocation should be adjusted to reflect the need associated with the socio-economic and environmental changes within the region.

The following strategies should be implemented in order to maintain the regional sustainable development and avoid water conflicts in a trans-boundary river basin. Firstly, the water resource allocation should make according to the changing socio-economic and environmental conditions, and the managers and stakeholders need to realize the unified dispatching and management in whole river basin [15]. Secondly, the river should be managed with a thorough understanding of the river dynamics so that the water distribution is coordinated to satisfactorily serve all the regions in the entire river basin. Thirdly, water conservation measures should be enforced to reduce the water demands of all affected regions. These measures include new technologies to improve water resources efficiency, enhancing water conservation, changing the water supply structure and expanding the use of recycled water [16]. Finally, sufficient mechanism should be established to ensure effective communications among all affected regions so that any potential conflicts can be avoid through cooperation.

## 5. Acknowledgments
This work was supported by the China Social Science Foundation (No.13BMZ057, No.14CMZ034), and The Natural Sciences and Engineering Research Council of Canada (NSERC, DG NO.443130).

## 6. Appendices A
Definitions: 1. UCRUP: the change rate of urban population of upstream regions; 2. UCRRP: the change rate of rural population of upstream regions; 3. UCUP: the change of urban population of upstream regions; 4. UCRP: the change of rural population of upstream regions; 5. UUP: urban population of upstream regions; 6. URP: rural population of upstream regions ; 7. URPCWC: rural per capita water consumption of upstream regions; 8. UUPCWC: urban per capita water consumption of upstream regions; 9. URHWD: rural household water demand of upstream regions; 10. UUHWD: urban household water demand of upstream regions; 11. UTHWD: total household water demand of upstream regions; 12. UTFPCIVAWD: table function of water demand of per unit of industrial value-added of upstream regions; 13. UPCIVAWD: water demand of per unit of industrial value-added of

upstream regions; 14. UCRIVA: the change rate of industrial value-added of upstream regions; 15. UCTIVA: the change of total industrial value-added of upstream regions; 16. UIVA: industrial value-added of upstream regions; 17. UIVAWD: water demand of industrial value-added of upstream regions; 18. UTFIFWD: table function of water demand of per thousand hectares irrigated farmland in upstream regions; 19. UQIFWD: quote of water demand of irrigated farmland in upstream regions; 20. UCRAIF: the change rate of irrigated farmland area of upstream regions; 21. UCAIF: the change of irrigated farmland area of upstream regions; 22. UAIF: irrigated farmland area of upstream regions; 23. UIFWD: water demand of irrigated farmland in upstream regions; 24. UTFEFWD: table function of water demand of ecological forestry in upstream regions; 25. UQEFWD: quote of water demand of ecological forestry in upstream regions; 26. UCRAEF: the change rate of ecological forestry area of upstream regions; 27. UCAEF: the change of ecological forestry area of upstream regions; 28. UAEF: ecological forestry area of upstream regions; 29. UEWD: ecological water demand of upstream region 30. UTWD: total water demand of upstream regions for Zhanghe River; 31. UWDDZ: the ratio of upstream regions dependent on Zhanghe River water resources; 32. ACRUP: the change rate of urban population of Anyang City; 33. ACUP: the change of urban population of Anyang City; 34. AUP: urban population of Anyang City; 35. AUUPCWC: urban per capita water consumption of Anyang City; 36. AUHWD: urban household water demand of Anyang City; 37. ACRRP: the change rate of rural population of Anyang City; 38. ACRP: the change of rural population of Anyang City; 39. ARP: rural population of Anyang City; 40. ARPCWC: rural per capita water consumption of Anyang City; 41. ARHWD: rural household water demand of Anyang City; 42. ATHWD: total household water demand of Anyang City; 43. ATFPCIVAWD: table function of water demand of per unit of industrial value-added in Anyang City; 44. APCIVAWD: water demand of per unit of industrial value-added of Anyang City; 45. ACRIVA: the change rate of industrial value-added of Anyang City; 46. ACTIVA: the change of total industrial value-added of Anyang City; 47. AIVA: industrial value-added of Anyang City; 48. AIVAWD: water demand of industrial value-added of Anyang City; 49. ATFIFWD: table function of water demand of per thousand hectares irrigated farmland in Anyang City; 50. AQIFWD: quote of water demand of irrigated farmland in Anyang City; 51. ACRAIF: the change rate of irrigated farmland area of Anyang City; 52. ACAIF: the change of irrigated farmland area of Anyang City; 53. AAIF: irrigated farmland area of Anyang City; 54. AIFWD: water demand of irrigated farmland in Anyang City; 55. ATFEFWD: table function of water demand of ecological forestry in Anyang City; 56. AQEFWD: quote of water demand of ecological forestry in Anyang City; 57. ACRAEF: the change rate of ecological forestry area of Anyang City; 58. AAEF: ecological forestry area of Anyang City;59. ACAEF: the change of ecological forestry area of Anyang City; 60. AEWD: ecological water demand of Anyang City;61. AWDDZ: the ratio of Anyang City dependent on Zhanghe River water resources; 62. ATWD= total water demand of Anyang city for Zhanghe River; 63. HCRUP: the change rate of urban population of Handan City; 64. HCUP: the change of urban population of Handan City; 65. HUP: urban population of Handan City; 66. HUUPCWC: urban per capita water consumption of Handan City; 67. HUHWD: urban household water demand of Handan City; 68. HCRRP: the change rate of rural population of Handan City; 69. HCRP: the change of rural population of Handan City; 70. HRP: rural population of Handan City; 71. HRPCWC: rural per capita water consumption of Handan City; 72. HRHWD: rural household water demand of Handan City; 73. HTHWD: total household water demand of Handan City; 74. HTFPCIVAWD: table function of water demand of per unit of industrial value-added in Handan City; 75. HPCIVAWD: water demand of per unit of industrial value-added of Handan City; 76. HCRIVA: the change rate of industrial value-added of Handan City; 77. HCTIVA: the change of total industrial value-added of Handan City; 78. HIVA: industrial value-added of Handan City; 79. HIVAWD: water demand of industrial value-added of Handan City; 80. HTFIFWD: table function of water demand of per thousand hectares irrigated farmland in Handan City; 81. HQIFWD: quote of water demand of irrigated farmland in Handan City; 82. HCRAIF: the change rate of irrigated farmland area of Handan City; 83. HCAIF: the change of irrigated farmland area of Handan City; 84. HAIF: irrigated farmland area of Handan City; 85. HIFWD: water demand of irrigated farmland in Handan City; 86. HTFEFWD: table function of water demand of ecological forestry in Handan City; 87. HQEFWD: quote of water demand of ecological forestry in Handan City; 88. HCRAEF: the change rate of ecological forestry area of Anyang City; 89. HCAEF: the change of ecological forestry area of Handan City; 90. HAEF: ecological forestry area of Handan City; 91. HEWD: ecological water demand of Handan City; 92. HTWD: total water demand of Handan City for Zhanghe River; 93. HWDDZ: the ratio of Handan City dependent on Zhanghe River water resources; 94. CER: the effect of climate on runoff; 95. ZCR: the change of runoff of Zhanghe River; 96. ZTR: total runoff of Zhanghe River; 97. RWW: the ratio of available water supply of Zhanghe River; 98. ZAWS: available water supply of Zhanghe River; 99. UWRA: water resources allocate to upstream Region; 100 UWRR: ratio of water resources allocate to upstream regions; 101. DWRA: water resources allocate t downstream regions; 102. AWWR: ratio of the downstream water resources allocate to Anyang City; 103. UWA: water adequacy of upstream regions; 104. AWWA: water resources allocate to Anyang City; 105. HWWA: water resources allocate to Handan City; 106. AWA: water adequacy of Anyang City; 107. HWA: water adequacy of Handan City; 108. ZIWRA: the instability of water resources allocation in downstream region; 109. DIWRA: the instability of water resources allocation in Zhanghe river basin

## 7. References

[1] Houba H, van der Laan G and Zeng Y 2014 Asymmetric nash solutions in the river sharing problem *Strat. Behavior Environ.* **4** 321–360

[2] Reuveny R 2007 Climate change-induced migration and violent conflict *Politic.Geo.* **6** 656-673.

[3] HanjraM A and Qureshi M E 2010 Global water crisis and future food security in an era of climate change *Food Policy* **5** 365-377

[4] Collins R D, de Neufville R, Claro J, *et al* 2013 Forest fire management to avoid unintended consequences: a case study of Portugal using system dynamics *J. Enviro.Manage.* **130** 1-9.

[5] Forrester J W 1958 Industrial dynamics: a major breakthrough for decision makers *Harvard Busine. Revi.* **4** 37–66

[6] Sun Y, Liu N, Shang J, et al 2017 Sustainable utilization of water resources in China: a system dynamics model *J. Cleaner Produc.* **142** 613-625

[7] Liu C L 2009 Study on the sustainable utilization of water resources in Zhang river upper reaches School of hydropower engineering(Handan: Hebei University of Engineering)

[8] Xiong X Y 2005 Study on Water Conflict in Upper Boundary of Zhang River(Beijing: Renmin University of China)

[9] Elsawah S 2011 Water resources planning and management based on system dynamics: a case study of Yulin city *Enviro.Develo. Sustain.* **2** 331-351

[10] Wang Q F 1995 New progress in theory and methodology of system dynamics System.

Engineer *Theo.Metho. Appli.* **2** 6–12

[11] Haddeland I, Heinke J, Biemans H, *et al* 2014 Global water resources affected by human interventions and climate change. *Proc. National Acade. Sci.* **9** 3251-3256

[12] Xu Z X, Takeuchi K., Ishidaira H., *et al* 2002 Sustainability analysis for Yellow River water resources using the system dynamics approach. *Wate. Res. Manage.* **36** 239-261.

[13] Niu F and Li Z W 2015 Analysis of the Present Situation of Water Resources in Upper Zhanghe River, *Haihe Wate. Res.* **4** 4-6

[14] Degefu D M, He W J, Yuan L, et al 2016 Water allocation in transboundary river basins under water scarcity: a cooperative bargaining approach *Wate. Resour. Manage.* **30** 4451-4466

[15] Jia N S 2003 Problems and countermeasures of solving water disputes in the upper reaches of Zhang River *Haihe Wate. Res.* **6** 23-25

[16] Jin L C and Wang L 2016 Reflections on the construction of water law enforcement ability at the basic level in the river basin: a case study of Zhanghe River *Haihe Wate. Res.* **2** 26-27

IOP Conf. Series: Earth and Environmental Science **111** (2018) 012002 doi:10.1088/1755-1315/111/1/012002

# Scenario Analysis of Soil and Water Conservation in Xiejia Watershed Based on Improved CSLE Model

**Jieying Liu, Ming Yu*, Yong Wu, Yao Huang and Yawen Nie.**

School of Geographical Science, Fujian Normal University, Fuzhou, China
E-mail: gisyuming@163.com

**Abstract**. According to the existing research results and related data, use the scenario analysis method, to evaluate the effects of different soil and water conservation measures on soil erosion in a small watershed. Based on the analysis of soil erosion scenarios and model simulation budgets in the study area, it is found that all scenarios simulated soil erosion rates are lower than the present situation of soil erosion in 2013. Soil and water conservation measures are more effective in reducing soil erosion than soil and water conservation biological measures and soil and water conservation tillage measures.

## 1. Introduction

Soil erosion has both natural and man-made causes, and the increase of soil erosion can cause the ecological environment to deteriorate, which is the common concerned of the ecological and environmental problems. Soil erosion is mainly affected by such factors as topographic factors, soil types, precipitation characteristics, and land cover. Topography, soil type and precipitation characteristics will not change very much in the short run for the impact of human activities, while surface coverings change easily [1]. Through the implementation of comprehensive control of soil erosion, and artificially change land cover it can reduce soil erosion and control its development, so as to achieve the goal of ensuring the safety of the ecological environment. This paper takes Xiejia watershed as the study area, on the basis of existing studies, and analyses the soil erosion in different scenarios. Then compare with the soil erosion status in the basin, evaluate and analyse the change of soil erosion intensity under different scenarios, to provide scientific basis for effective control of soil erosion and assessment of watershed ecological security.

## 2. Overview of Research Area, Research Data and Research Methods

### *2.1. Overview of Research Area*

Xiejia watershed is located in the northwest of Fangtian township, longitude 116 °33 'E and latitude 26 °12'N (Figure 1). Watershed lie in the east of Wuyi Mountain, which relative height is generally between 100m. Xiejia watershed belong to subtropical monsoon climate zone, which has abundant rainfall, mild climate and obvious monsoon. The forest resources in the study area are unevenly distributed that most of the tree species are coniferous forests with single stand structure, and the proportion of sparse forest is large and site conditions are poor. In addition, human activities reduced the area of native forest vegetation, resulting in bare land surface and weakened forest capacity for regulating surface water storage and runoff that aggravated soil erosion.

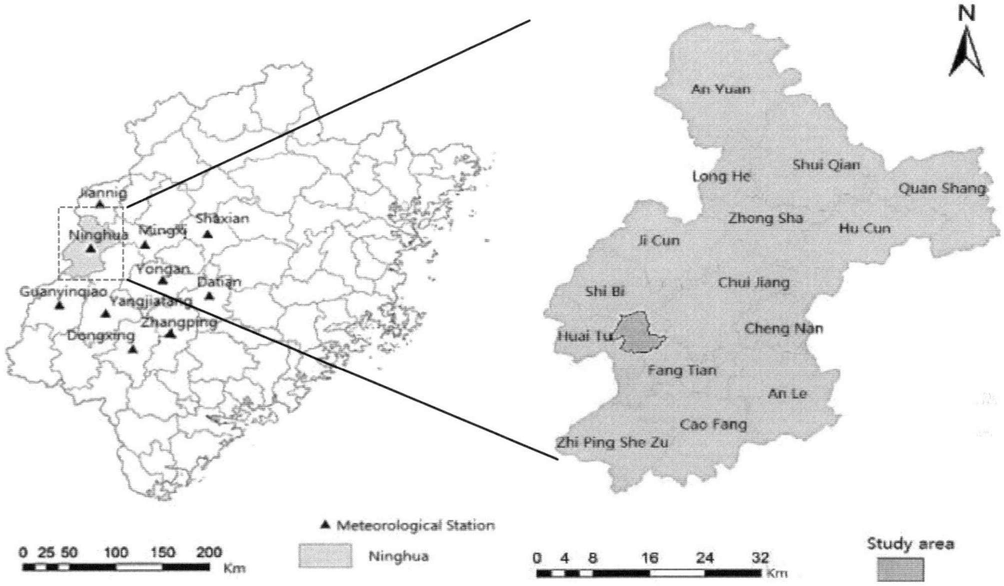


**Figure 1**. Geographic locations of the study area and meteorological stations.

*2.2. Research Data*

Data Source:

(1)DEM data with a resolution of 30m of Xiejia watershed is available for free download via ASTER GDEM.

(2)Remote sensing image data: Landset-8OLI image data with imaging time on August 11, 2013.

(3)Meteorological data: daily rainfall data provided by China Meteorological Science Data Sharing Service Network.

(4)Soil data: 1: 250000 soil type database obtained by Ninghua County Land Bureau.

(5)Other data: Ninghua County Statistical Yearbook from the Ninghua County Bureau of Statistics; The present situation of soil and water loss in Ninghua County was obtained from the Fujian Provincial Water Resources Bulletin in 2013; The required data are pretreated according to model requirements and research purposes.

*2.3. Research Methods*

*2.3.1. Scenario Analysis.* Scenario analysis, also known as prospects description methods, is a method of predicting the consequences of a prediction phenomenon or a situation that may occur if a trend or a phenomenon is assumed to be sustained to the future, which is considered to be an extension of the Multi - objective Planning and Screening Method for System Engineering [2]. By setting different scenarios, quantitatively calculate the soil erosion levels under the different soil erosion scenarios.

Then compare the advantages and disadvantages of each scenario, in order to provide suggestions and reference for the calculation of soil erosion modulus and soil erosion control [3-4].

*2.3.2. Scenarios Design of Xiejia Watershed.* At present, the main ways of comprehensive management of soil erosion are to adjust the land use, rational allocation and implementation of soil and water conservation measures. The natural factors such as soil, rainfall and slope will not change greatly with human activities in a short period of time [5-6]. Therefore, the most important consideration in scenario setting is the factor of soil and water conservation under human action. The study based on the present situation of soil erosion in Xiejia watershed and the key control decisions of soil erosion, worked out the scenario programs of soil erosion in Xiejia watershed (Table.1)

**Table1.**Setting of scenarios for soil erosion control in Xiejia watershed.

| Situational condition | Scenario setting |
|---|---|
| soil and water conservation biological measures | ($a_1$) maintaining the status in 2013 |
| | ($a_2$) the naked land was changed into irrigated grassland |
| | ($a_3$) bare land was changed to forest land in the watershed |
| soil and water conservation tillage measures | ($b_1$) maintaining the status in 2013 |
| | ($b_2$) change the cultivated land with a slope of 15 degrees or more to grassland |
| | ($b_3$) change the cultivated land that slope is above 25 degrees into forest land |
| soil and water conservation engineering measures | ($c_1$) maintaining the status in 2013 |
| | ($c_2$) measures for soil and water conservation projects will be implemented in the whole watershed according to local conditions. |

The above scenarios are combined to obtain 10 kinds of soil erosion scenarios in Xiejia watershed. On the basis of existing studies, changing the factor values of the B, E, T and estimate soil erosion value in study area under different soil and water conservation measures. The current situation of soil erosion in 2013 is set as $a_1b_1c_1$, which is the background value of soil erosion. Then compare and analyze the degree of soil erosion under the combination of scenarios, so as to evaluate the influence of soil and water conservation measures on soil erosion, and to provide scientific basis for soil erosion control in the future.

## 3. Scenario Analysis of Soil Erosion in Xiejia Watershed

In this paper, the CSLE model is used to calculate the soil erosion modulus [7-8], and the model of soil erosion in China is expressed as:

$$A=R\times K\times L\times S\times B\times E\times T \tag{1}$$

where A is the amount of soil loss *(t/ hm²•a)*; *R* is rainfall erosive factor *(t/ hm²•a)*; *K* is soil erodibility factor *((t•hm²•h) /(hm²•M J. mm ) )*; *L is* slope length factor; *S* is slope factor; *B* is biological measures factor; *E* is engineering measure factor; *T* is tillage measures factor. According to the design of soil erosion scenarios in Xiejia watershed in 1.2.2 section, applied formulas (1) to estimate soil erosion modulus under different soil erosion scenarios. Then compare the degree of soil erosion under different situations, and evaluate the impact of soil and water conservation measures on soil erosion [9-15].

*3.1. Effects of Soil and Water Conservation Biological Measures on Soil Erosion.*

Keeping the E and T factor unchanging, and changing the B factor, obtain different biological conservation measures scenario. CSLE equation is used to estimate the total amount of soil erosion under each scenario. According to the "Soil Erosion Classification Criteria" (*SLI90-2007*) in the southern soil erosion grading standards, grade the soil erosion amount and calculate the area of soil erosion type under different scenarios by using GIS software.

It can be seen from Table 2 that when the soil and water conservation biological measures are simulated in bare land, the soil erosion rate of Xiejia watershed has declined. In 2013, 24.13% decreased to 23.71% and 23.54%, the decrease rate is not obvious, because the total land area of the bare in Xiejia watershed is small, only 0.15 km$^2$ and the contrast is not significant. As can be seen from Table 1, scenario scenarios $a_2b_1c_1$ and $a_3b_1c_1$ have improved in all types of erosion compared to the 2013 situation, and the area of mild erosion type accounted for a larger proportion. Contrasting scenario scenarios $a_2b_1c_1$ and $a_3b_1c_1$, $a_3b_1c_1$ are lower than the $a_2b_1c_1$ in the other types of erosion types, except the intensity erosion type area. It shows that compared with planting grassland in bare land, afforestation in soil erosion area is more effective for soil erosion improvement.

**Table 2.** The statics of soil erosion in different soil erosion conservation biological measures (m$^2$).

| Project | Scenarios | Erosion rate (%) | Slight | Moderate | Strength | Extreme | Violent |
|---|---|---|---|---|---|---|---|
| 2013 | $a_1b_1c_1$ | 24.13 | 5783400 | 867600 | 277200 | 164700 | 41400 |
| S1 | $a_2b_1c_1$ | 23.71 | 5767200 | 843300 | 244800 | 122400 | 30600 |
| S2 | $a_3b_1c_1$ | 23.54 | 5734800 | 841500 | 244800 | 110700 | 28800 |

### *3.2. Effects of Soil and Water Conservation Tillage Measures on Soil Erosion.*

In the land use types of Xiejia watershed, the cultivated land accounts for a larger proportion of the total land area, therefore, the implementation of reasonable soil and water conservation tillage measures can effectively improve soil erosion in the study area. In combination with the actual situation, under the condition that B and E factor remain unchanged and T factor are changed, two soil and water conservation tillage measures, namely $a_1b_2c_1$ and $a_1b_3c_1$, are obtained (Table 3). According to the above mentioned method, calculate the soil erosion rate of each scenario. It can be seen from Table 3 that the soil erosion rate decreased after the two soil and water conservation tillage measures were carried out in the watershed. The decrease was small, because most of the cultivated land in the valley was located in the mountain depression which slope below 5°. However, the soil erosion is still severe in the cultivated area above 15 ° where human activities are more frequent. There is little difference in soil erosion rate and each erosion type between scheme $a_1b_2c_1$ and scheme $a_1b_3c_1$. In addition, the area of slight erosion have decreased greatly, both decreased more than 400000m$^2$. This shows that the implementation of returning farmland to cropland or returning farmland to forest in the cultivated area can achieve better effects of soil erosion control.

**Table 3.** The statics of soil erosion in different soil erosion conservation tillage measures (m$^2$).

| Project | Scenarios | Erosion rate (%) | Slight | Moderate | Strength | Extreme | Violent |
|---|---|---|---|---|---|---|---|
| 2013 | $a_1b_1c_1$ | 24.13 | 5783400 | 867600 | 277200 | 164700 | 41400 |
| G1 | $a_1b_2c_1$ | 22.66 | 5355900 | 864900 | 274500 | 162900 | 41400 |
| G2 | $a_1b_3c_1$ | 22.60 | 5366700 | 851400 | 268200 | 158400 | 37800 |

### *3.3. Effects of Soil and Water Conservation Engineering Measures on Soil Erosion.*

The common measures such as tractor road construction, drainage and water reservoir are classified as slope change ladder and other surface treatment which use to harness the small watershed, therefore, select slope change ladder as soil and water conservation engineering measures. The forest land and grassland which no engineering measures are implemented and the vegetation coverage is below 60%, are managed by slope and ladder. In the same situation as other soil and water conservation measures, change the soil and water conservation engineering measures, then calculate the soil erosion rate under different scenarios with the method mentioned above, and count the area of different types of soil erosion under different schemes(Table 4). Compared with the data in the comparison table, the soil and water conservation governance effect is obvious after the soil and water conservation project measures are implemented. Soil erosion rate decreased by 7.85%, among the soil erosion types, the area of slight erosion, moderate erosion and severe erosion decreased by nearly 50%. The reason is that the total area of forest land and grassland in the watershed reaches 62.84%, therefore, the implementation of a large area of water and soil conservation measures in the watershed has positive effects on soil erosion improvement in the study area.

**Table 4.** The statics of soil erosion in different soil erosion conservation engineering measures ($m^2$).

| Project | Scenarios | Erosion rate (%) | Slight | Moderate | Strength | Extreme | Violent |
|---|---|---|---|---|---|---|---|
| 2013 | $a_1b_1c_1$ | 24.13 | 5783400 | 867600 | 277200 | 164700 | 41400 |
| C1 | $a_1b_1c_2$ | 16.28 | 3984300 | 432000 | 244800 | 122400 | 28800 |

### *3.4. Effects of Soil and Water Conservation Measures on Soil Erosion.*

Combining the soil and water conservation biological measures, the soil and water conservation tillage measures and the soil and water conservation engineering measures, five scenarios are got (Table 5). Then calculate soil erosion type area and a comparison of soil erosion rates under eleven soil erosion scenarios was obtained at last (Figure 2). As can be seen from the table, compared with the area of soil erosion type in 2013, the soil erosion types and areas calculated under the five combinations are obviously reduced in different degrees, except the scenario $a_2b_2c_2$ and $a_3b_2c_2$ which the area of severe erosion is zero, and the area of extreme strength erosion decreases greatly. From the soil erosion rates of all scenarios, scenario $a_3b_2c_2$ has the lowest soil erosion rate which assumed that the naked land in the watershed is changed to the grassland (grassland coverage >80%), and the cultivated land with the slope of 15° or more is transformed into the grassland, besides the engineering measures are not implemented in the watershed and Vegetation coverage below 60% of the woodland and shrubs to implement slope change ladder and other treatments to improve the effect of soil and water conservation measures in the model, so as to control the soil erosion rate and achieve the effect of soil erosion control. Although the soil erosion rate of bare land is above 90%, the proportion of bare land in Xiejia watershed is smaller, only 0.5% of the total area of soil. Compared with the single soil and water conservation measures, the combination of soil and water conservation comprehensive measures are more effective not only in the soil erosion rate but also in the soil erosion type area.

**Table 5.** The statics of soil erosion in different scenario ($m^2$).

| Project | Scenarios | Erosion rate (%) | Slight | Moderate | Strength | Extreme | Violent |
|---|---|---|---|---|---|---|---|
| 2013 | $a_1b_1c_1$ | 24.13 | 5783400 | 867600 | 277200 | 164700 | 41400 |
| Z1 | $a_1b_2c_2$ | 7.33 | 1925100 | 119700 | 51300 | 49500 | 20700 |
| Z2 | $a_2b_2c_1$ | 14.05 | 3864600 | 157500 | 49100 | 55800 | 26100 |
| Z3 | $a_2b_2c_2$ | 4.69 | 1288800 | 72000 | 13500 | 10800 | 0 |
| Z4 | $a_3b_2c_1$ | 11.88 | 3234600 | 148500 | 48600 | 54000 | 26100 |
| Z5 | $a_3b_2c_2$ | 4.56 | 1255500 | 68400 | 14400 | 9000 | 0 |

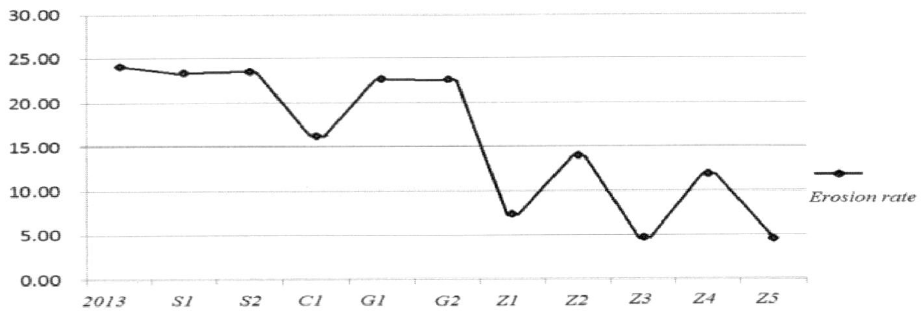


**Figure 2.** Comparison of soil erosion rate under scenario design.

### *3.5. Summary*
Through the study of the soil erosion situation setting and model simulation budget in the study area, it is found that the soil erosion rate simulated under all scenarios are lower than the soil erosion status in 2013, and combining scenarios of soil and water conservation measures are more effective in reducing soil erosion. As the bare area accounts for less total land area, and woodland and shrub land accounted for more than half of the total area of the watershed, therefore, compared with soil and water conservation biological measures and soil and water conservation tillage measures, the implementation of soil and water conservation engineering measures in the watershed will contribute more to the reduction of soil erosion. In addition, the soil erosion rate of bare land has reached more than 90%, which is the type of land use that must be controlled, so it is necessary to plant trees on bare land (shrub grass) when carried out the implementation of returning farmland to forests (shrubs), and the implementation of slope change ladder in forest land and grassland in Xiejia watershed, so as to achieve the effect of reducing and controlling soil erosion and ensuring the safety of the river watershed.

## 4. Conclusion
Based on the combination of RS/GIS and technology, this paper takes the Xiejia watershed in Ninghua County as the research area, extracts the factors affecting soil erosion and sets up the spatial data sets of environmental factors. The distribution of R factor in the rainfall erosivity of Xiejia watershed in Ninghua County is relatively uniform and the spatial difference is small. The soil erodibility factor is in the range of 0.2357 to 0.3391, and the K value of the paddy soil is the largest and its erosion resistance is the weakest. Compared with the slope length factor, the slope factor has a great influence on soil erosion. Based on the present situation of soil erosion in Xiejia watershed, formulate a scenario of soil erosion scenarios and through analysis, it is found that the soil erosion rate under different scenarios is lower than that of soil erosion in 2013. Compared with soil and water conservation measures and soil and water conservation tillage measures, water and soil conservation measures in the watershed contribute more to the reduction of soil erosion. This paper, based on the collection of large amounts of data and data, try to choose the most reasonable and scientific research method for the actual situation of the study area. However, soil erosion is uncertain, but scenario simulation is in the ideal state, and some scenarios may not be realistic enough. Therefore, how to simulate the soil erosion in the study area according to the existing research data, it will be further explored in the future research.

## 5. Acknowledgments

The study is supported by the National Science Foundation of China (No. 41171232) and the National Science Foundation of Fujian Province (2014J01149 and 2017R1034-4).

## 6. References

[1] W H Wischmeier, D D Smith 1978 *Predicting Rainfall Erosionesses A Guide to Conservation Planning*(Washington D C:USDA) p1-58

[2] F Y Wei 2007 *Statistical Diagnosis and Prediction Technology of Modern Climate* (Beijing: Meteorological Press) p29

[3] G Q Fang, F S Ruan, X H Wu and Z M Guo 1997 Preliminary study on erodibility characteristics of main soil in Fujian Province *J.Soil Water Conserv. Fujian.* **2** 19-23

[4] X X Lv and R M Shen 1992 Preliminary study on soil erodibility factor K *J. Soil Water Conserv.* **6** 63-70

[5] H M Xu, Q Z Gao and Y Jiang 2008 Scenario analysis of soil erosion in the middle reaches of the Yellow River sandstone area in the watershed *J. Ecol. Rural Environ.* **01** 10-4

[6] J H Pan 2009 Simulation of Soil Erosion in Small Watershed of Loess Hilly and Gully Region - Take Qingshui watershed in Jingning County of Gansu Province as an example *J. Nat. Resour.* **44** 577

[7] Z C Zhou and Z P Shangguan 2004 Study on Soil Erosion Model *J.Sci. Soil Water Conserv.* **2** 52-6

[8] Z S Jiang, Z Q Wang and Z LIU 1996 Quantitative study on spatial variability of soil erosion in small watershed of Loess Hilly Region *J. Soil Water Conserv.* **2** 1-9

[9] L X Cao and X H Fu 2007 Comparison and analysis of slope length calculation methods based on DEM *J. Bull. Soil Water Conserv.* **27** 58-62

[10] Y Zhang, B Y Liu P J Shi and Z S Jiang 2001 Calculation of soil erosion crop coverage factor on the Loess Plateau*J.Acta Ecol.Sin.***21** 1050-6

[11] J L Xu and X Zha 2009 Calculation of Soil Erosion Risk and Analysis of Influencing Factors—Taking red soil erosion area in Fujian, Anxi as an example *J.Geo-Inf.Sci.* **11** 577-84

[12] Q K Guo, B Y Liu, S B Zhu, G Y Wang, Y N Liu and A J Wang 2013 Major tillage measures for soil and water conservation in China *J.Soil Water Conserv. China.* **10** 22-6

[13] Y Sun Soil erosion modulus monitoring based on GIS, Master, Northeast Normal University, Jilin Changchun China 2012.

[14] B Y,Liu Y N Liu, K L Zhang and Y Xie 2013 Classification of soil and water conservation measures *J Soil Water Conserv.* **27** 80-4

[15] K Wang, Q Y Xia, J H Ma and Qu 2015 Quantitative evaluation of soil erosion in Gushan River Watershed Based on CSLE and high resolution aerial images *J.Res.Soil Water Conserv.* **22** 26-32

# Concept and Connotation of Water Resources Carrying Capacity in Water Ecological Civilization Construction

**Zhilong Chao [1], Xiaoyu Song [2] and Xianghua Feng [2]**

[1] Shaanxi Hydrology and Water Resources Survey Bureau, Xi'an, Shaanxi 710068, China

[2] State Key Laboratory of Eco-hydraulic Engineering in Arid Area, Xi'an University of Technology, Xi'an, Shaanxi 710048, China.

Email: 1042067132@qq.com

**Abstract**: Water ecological civilization construction is based on the water resources carrying capacity, guided by the sustainable development concept, adhered to the human-water harmony thoughts. This paper has comprehensive analyzed the concept and characteristics of the carrying capacity of water resources in the water ecological civilization construction, and discussed the research methods and evaluation index system of water carrying capacity in the water ecological civilization construction, finally pointed out that the problems and solutions of water carrying capacity in the water ecological civilization construction and put forward the future research prospect.

## 1. Introduction

The water resources are not only basic natural resources of social and economic development, but also the basic elements of comprise the whole ecological environment system. It has played an indispensable role in ensuring the healthy development of human society and the benign circulation of ecosystem. It's importance even more pronounced to realizing sustainable development in water shortage regions particularly[1]. Water resources will become an important factor to constraint the comprehensive development and development scale of a country or region, and has attracted wide attention of scholars both at home and abroad. With the growth of population in China, and comprehensively promoting the rapid development of social economy and urbanization advancing, water resources shortage has become increasingly prominent for the increase of city water demands, and water quality is not up to standard for the serious environmental pollution caused by urbanization and industrialization. Therefore, it's imperative to rational allocate the regional water resources, comprehensive coordinate the urban water use in all sectors, and actively build water ecological civilization city [2-10].

With the diversification of the water ecological system problem, the institutionalization of general management in river basin water ecological system, together with the number and diversity of aquatic creatures are threatened, which requires that people have to from a new angle clarify the carrying capacity of the water ecosystem. The requirement in the future research on water resources carrying capacity, must be reflected the current water ecological problems and taken into account the population, ecological environment, social and economic. The concept of water ecological carrying capacity is put forward to overcome the deficiency of the water resources carrying capacity and water environment carrying capacity study, promoted to regional water resources development and water environment protection and understanded the level of the relationship between regional economic development, which can be better to coordinate the water ecological system health and provide a scientific basis for social and economic development. In recent years, with the national and local ecological civilization construction policies promoted vigorously, the protection of water resources and ecological environment has been intensified, and the water ecological environment has been effectively improved. This paper aims to integrate the concept and connotation research of the water resources carrying capacity in recent years, and analyze the water resources carrying capacity concept and connotation in the process of the water ecological civilization construction, to promote the further development of water ecological civilization construction.

## 2. The Water Resources Carrying Capacity Concepts and Characteristics in Water Ecological Civilization Construction

### *2.1. Concept of Water Carrying Capacity*

*2.1.1. The Origin of Water Resource Carrying Capacity Concept.* The term "bearing capacity" was derived from the concept of physics, which refered to the maximum pressure that an object bears when it was subjected to external pressure without damaging it. The earliest concept was applied to the study of human ecology by Park and Burgess [11] in 1921, and was defined for the first time in the field of ecology, namely according to the food resources in the region to determine the capacity of the population. Until the early 1980s, the resources carrying research began by the United Nations educational, scientific and cultural organization and put forward the concept of resources carrying capacity, which included the water, mineral, land resources, and other areas of the main resources [12]. With the rapid development of human society, the problems of environmental pollution and resource shortage have become very serious, and the concept of carrying capacity has gradually attracted wide attention. The research about water resources carrying capacity was to realize the sustainable development and utilization of water resources, and to ensure the sustainable development of social economy. Today the water related research on carrying capacity is used more in two aspects of the carrying capacity of water resources and water environment, that is, "when the regional social and economic development to a certain historical stage, on the premise of sustainable development and utilization of water resources and the maximum available water quantity can meet the maximum population and social economic development scale". The water resources carrying capacity included two parts: the supply capacity of water resources and water environment capacity. Water environmental capacity refered to the water functional areas pollutant-holding capacity, namely under the target water quality that met the requirements of the water function area, the maximum number of waters area to allow acceptance of all kinds of pollutants, the environmental capacity's support ability to human activity would directly affect the size of the water resources carrying capacity, and the water environmental capacity of a certain period was analyzed, which should be carried out after the combination of measures of water quality investigation and analysis, water quality planning, water supply project and sewage treatment; Water supply ability was mainly refered to the part which could be utilized by the human production and life, and the size of the water supply capacity was constrained by the ecological balance problem, namely the regional water resources supply capacity was equal to the total amount of water resources minus ecological water demand, the total amount of water resources of regional was generally expressed in a multi-year average minus ecological water demand, there was no exact and mature method for calculating the ecological water demand at present, mainly depended on the ecosystem scope definition and ecological protection criteria in different regions (how to keep the ecological scale and quality).

Therefore, the understanding of water resources carrying capacity needs to be conditioned: the clear research area, the level of science and technology and economy under specific conditions, the principle of sustainable development, the supporting target of the water resources. So the water resources carrying capacity can be defined as: Under the established economic level and technical conditions, the population, socio-economic and ecological environment that water resources can carry in a particular research area can attain the largest scale of inter-coordination sustainable development.

*2.1.2. Connotation of Water Resources Carrying Capacity.* Through integration of water resources carrying capacity research, the characteristics of water resources carrying capacity in the water ecological civilization construction, In general, it can from the following four aspects to understand the connotation of water resources carrying capacity in the water ecological civilization construction:

Research scope. Firstly, the water resources carrying capacity should analyze the load of population, social economy and resource environment carrying capacity standing in the perspective of the amount of water resources and consider the quantity, quality and distribution of water resources. Secondly, it is necessary to evaluate the contradiction between water supply and demand from the perspective of water resources management, technology, economy and reasonable allocation, and analyze the water resources, society, economy, ecological environment within each subsystem characteristic and relationship each other.

Carrying objects. As it is well known, water is the supporting body in the study of water resources carrying capacity, so the water resources carrying capacity of water ecological civilization construction will be different from the previous research on the water resources carrying capacity. The water resources carrying capacity through the ecological environment is regarded as a subsystem together with population and social economy system as carrying body research. But at present the research of ecological environmental water demand is not enough to accurately grasp the quantitative relationship between eco-environment and water resource, which brought difficulties to the study of water resources carrying capacity in water ecological civilization construction. As a measure of regional social and economic development degree index of water resources carrying capacity, should reflect the requirements, configuration relationship between the supporting and carrying body. So only understanding and analyzing the relationship between the supporting and carrying body, between the each carrying body and the carrying body inscape the relationship of in-depth, it can be better clarify the connotation of the water resources carrying capacity, such as the development and change of the population, which is not only bring the change of quantity of supply and demand, the regional economic development and ecological environment will be affected.

Carrying conditions. Because of the water resources carrying capacity research is a complicated system which is multi-level by the supporting body and the carrying body is coupled, and must be reflected the different time and different life level and the development level of science and technology of water resources to be bearing body support ability, it puts forward a new challenges to the research method of water resources carrying capacity, which has a clear study area, the relatively stable boundary conditions, the regular hydrometeorological environment, water resources quantity is relatively stable and easy to measure, to use the existing mature research method to evaluate the water resources carrying capacity.

The thoughts of water resources sustainable development and utilization to ensure economic and social sustainable development, which were derived from water resources in different river basin within the scope of ecological environment and social economy carrying capacity. Because science and technology condition and the corresponding economic life level in the different periods are constant development and change by the ecological environment and social economic development scale.

The impact of the science and technology development on the water resources carrying capacity is reflected in the scientific and technological development level determines the level of development, utilization and management of water resources at each stage, which will affect the development scale of the social economy.

Expression form. First of all, water resources are the dependent and irreplaceable resources of all living beings in the world, and are also the key factors of population, ecological environment, society and economy in the support region. But the progress of human society have mineral, land and other natural resources as the support, and the development conditions of a certain period of the human society is often accompanied with a long history of background, which can't be free from the influence of factors such as politics, culture, geography, so the single factor or condition changes is very difficult to suddenly affect the society development process. Second, the nature of water resources carrying capacity is used to quantitative the scale of social and economic development that regional water resources can support, so people often use a single index or as little as possible index in hopes of simple, quantitative reflects the current carrying capacity of regional water resources. But water resources carrying capacity is a typical multi-objective system, which needs to reflect the relationship between the three aspects of water resources, ecological environment and social economy. so the research of the water resources carrying capacity is inevitable to be taken into multi-index system evaluation or multi-objective decision making direction. Therefore, the focus of current research is how to express this kind of measurement which can highlight both simple and quantitative problems, and reflect the whole connotation of water resources carrying capacity.

### 2.2. *Water Resources Carrying Capacity Characteristics*

*2.2.1. Space-Time.* The carrying capacity of water resources has temporal and regional characteristics. From the time perspective, due to the different periods of social and economic development level difference obvious, water resources development and utilization of capacity are different, regional water resources use efficiency are variable, namely water resources carrying capacity in different periods are developing and changing. From the spatial perspective, even in the same period, due to different regional economic base and technical level are different, regional resource can carry the population, ecological environment level and economic and social development scale differently.

*2.2.2. Social Economy.* The social economic characteristics of water resources carrying capacity are reflected in the human development and utilization capacity of water resources, the structure of water use and the allocation and management of water resources by the government, which can be improving water resources carrying capacity through the measures adjusting industrial structure and using science and technology and other means achieving the changing objects of regional water resources carrying condition.

*2.2.3. Sustainability.* On the one hand, sustainability is to achieve sustained supply of water resources to the social economy, namely to take appropriate management mode to realize the sustainable development and utilization of water resources in order to achieve the purpose of the social and economic sustainable development, should be formed "define demand by supply" relationship between social economic development and water resources carrying capacity.

On the other hand, sustainability is also reflected in the upward trend of water resources carrying capacity, which should be along with the social economy development, but this is not to enhance economic growth for the purpose. Only ensuring the zero growth of water demand and even downward trend, to achieve social and economic sustainable development, namely by raising water utilization efficiency, the regional social economy development scale can be matched with the regional water resources carrying capacity.

## 3. Establishment of Water Resources Carrying Capacity Evaluation Index System in Water Ecological Civilization Construction

In the aforementioned 2.1.1 section is preliminary already defined the concept of water resources carrying capacity, namely under certain economic and technological conditions, on the basis of water resources evaluation, closely linked optimized allocation of water resources, and discussed whether the water resources that could be developed and utilized in the region met the needs of the current and foreseeable future social and economic development. The key to evaluate the carrying capacity of water resources is to establish a reasonable and feasible comprehensive evaluation index system. But because of the particularity and complexity of the system, and the influence of many factors, involves the system all aspects of the social, economic and ecology, so if we can establish a reasonable and feasible comprehensive evaluation index system which requires fully and accurately mastering the water resources conditions, eco-environmental quality, development and utilization situation and supply demand relationship of the study area; So the establishment of water resources carrying capacity comprehensive evaluation index system should reflect regional water resource conditions, the regional ecological environmental quality, the status of regional social and economic development, the region demand water structure and the level of water-saving and the regional supply water structure and water resources development and utilization situation other content[13].

To establish evaluation index system as the key part of water resources carrying capacity evaluation, the requirements of its core is to based on features of the study area to choose the feasible evaluation index, because the method itself has been used in the development of a mature, so in the water resources carrying capacity of specific evaluation, it must be combined with the actual, the different categories of several representative index to build index evaluation system, and adopt appropriate methods to evaluate.

## 4. Evaluation Method and Applicability Analysis of Water Resources Carrying Capacity in Water Ecological Civilization Construction

With the gradual deepening of the concept of water resources carrying capacity, the research methods are becoming more and more perfect. The main research methods of water resources carrying capacity in water ecological civilization construction included:

### *4.1. Background Analysis.*

On a certain time scale, by comparing the natural conditions and social backgrounds of the similar study area, the possible carrying capacity of the study area is obtained by using the kind of ratio. Because this method is mainly for small and independent of the load factor analysis, the analysis becomes simple, the disadvantage is that this method is limited to a certain historical background, without considering the water resources, ecological environment, social and economic relationships between each subsystem, which can not fully show the water resources carrying capacity as the characteristics of multi-level and multi-objective complex large system.

### *4.2. Conventional Trend Method.*

This method is introduced the idea of "Determine production by quantity", namely on the basis of regional available water resources, the premise is the guarantee basic ecological environmental water requirements and reasonable allocation of various national economy departments under the proportion of water, water-saving agriculture and water-saving society construction by calculating the social and economic scale that the region's water resources can support. This method is simple and operable, and the development of single factor is more considered, so this method can be used to estimate the potential of some bearing factors. And water resources carrying capacity is made up of water resources, ecological environment, social and economic system of each subsystem, involves many factors, this method did not consider the various factors each other mutual promotion and mutual restriction relationship each other. Therefore, only when the boundary conditions are determined, the supply and demand water can be obtained, and it is difficult to truly reflect the carrying situation of regional water resources by simple supply and demand balance analysis.

### *4.3. Fuzzy Comprehensive Evaluation Method.*

In this method, the comprehensive evaluation index system is established by selecting the carrying object which affects the carrying capacity of water resources, and it can realize the evaluation of single factor and realize the integration of various factors through the comprehensive evaluation matrix, so that the system can draw a more clear results for water resources carrying capacity of multi-factor comprehensive evaluation. This avoids the fact that the two methods described above allow each other to be independent, so it can more comprehensively determine the water resources carrying capacity of the region, not only as a multi- level of large-scale system, but also through this quantitative conclusion to measure the water resources on the socio-economic development of carrying capacity. However, the fuzzy comprehensive evaluation method is based on the subjective "discrete" process to be integrated, so the results vary from person to person; on the other hand fuzzy comprehensive evaluation method to lose large algorithm which will lose a lot of information, not only reduced the amount of information available to the model ,also result in miscarriage of justice, and with the increase the selection of evaluation indicators, the amount of information lost, which is more likely to cause misjudgment [14]. Therefore, in the use of fuzzy comprehensive evaluation method for water resources carrying capacity evaluation, it is necessary to break through the factors in the bearing capacity of the degree of influence and the selection of factors on the limitations [15].

### 4.4. *Principal Component Analysis.*

In this method, the multidimensional variable is replaced by the linear variable transformation and the small part of the information, and the multivariate variables are replaced by the processed variable. This makes the data loss less, which makes up the fuzzy comprehensive evaluation method some of the defects [16]. The innovation of this method is that after the best synthesis and simplification of the high-dimensional variable system, the objective weight is obtained and the subjective assumption is obtained. In addition, it is only applicable to the study of the spatial difference of water resources carrying capacity in the case of clear regional conditions, which is unsatisfactory in the future researching on water capacity carrying capacity.

### 4.5. *System Dynamic Method (SD).*

This is a method to study the complex-social economic system. Based on the feedback control theory, the evaluation of water resources carrying capacity is completed by the measure of computer simulation technology. This method was founded by Prof. J Forrester [17] of the Massachusetts Institute of Technology in the mid 1950s. Through the combination of qualitative and quantitative, the system, analysis, synthesis and reasoning integration method, and with a special DYNAMO software. This method brings model simulation, policy simulation great convenience, and it can better grasp the system of various feedback. Moreover, it is an ideal method to study the theory of complex large-scale systems, which is suitable for systems with high order, nonlinear, multiple feedback, complex mechanism and time-varying characteristics. On the other hand, the disadvantage of this method is that it is difficult to grasp the parameters accurately so that the results can not be obtained to some extent when the long-term development planning simulation is carried out. Therefore, the result of the SD method is more reliable when simulating the short-term development of the region.

### *4.6. Multi-Objective Decision-Making Method.*

The water resources system and other carriers are composed of a multi-objective, multi-level complex large system, that can not only consider a single target optimization, but through the coordinated development of the overall optimization; as the principle of sustainable development, the method needs choose several targets from the index of water resources carrying capacity, and give the constraints of the target according to the regional characteristics. The system analysis and dynamic analysis method are used to solve the different scenarios. The socio-economic development scale and ecological environment of water resources carried by different levels situation. The advantage of this method is that through the multi-objective decision-making model, the natural system is connected with the social system, and the theory of water resources system is improved, which provides a new solution to the idea and the technical means for water resources planning and decision-making. In the same way, the shortcomings of this method lies in the existence of interdependence and mutual restraint between the objectives, and the main influencing factors of the selected targets are related to each other, and the complexity is self-evident. In addition, the method of determining the weight of each influencing factor (such as the ancient forest method, ABP method, Delphi method) is the subjective judgment method, which affects the whole evaluation process.

In view of the complexity, randomness and ambiguity of the socio-economic system of water resources, the influencing factors of water resources carrying capacity are multifaceted and multi-level, which makes these methods seem to be a bit stretched in the evaluation of water resources carrying capacity and can not more fully or accurately reflect the regional water resources carrying capacity. Therefore, It is imperative to strengthen the research on water resources carrying capacity.

## 5. Problems and Prospects of Water Resources Carrying Capacity in Water Ecological Civilization Construction

Water ecological civilization construction is an extension of the ecological civilization construction, its core idea is to realize people and water in harmony, and the water pollution, water shortage, water ecological degradation and other various pressures exist at the same time, so to strengthen the water ecological civilization construction is not only need of era, and which has far-reaching influence for the economic and social sustainable development. Therefore in the process of water ecology civilization construction to the water resources carrying capacity research is different from the past, one is to study the method and technology to follow the pace of era development, to apply the latest achievements in the field of water resources carrying capacity research area, created a new research direction, and form a new theoretical basis, to make up the insufficient in the study of the past. Two is to broaden the research area, set up the connection between the water ecological system and other subsystem, to realize the effective combination of the human activities and natural system feedback mechanism.

### *5.1. Strengthening the Theoretical System Research of Water Resources Carrying Capacity in Water Ecological Civilization Construction*

The concept and basic theory system of water resources carrying capacity in water ecological civilization construction is improved, such as, the research purpose, service object, discriminant criterion and research scope of water resource bearing capacity are clarified; it also need to perfect the connotation of water resources carrying capacity and its relationship with water resources evaluation, water resources allocation and sustainable development and the support theory of water resource carrying capacity, including the water environment capacity and ecological water demand with other problems

### *5.2. The Model and Dynamics of Analysis Methods and the Expansion of Evaluation Methods*

The use of computer technology in the field of water resources carrying capacity research in-depth, ignoring the dynamic characteristics of the water resources carrying capacity, and to change the pattern research, the idea is to establish a set of the social economic system of evaluation index system reflecting the water ecosystem health condition and the water resources bearing capacity , and realize the dynamic simulation research of the water resources carrying capacity which can complete the dynamic process prediction and estimation of water resources carrying capacity. In the process, it should add natural factors such as biological ecological index in the research and evaluation, and not make social factors dominate. In addition, due to the research methods have some shortcomings and the differences of the study area, so it must weigh the advantages and disadvantages of each evaluation method in different study areas for water resources carrying capacity evaluation.

### *5.3. Pay Attention to Ecological Water Demand Research*

Sustainable development is the premise of ensuring the benign development of the ecological environment, the request we must pay attention to the theory and practice research of ecological environment water demand in the study of water resources carrying capacity, through summarizing and analyzing the theoretical system of ecological environment water demand in the past, realize the ecological water demand scientific computing, make the optimal allocation of water resources is more reasonable.

### *5.4. Strengthening the Study of Water Environmental Capacity of Quantitative Standard*

Water environmental capacity (pollutant-holding capacity), as a natural factors, the most important evaluation index of the water resources carrying capacity which refers to the aquatic environment under specified environmental goals can hold the pollutant quantity, size and water features, water quality target and pollutant characteristics, as well as and the ways of emissions of pollutants and emissions time and space distribution is a close relationship. This requires that it must be strengthen overall water quality monitoring, however, on the one hand because of water pollution control started relatively late in China, compared with hydrological data, most of the lack of a long series of water quality monitoring data; On the other hand, due to the outstanding problems in the process of water quality monitoring, the monitoring result credibility is not high, this determines the foreign advanced methods cannot be directly applied in China's water environmental capacity calculation and distribution system. And the most strict water resources management system is the implementation of the requirements of water environmental capacity theory to implement the management practice, so it has to be realized the conformity of the quantitative calculation method through scientifically selecting appropriate evaluation method to make quantitative results accord with actual.

### 5.5. *Introduce New Technology and New Method*

The water resources carrying capacity research should draw the latest research theory on other areas on the basis of interdisciplinarity, through the introduction of new technologies and new methods in the use of the subject, inject new vitality for the water resources carrying capacity research, and bring it to a new height. Such as on the basis of previous theoretical studies, it is applied to the water resources carrying capacity research field through the advanced technology of remote sensing and geographic information system, which can not only make up for the shortcomings of traditional methods in the past, but also can make the research results closer to reality.

The purpose of carrying capacity of water resources is to achieve the strategic goal of sustainable development of human society by realizing the benign cycle of regional ecological environment. This request is based on the water cycle theory, regional economic development theory and system theory research, first of all, it must be made clear the balance between supply and demand of regional water resources system and the level of the ecological environment, to determine the regional water resources carrying social and economic scale; Second, on the one hand because of the change of regional climate and the strengthening of environmental protection consciousness, which makes the quantity and quality of regional water resources dynamic changes, on the other hand, the study area of social and economic system development level, industrial structure and scale of development are not immutable, therefore the influence of dynamics on water resources carrying capacity can not be neglected. It should be based on the integration of various disciplines through the introduction of innovation in other fields of theoretical results, to achieve breakthroughs in the field of water resources carrying capacity and better solve the existing problems in the study. Looking back the development history of water resources carrying capacity research, it makes the contradiction between social economic development and water issues highlighted through further research of the water resources carrying capacity, which provides a train of thought to solve the problems encountered on the way to insist on sustainable development. Today the water problem is diversified and many problems cross coexist, therefore it is urgent to carry out many aspects, multi-level research of water resources carrying capacity and protect the sustainable development of the economy under the extremely severe social background.

## 6. Acknowledgements

The research was financially supported by the National R & D program(2016YFC0400301) and Shaanxi Provincial Water Conservancy Science and Technology Project (2016slkj-11).

## 7. References

[1] Wang X Q, Gao W, He F and Peng W Q 2011 Discussion on concept and connotation of water ecological carrying capacity *Journal of China Institute of Water Resources and Hydropower Research.* **9** 41-46

[2] Cui J F 1998 The carrying capacity of municipal water environment and its case study 1998 *Journal of Natural Resources.* **13** 58-62

[3] Jia R, Jiang X H, Xue H F and Shen B 2000 Study on carrying capacity of regional water resources 2000 *Journal of Lanzhou University(Natural Sciences).* **36** 114-121

[4] Wei Y H, Jiang X H, Huang Q and Xue X J 2001 On system dynamic simulation model of water

resources bearing capacity in duality mode *Geographical Research.* **20** 191-198

[5] Xia J and Zhu Y Z 2002 The measurewent of water resources security: A study and challenge on water resources carrying capacity *Journal of Natural Resources.* **17** 262-269

[6] Long T R, Jiang W C and He Q 2004 New understanding of the connotation of water resources carrying capacity *Journal of Hydraulic Engineering.* **35** 38-45

[7] Xia J, Zhang Y Y, Wang Z G and Li H 2006 Water carrying capacity of urbanized area *Journal of Hydraulic Engineering.* **37** 1482-88

[8] Zhang Y Y and Fan X Y 2011 The construction of ecological civilization and the carrying capacity of resource and environment *Natural Resource Economics of China.* **24** 9-11+8+54

[9] Yuan G H, Zhen J E, Jia L B, Wang S H, Luo S X and Xi X 2014 Evaluation Monitoring and the Idea of Warning concerning Carrying Capacity on Resource and Environment *Natural Resource Economics of China.* **27** 20-24

[10] Cao L, Liu J H, Qin D Y, Sang X F, Wang M N and Liu M 2011 The construction of water construction in shaanxi based on the regional water resources carrying capacity *China Rural Water and Hydropower.* **52** 1-4

[11] Park R E and Burgess E W 1921 *Introduction to the Science of Sociology* Chicago: University of Chicago Press

[12] UNESCO F 1985 *Carrying Capacity Assessment with a Pilot Study of Kenya: Population-Resources-Environment-Development: A Resource Accounting Methodology for Exploring National Options for Sustainable Development* Rome: Isric Library.

[13] Gong L 2007 Research on water resources carrying capacity in lanzhou *ShaanXi: Northwest A & F University.*

[14] Zai S M, Wen J, Wu F and Sun H 2011 Evaluation of water resources carrying capacity in Xinxiang City *Journal of Hydraulic Engineering.* **42** 783-8

[15] Wang Y Z 2005 Study on evaluation of regional water resources carrying capacity *JiangSu: HoHai University*

[16] Fu X and Ji C M 1999 A comprehensive evaluation of the regional water resources carrying capacity—Application of main component analysis method *Resources and Environment in the Yangtze Basin.* **8** 168-173

[17] Forrester J W 1958 Industrial dynamics: a major breakthrough for decision makers *Harvard business review.* **36** 37-66.

IOP Conf. Series: Earth and Environmental Science 111 (2018) 012004 doi:10.1088/1755-1315/111/1/012004

# Land Ecological Security Evaluation of Underground Iron Mine Based on PSR Model

**Xiao Xiao [1], Yong Chen *[1,2], Jinghua Ruan[1], Qiang Hong[1,2] and Yong Gan[1]**

[1] School of Resources and Environmental Engineering, Wuhan University of Science and Technology

[2] Hubei Key Laboratory for Efficient Utilization and Agglomeration of Metallurgic Mineral Resources

E-mail: yongchen1968@163.com

**Abstract.** Iron ore mine provides an important strategic resource to the national economy while it also causes many serious ecological problems to the environment. The study summed up the characteristics of ecological environment problems of underground iron mine. Considering the mining process of underground iron mine, we analysis connections between mining production, resource, environment and economical background. The paper proposed a land ecological security evaluation system and method of underground iron mine based on Pressure-State-Response model. Our application in Chengchao iron mine proves its efficiency and promising guide on land ecological security evaluation.

## 1. Introduction

Land ecological safety is the core of the sustainable land use and a new research field in environmental impact assessment, which is highly connected with human's survival and development [1]. Foreign studies mainly focus on ecological risk and ecosystem health while many domestic scholars aim at mechanism, warning and evaluation method of land ecological safety in special areas especially the ecological sensitive district [2]. They all largely enrich evaluation means and methods as well as enhance the awareness of land ecological security issues.

Steel is the most consumed metal in the world and Iron ore is the main raw materials in its production. With domestic open-pit mining facing an depletion, underground iron mining is becoming more and more popular, so it's high time we carry out study aims at land ecological safety issues of underground iron mine. They are typical ecological sensitive districts, researchers have obtained much achievements on some aspects: the vulnerability, stability and health assessment of ecosystem, the ecological civilization and the ecological risk assessment and etc [3-6]. But the deficiencies are as follows: (1) Different types of mine own unique resources, mining technology, characteristics and social background with relatively great difference in ecological safety issues. Current studies are mainly generalized evaluation or only suitable for coal mine. (2) The evaluation result is not comparable for its evaluation indexes and standards are limited to a special location thus restricting its application.

## 2. Land Ecological Safety Problems of Underground Iron Mine

### *2.1. Characteristic*

A fuzzy indication between risk source and damage. There are many kinds of risks during mining process jeopardizing ecological system via different ways, which results that one kind of problem derives from varied sources and one source accompanies more than one problems [7].

Concealment and hysteresis. The underground producing activity of mining has less obvious effect and damage on ecological environment compared with open-pit mining. In addition, risks such as surface subsidence resulted from roadway and empty area collapse belongs to the kind of accumulative and permanent damage, which will progress as time goes by.

Mutual interaction and feedback effect. On the one hand, The mining force is assumed to be source of ecological safety issues, whose ecological effects are subjected to local environment. On the other hand, the destructed geological environment will affect the choice of mining technique and process in underground iron ore, which may influence investment and social cost during production.

### *2.2. Associated Patterns*

Considering the mining process of underground iron mine, we analysis the emergence and development of ecological security problems, then put up with a causal association patterns between ecological safety, mining process, resource, environment and economical background (figure 1).

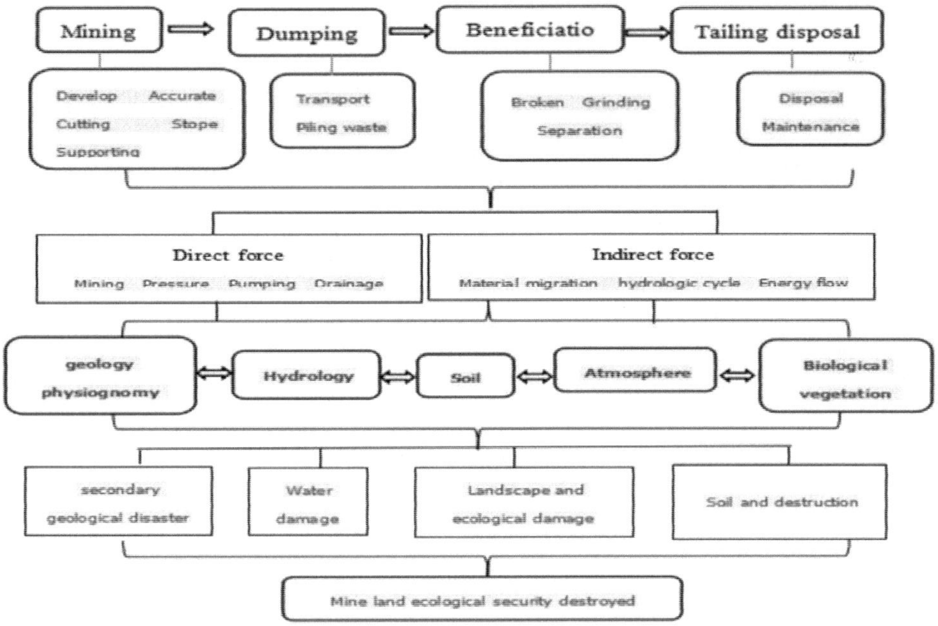


**Figure 1.** The causal association pattern based on mining process

## 3. Evaluation Process and the Method

### *3.1. Evaluation Index System*

The PSR model is a widely used model based on "pressure - state - response" index system and was established by the United Nations agency for economic cooperation development. Pressure index reflects the load caused by human activities to the system. State index shows natural resources and the environment quality condition of ecosystem. Response index characterizes human countermeasures on ecological safety issues. So PSR model has a clear causal relationship with systematicness [8]. So this paper presents a modified PSR model(table 1), considering the natural and human pressure rather than human pressure alone.

**Table 1.** The evaluation index system of the land ecological security

| First level index | Second level index | Third level index |
|---|---|---|
| Pressure(P) | Natural pressure($P_1$) | Geomorphological and geological conditions:$P_{11}$ Climate condition:$P_{12}$ Ore grade:$P_{13}$ Occurrence of ore body:$P_{14}$ |
| | Mining pressure($P_2$) | Mining intensity:$P_{21}$ Technology modernization:$P_{22}$ Condition of equipment:$P_{23}$ |
| | Economical pressure($P_3$) | Per capita output:$P_{31}$ Roe:$P_{32}$ Ton ore mining costs:$P_{33}$ Per capita income:$P_{34}$ |
| State(S) | State of geological hazard ($S_1$) | Disaster area ratio:$S_{11}$ Risk level of tailings:$S_{12}$ Comprehensive risk in mine-out area:$S_{13}$ Comprehensive risk of waste rock:$S_{14}$ |
| | State of Water Resources($S_2$) | Environmental quality of surface water:$S_{21}$ Environmental quality of underground water:$S_{22}$ Groundwater level:$S_{23}$ |
| | Landscape ecological state ($S_3$) | Ratio of soil erosion area:$S_{31}$ Biological diversity:$S_{32}$ Landscape fragmentation:$S_{33}$ |
| | Soil environmental state($S_4$) | Green coverage:$S_{41}$ Land reclamation rate:$S_{42}$ Soil environmental quality:$S_{43}$ |
| Response (R) | Management technology response($R_1$) | "Three wastes" emission compliance rate:$R_{11}$ Comprehensive utilization rate of waste rock:$R_{12}$ "Three simultaneous" implementation:$R_{13}$ |
| | Institutional policy response($R_2$) | Governance funds proportion:$R_{21}$ Science and technology investment:$R_{22}$ |

### *3.2. Evaluation Methods and Steps*

This paper employ AHP to determine the weight of each index to set up fuzzy comprehensive evaluation mathematics model so as to study this complex ecotope issues.

Weighing index: a judgment matrix is used to identify importance with other indexes in the same level, importance numbers are judged by a certain criterion comes from the standard of its dominant level.

$$(A\,ij)n \times n = \begin{bmatrix} \delta_{11} & \cdots & \delta_{1n} \\ \vdots & \ddots & \vdots \\ \delta_{n1} & \cdots & \delta_{nn} \end{bmatrix} \tag{1}$$

The weight of each index is obtained by the greatest characteristic root and characteristic vector of judgment matrix.

Evaluations: we grade the land ecological safety into four ranks: extremely low, low, medium and high and the evaluation set are built as V={extremely low, low, medium , high}.

Index classification standard: the quantifiable indexes are referred to national standards and criteria as table 2.

**Table 2.** The classification standards of quantitative indexes

| **Index** | | **Standard and reference** | | | | **Index** | | **Standard and reference** | | | |
|---|---|---|---|---|---|---|---|---|---|---|---|
| **Index** | Unit | $V_1$ 0-2.5 | $V_2$ 2.5-5 | $V_3$ 5-7.5 | $V_4$ 7.5-10 | **Index** | Unit | $V_1$ 0-2.5 | $V_2$ 2.5-5 | $V_3$ 5-7.5 | $V_4$ 7.5-10 |
| Ore grade | % | <25 | 25-35 | 35-45 | >45 | Biological diversity* | / | <0.5 | 0.5-1 | 1-1.5 | >1.5 |
| Mining intensity | 10t/a | >50 | 30-50 | 20-30 | <20 | Landscape fragmentation* | / | <0.3 | 0.3-0.5 | 0.5-0.7 | >0.7 |
| Per capita output | 10t | <15 | 15-25 | 25-30 | >30 | Green coverage | % | <30 | 30-45 | 45-60 | >60 |
| Roe | % | <0 | 0-6 | 6-15 | >15 | Land reclamation rate | % | <20 | 20-50 | 50-80 | >80 |
| Ton ore mining costs | yuan/t | >200 | 160-200 | 120-160 | <120 | Soil environmental quality* | / | >2 | 1-2 | 0.7-1 | <0.7 |
| Per capita income | 10t | <1.5 | 1.5-2.5 | 2.5-3.5 | >3.5 | "Three wastes" emission compliance rate* | % | <60 | 60-75 | 75-90 | >90 |
| Disaster area ratio | % | >12 | 8-12 | 4-8 | <4 | Comprehensive utilization rate of waste rock | % | <10 | 10-20 | 20-30 | >30 |
| Environmental quality of surface water* | / | V | IV | III | aboveII | "Three simultaneous" implementation | / | <60 | 60-75 | 75-90 | >90 |
| Environmental quality of underground water* | / | V | IV | III | aboveII | Governance funds proportion | % | <1 | 1-2 | 2-3 | >3 |
| Ratio of soil erosion area | % | >30 | 20-30 | 10-20 | <10 | Science and technology investment | % | <5 | 5-10 | 10-15 | >15 |

For other qualitative indexes, we establish a index classification basis(table 3)based on its major

IOP Conf. Series: Earth and Environmental Science **111** (2018) 012004 doi:10.1088/1755-1315/111/1/012004

influence and characteristics on ecological safety issues. So different descriptive definitions is used to provide reference for the expert scoring that are matched with evaluation set V={extremely low, low, medium , high}.

**Table 3.** The classification basis

| Index | Basis |
|---|---|
| Geomorphological geological and conditions | Degree of tectonic activities; Joint fracture development; Hydrogeological conditions; Geomorphic type; Elevation and cutting conditions |
| Climatic conditions | Annual rainfall; Intensity of heavy rain; Matching condition of light, warmth and water |
| Occurrence of ore body | Depth, thickness and dip angle of the orebodies; Continuity degree of the orebodies; Rock types and strata conditions of the roof rock |
| Technology modernization | Scientificity in development and utilization; Green method in mining; Environmental protection in ore dressing; Completeness in six system |
| Condition of equipment | Advanced level of production equipment; Exist of environmental protection equipment |
| Risk level of tailings | Land use; Engineering geological conditions; Condition of the dam; Damage if collapse |
| Comprehensive risk in mined - out area | Accumulation degree and scale; Geological condition of surrounding rock; Stability of its structure and support; Ecological damage |
| Comprehensive risk of waste rock | Land use; Terrain and formation characteristics; Physical properties of waste rock; Impacted objects and scope |
| Groundwater level | Development and burial depth of underground water; Water intake and water gushing in mining area; Influence of ground disaster on groundwater recharge channels |

Membership function: The membership function is the basis and a key point in a reasonable fuzzy evaluation. Quantifiable indexes in table 2 are divided into classification index and numerical index. Classification indexes are determined by the lowest category amid monitoring samples. As for numerical index, we apply a improved distribution function to a avoid subjectivity in evaluation. Membership function is as follows:

$$\mu(c)=\begin{cases}1-\dfrac{|c-(a_1+a_2)/2|}{a_2-a_1}\\ \dfrac{|c-(a_1+a_2)/2|}{a_2-a_1}\end{cases} \tag{2}$$

C represents real value of the quantitative index, $a_1$ is the lower threshold of grade interval where c belongs to and $a_2$ is the upper threshold. When there is no concrete upper threshold or lower one, we calculate the interval span to add or plus. For other qualitative indexes in table 3, we take a average of expert scoring as c to calculate.

**Table 4.** The standard value and range of score

| | $V_1$(extremely low) | | $V_2$(low) | | $V_3$(medium) | | $V_4$(high) | |
|---|---|---|---|---|---|---|---|---|
| | Standard values | Score interval | Standard values | Score interval | Standard values | Score interval | Standard values | Score interval |
| c | 1 | 0-2 | 2 | 1-3 | 3 | 2-4 | 4 | 3-5 |

Membership of c in table3 is derived from a trapezoidal distribution function.

$$\mu_{1c}=\begin{cases}c(0<c<1)\\1(c=1)\\-c+2(1<c<2)\end{cases}\quad \mu_{2c}=\begin{cases}c-1(1<c<2)\\1(c=2)\\-c+3(2<c<3)\end{cases}\quad \mu_{3c}=\begin{cases}c-2(2<c<3)\\1(c=3)\\-c+4(3<c<4)\end{cases}\quad \mu_{4c}=\begin{cases}c-3(3<c<4)\\1(c=4)\\-c+5(4<c<5)\end{cases} \tag{3}$$

Fuzzy comprehensive evaluation: we aim at getting a membership vector of four evaluations for land ecological safety in studying area. Calculation starts from lowest index level with function (2) and (3) to get the membership degree to the four evaluations. Then the fuzzy relation matrices of upper level index is built by calculated membership, we multiply the fuzzy relation matrices by its weight to get another fuzzy relation matrices of further upper level. In this way, we can have a final membership of four evaluations.

A fuzzy relation matrix :

$$R_{mn}=\begin{bmatrix}r_{11} & \cdots & r_{1n}\\ \vdots & \ddots & \vdots\\ r_{m1} & \cdots & r_{mn}\end{bmatrix} \tag{4}$$

M is index numbers in next level, n is 4 (number of evaluations), r is the membership value.

## 4. Empirical Test

Chengchao iron mine is a integrated mining founded in 1958 with mainly pillarless caving method whose production scale ranks third among iron mine around China, thus being an important source of Wuhan Iron and Steel. The evaluation data mainly come from geological topographic map of Chengchao iron ore, geological report of Chengchao iron ore, statistical report of Chengchao iron ore, environmental quality report of Ezhou, remote sensing image of mining area and field survey data. Here are the results:

The general score of land ecological security in Chengchao Iron Mine is (0.158, 0.385, 0.354, 0.103), mainly lies in the range between "low" and "medium", whose final result is lower in land ecological security.

Score in stress is (0.126, 0.302, 0.425, 0.147), close to "medium", which due to the complex geological conditions, strong tectonic movement, high mining intensity and caving mining that brings about great ecological security pressure. Meanwhile, the advanced mining equipment with lower cost favorable income of workers reduce the mine land ecological security pressure in turn.

Score in state is (0.276, 0.412, 0.229, 0.083), belongs to "low" and had a tendency to "extremely low", which is related to the frequent occurrence of geological disaster, massive large mine-out areas with poor support, serious soil pollution, low biodiversity, and landscape fragmentation in Chengchao iron mine. Although the greening coverage of mining area is high, soil and water loss is managed to a certain extent, but it is still not enough to alter the "low" state in land ecological security.

Score in response is (0.033, 0.432, 0.450, 0.085), lies in the range between "low" and "medium" and prior to "medium", which is connected with the economic benefits, increased governance funds, science and technical investment.

The whole evaluation results reflect the actual situation and main contradiction of land ecological security in Chengchao iron mine, which point the direction in management and regulation for mine land ecological safety issues.

## 5. Discussions

This paper aims at underground iron mine and based on the analysis of potential causal connections between safety and mining activities, resources, environment and economic background, attempting to establish a universal evaluation index system and method of land ecological safety for underground mines. We apply it in Chengchao iron ore, then achieve a good result.

Land ecological safety is a relative concept consists of not only the relativity of region (space), but also the relativity of development stage (time). Therefore, the evaluation needs reflecting the relative difference of land ecological security among the same type of mines to make comparisons, and provide the basis for further classification management of a certain type mine in the country or region. Besides, the evaluation on pixel or plaque level based on the special local and spatial heterogeneity of land ecological security reserves further research.

## 6. References

[1] Yu J, Fang Li and Cang D B 2012 Evaluation of land eco-security in Wanjiang district base on entropy weight and matter element mode. *Transactions of the Chinese Society of Agricultural Engineering.* **28(5)** 260—266

[2] Chu P P and Fu M C 2014 Research progress on ecological security and evaluation of regional land in China. *Chinese Agricultural Science Bulletin.* **30(11)** 160-164

[3] Xu J X, Li G and Chen G L 2013 Dynamic changes and evaluation of land ecological quality in coal mining area. *Journal of China Coal Society.* **38(S1)** 180-5

[4] Quan Z J, Li Y and Li J S 2013 Ecological vulnerability of coal mining area:A case study of Shengli Coalfield in Xilingguole of Inner Mongolia,China. *Chinese Journal of Applied Ecology.* **24(6)** 1729-38

[5] Cheng J H, Chen J and Li J S 2013 Research on evaluation indicators system of ecological civilization in mining area. *China Popolation Resources and Environment.* **30(2)** 1-10

[6] Chang Q,Qiu Y and Xie M M 2012 Theory and method of ecological risk assessment for mining areas based on the land destruction. *Acta Ecologica Sinica.* **32(16)** 5164-74

[7] Liu X Y, Chen Y and Lv L2012 Anti-disturbance capacity evaluation of metal mining geological environment: a case study on High-phosphorus iron area in western Hubei. *Metal Mine.***(11)** 142-5

[8] Li Z C, Liu L D and Sun Y F2010 Assessment method of regional ecological security based on Pressure-State-Response model. *Acta Ecologica Sinica.* **30(23)** 6495-503

# Fuzzy Comprehensive Evaluation of Ecological Risk Based on Cloud Model: Taking Chengchao Iron Mine as Example

**Jinghua Ruan[1], Yong Chen* [1,2], Xiao Xiao[1], Gan Yong[1], Ranran Huang[1] and Zuohua Miao[1,2]**

[1] School of Resources and Environmental Engineering, Wuhan University of Science and Technology, Wuhan, 430081, China
[2] Hubei Key Laboratory for Efficient Utilization and Agglomeration of Metallurgic Mineral Resources, Wuhan, 430081, China
E-mail: yongchen1968@163.com

**Abstract.** Aimed at the fuzziness and randomness during the evaluation process, this paper constructed a fuzzy comprehensive evaluation method based on cloud model. The evaluation index system was established based on the inherent risk, present level and control situation, which had been proved to be able to convey the main contradictions of ecological risk in mine on the macro level, and be advantageous for comparison among mines. The comment sets and membership functions improved by cloud model could reflect the uniformity of ambiguity and randomness effectively. In addition, the concept of fuzzy entropy was introduced to further characterize the fuzziness of assessments results and the complexities of ecological problems in target mine. A practical example in Chengchao Iron Mine evidenced that, the assessments results can reflect actual situations appropriately and provide a new theoretic guidance for comprehensive ecological risk evaluation of underground iron mine.

## 1. Introduction

The mining area belongs to the typical ecological fragile area, the ecological and environmental problems of which have aroused the concern of researchers who have made a series of studies in environmental quality, the stability and vulnerability of ecosystem, ecological civilization and ecological security [1]. In recent years, the ecological risk assessments of mining area have been transformed from special to comprehensive [2], resulting in appearing various research perspectives such as landscape [3], land destruction [4], ecological gradient [5] and ecological carrying capacity [6]. However, the ecological risk assessment of mining area still faces many problems, which are mainly manifested in the representation of uncertainty in evaluation process. The formation mechanism of mining ecological environment is complex, which leads to the ambiguity and randomness in the process of ecological risk assessment. In order to solve the problem, a cloud model based on fuzzy comprehensive evaluation is proposed to solve fuzziness and randomness during the evaluation.

## 2. The Evaluation Index System

The comprehensive evaluation of ecological risk is introduced to reflect the differences of the ecological risk degree among the same types of mine. Therefore, the evaluation index system is supposed to represent the main contradiction of the production technology, technical characteristics and the ecological environment.

At present, the ecological risk evaluation generally follows the analysis framework of "risk source — risk receptor — exposure path — ecological endpoint", which is more effective for special ecological risk assessment of single risk source. It is inappropriate to adopt traditional method when comprehensive evaluation of ecological risk involves multiple sources, multiple receptors and multiple ecological endpoints. So, this paper built an index system from three aspects of inherent risk, present level and control situation to reflect the comprehensive risk degree of mines, to serve on the government decision-making management. The selected indexes are shown in table 1.

**Table 1.** The evaluation index system of the land ecological risk

| First-class index | Second-class index | Third-class index | |
|---|---|---|---|
| Inherent risk (D) | Natural occurrence (D1) | Geology and geomorphology D11 | Climate and hydrology D12 |
| | | Occurrence of ore D13 | |
| | Mining activities (D2) | Mining intensity D21 | Technological modernization D22 |
| | | Advanced level of equipment D23 | |
| Present level (R) | Geological disaster (R1) | Damage R11 | Tailing risk R12 |
| | | Goaf risk R13 | Waste dump risk R14 |
| | Water resources (R2) | Surface water R21 | Ground water R22 |
| | Landscape ecology (R3) | Biodiversity R31 | Landscape fragmentation R32 |
| | Soil environment (R4) | Soil erosion R41 | Soil pollution R43 |
| Control situation (C) | Technical means (C1) | "Three wastes" emission C11 | Waste stone utilization C12 |
| | | Monitoring and forecasting facilities C13 | |
| | Institutional policy (C2) | Land reclamation C21 | Investment in science and technology C22 |

## 3. The Evaluation Method

As mentioned above, there are various factors influencing the ecological environment, whose mechanism and magnitude in the evolution of the ecological environment have not yet been fully revealed. Therefore, it is inappropriate to describe them as definite states. In this paper, a fuzzy comprehensive evaluation model based on cloud model is established to improve the traditional fuzzy comprehensive evaluation model which ignores the randomness.

### *3.1. Fuzzy Comprehensive Evaluation Model*

Fuzzy mathematics based on the "fuzzy set" theory is a kind of quantification method for dealing with uncertain and imprecise problems. It is applied in many fields such as fuzzy control, fuzzy recognition, fuzzy decision, information retrieval and fuzzy evaluation. And fuzzy comprehensive evaluation model is the most commonly used to evaluate fuzzy problems for better evaluation results. For the three-level evaluation index system involved in this study, the multilevel fuzzy comprehensive evaluation model can be used to evaluate, the general expression is:

$$P=\begin{pmatrix} & v_1 & v_2 & \cdots & v_j \\ u_1 & p_{11} & p_{12} & \cdots & p_{1j} \\ u_2 & p_{21} & p_{22} & \cdots & p_{2j} \\ \vdots & \vdots & \vdots & \vdots & \vdots \\ u_i & p_{i1} & p_{i2} & \cdots & p_{ij} \end{pmatrix}, \qquad Y=\bigoplus_{i=1}^{m}\omega_i \otimes p_{ij} \tag{1}$$

In which {u$_1$, u$_2$, ...u$_i$} and {v$_1$, v$_2$, ...v$_i$} are the evaluation factors set and comment set, respectively, P is relation matrix, $p_{ij}$ is the membership of participation factors u$_i$ in comment v$_j$. Y is evaluation result. The symbols $\oplus$ and $\otimes$ are fuzzy operators, which stands for cyclic addition and cyclic multiplication, respectively.

Among them, constructing a reasonable membership function is the key to fuzzy evaluation. But the membership used in most of the existing methods is given by subjective value or empirical formula. Once the membership function is determined, the fuzzy problem is replaced by the precise mathematics, which can not express the randomness.

### *3.2. Cloud Model Improvement Process*

Cloud model firstly proposed by Academician Li D Y [7] is an uncertainty conversion model between qualitative concepts and quantitative expression, which has been proved to be universally applicable.

The cloud model includes three digital features: expected value ( Ex ), entropy ( En ) and hyper entropy ( He ). Among them, Ex is the most representative point of a concept in the theoretical domain; En is used to measure the ambiguity and probability of the qualitative concept; He expresses the uncertainty of En, its value can be described by the dispersion and thickness of cloud.

The process of fuzzy comprehensive evaluation of ecological risk based on cloud model includes:

3.2.1. *Determining the Weights of the Evaluation Indexes.* Analytic hierarchy process (AHP) is a hierarchical and systematic analysis method; it can solve the complicated fuzzy problems effectively and conveniently. Therefore, AHP was introduced to determine the weights of evaluation indexes in this paper.

3.2.2. *Setting Comment Sets and Benchmark Cloud.* In this paper, ecological risk of underground iron mine was divided into four levels: high risk, medium risk, low risk and slight risk, whose evaluation scores are set to [0, 2.5], [2.5, 5], [5, 7.5) and [7.5, 10] correspondingly. And the normal cloud model is introduced to describe the risk status of these four levels respectively to form benchmark cloud.

The parameters of the benchmark cloud are determined by the formulas under bilateral constraints.

$$\begin{cases} E_x=(B_{\max}+B_{\min})/2 \\ E_n=(B_{\max}-B_{\min})/6 \\ H_e=k \end{cases} \tag{2}$$

In the above equations $B_{max}$ and $B_{min}$ are the upper and lower limits of the quantitative interval respectively, k is a constant, adjusted according to the fuzzy threshold of the variable itself. Since k equals 0, the model degenerates into the membership function. Referencing to the relevant research, k equals to 1 in this paper.

3.2.3. *Generating the Cloud Model*. This paper set up scoring basis and created descriptions firstly. Then, experts were invited and informed of scoring rules that scores from 0-10 indicate that the degree of risk is getting lower and lower.

According to the principle of inverse cloud generator, taking each score as cloud droplet generates membership clouds of 4 risk grade, then combined as the membership cloud model of each evaluation index. The parameters are calculated as follows:

$$\begin{cases} E_x = \bar{x} = \sum_{j=1}^{n} x_j \\ E_n = \sqrt{\frac{\pi}{2}} \times \frac{1}{n} \times \sum_{j=1}^{n} \left| x_j - E_x \right| \\ H_e = \sqrt{S^2 - E_n^{\ 2}} = \sqrt{\frac{1}{n-1} \times \sum_{j=1}^{n} (x_j - \bar{x}) - E_n^{\ 2}} \end{cases} \tag{3}$$

In equation ( 3 ) $x_j$ is the score assigned by the j-th expert to the description. Using the three characteristics ( Ex, En, He ) of cloud, we can express the randomness between the evaluation index and the membership of risk grade. When the membership model is established, the more the number of experts are, the higher the accuracy of the model will have.

3.2.4. *Comprehensive Evaluation of Ecological Risk.* The experts graded each index according to the scoring basis. Substituting the experts' scores into the membership cloud model can obtain $Ce_i$ which represents the membership of evaluation index to each risk level respectively. Eventually the cloud model can be obtained by assigning the membership cloud a weight under each risk level. The weighted synthesis method of cloud model is as follows [8]:

$$(E_x, E_n, H_e) = (E_{x_1}, E_{n_1}, H_{e_1}) \oplus (E_{x_2}, E_{n_2}, H_{e_2}) \oplus \cdots \oplus (E_{x_m}, E_{n_m}, H_{e_m}) \tag{4}$$

$$\begin{cases} E_x = \sum_{i=1}^{m} E_{x_i} \times \omega_i \\ E_n = \sqrt{\sum_{i=1}^{m} E_{n_i}^{\ 2} \times \omega_i} \\ H_e = \sum_{i=1}^{m} H_{e_i} \times \omega_i \end{cases} \tag{5}$$

In which $\omega_i$ is the weight, *i* is the number of evaluation indexes, ⊕ represents the weighted composition of the cloud.

*3.2.5. Characterization of Fuzzy Degree of Evaluation Results.* It is difficult for single grade evaluation results to reflect the details of the ecological risk degree of underground iron mine because of the hierarchical distribution inconsistency of each evaluation index. So the De Luca-Trimini definition of fuzzy entropy is used to characterize the ambiguity of the evaluation result of the cloud model.

$$E = -k_2 \sum_{j=1}^{n} [C_{e_i} \ln C_{e_i} + (1 - C_{e_i}) \ln(1 - C_{e_i})] \tag{6}$$

$$k_2 = -\ln(\frac{1}{n}) - (n-1)\ln\frac{n-1}{n} \tag{7}$$

In the above equations $E$ is fuzzy entropy, characterizing the fuzzy degree of evaluation results; $Ce_i$ is i-th element of the vector $Ce$ for cloud model's certainty under different risk grade, setting ln0 = 0; and $k_2$ is the standard coefficient. The relations are [0, 0.25], [0.25, 0.5], [0.5, 0.75], [0.75, 1] corresponding to "significant", "less significant", "less fuzzy" and "fuzzy" respectively.

## 4. Instance Verification and Result Analysis

### *4.1. Overview of the Study Area*

Located in Ezhou City, Hubei Province, Chengchao Iron Ore has the most abundant resources and the largest mining scale of underground iron in Central South Region of China. The main geological structure is affected by the two tectonic systems, the Huaiyang epsilon structural system and the Neocathaysian structural system. Faults and fractures are extremely developed. Marble is the main aquife with zonal distribution. There is no bulk surface water, and ground water is replenished by the infiltration of precipitation. The mining area is located in the subtropical monsoon climate zone, whose annual average temperature and rainfall are 17.2 degrees and 1306mm respectively.

### *4.2. Evaluation Results and Analysis*

Referring to the geological report, statistical report, environmental quality monitoring report and relevant studies in Chengchao Iron Ore, ten experts are invited to score each index according to comment sets and scoring rules. Then, according to the fuzzy comprehensive evaluation method based on cloud model, the ecological risk assessments are carried out under the Matlab 7.0 environment. The results are shown as figure 1.

The ecological risk degree in Chengchao Iron Ore is between "medium risk" and "low risk". It can be further determined as "medium risk" by the expected value of evaluation cloud which equals 0.476. Substituting the scores into the membership cloud model can obtain the membership's certainty of evaluation index to each risk level. By the formulas ( 6 ), ( 7 ), the fuzzy entropy of evaluation results is 0.481, which points to "less significant".

The ecological risk in Chengchao Iron Ore can be expressed as [medium risk, less significant]. In order to further examine the reliability of evaluation results, we carried out the fuzzy comprehensive evaluation of the 20 experts scoring results and the membership of four comments determined by using the trapezoidal distribution function. The evaluation results are ( 0.158, 0.385, 0.354, 0.103 ), which points to “medium risk”, the same as the cloud model.

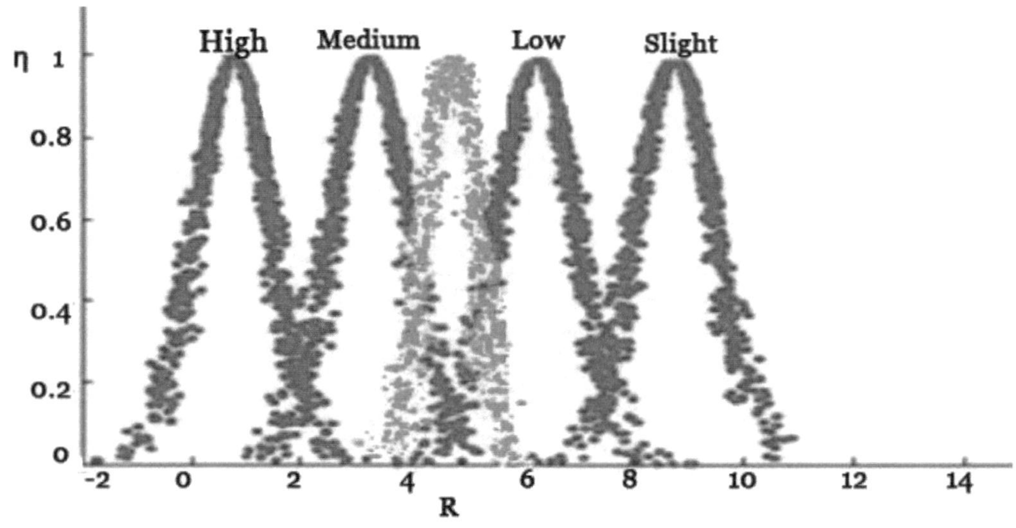


**Figure 1** The cloud chart of risk rank evaluation results

## 5. Conclusions

The establishment of comprehensive evaluation index system based on the inherent risk, present level and control situation is able to convey the main contradictions of ecological risk in mine and it is also beneficial to the comparison between different mines.

The empirical study of Chengchao iron ore shows that the fuzzy comprehensive evaluation model which is improved by cloud can reflect the actual situation of ecological risk better.

The introduction of cloud model can express the fuzziness and randomness during the evaluation process, but whether it will bring more uncertainty is still needed further researches.

## 6. Reference

[1] Hong Q, Chen Y, Liu Y Z, Zeng X Y and Gan Y 2015 Establishment and application of matter -element evaluation model of land ecological security in Chengchao iron mine *Industrial Safety and Environmental Protection* **41 (11)** 4-8

[2] Pan Y J, Wang Y L, Peng J and Han Y N 2012 Research progress in ecological risk assessment of mining area *Acta Ecol. Sin.* **32 (20)** 6566-74

[3] Wu J S , Qiao N , Peng J, Huang X L, Liu J Z and Pan Y J 2013 Spatial variation of landscape eco-risk in open mine area *Acta Ecol. Sin.* **33 (12)** 3816-24

[4] Chang Q, Qiu Y, Xie M M and Peng J 2012 Theory and method of ecological risk assessment for mining areas based on the land destruction Acta Ecol. Sin. 32 (16) 5164-74.

[5] Zhang S F, Zhang L and Zhang Y H 2012 Ecological risk assessment of the Yulin coal mining area based on Petar method *Resources Science* **33 (10)** 1914-23

[6] Jia Y and Cao L X 2011 Review on ecological risk assessment method for coal-mining area *Environ. Sci. A. Manage.* **36 (4)** 177-182

[7] Liu D F, Wang D, Ding H and Wang L C 2014 Eutrophication assessment by entropy-cloud model *J. Hydraul. Eng.***45 (10)** 1214-22

[8] Wei B, Wang J D, Zhang H W and Yu D K 2014 Service trust evaluation method based on weighted multiple attribute cloud *Journal of Computer Applications* **34 (3)** 678-682, 689

# The Evaluation of Land Ecological Safety of Chengchao Iron Mine Based on PSR and MEM

**Xiangdong Jin[1] and Yong Chen* [1, 2]**

[1]School of Resources and Environmental Engineering,Wuhan University of Science and Technology, Wuhan, 430081,China;
[2]Hubei Key Laboratory for Efficient Utilization and Agglomeration of Metallurgic Mineral Resources, Wuhan, 430081, China;
E-mail:yongchen1968@163.com

**Abstract**. Land ecological security is of vital importance to local security and sustainable development of mining activities. The study has analyzed the potential causal chains between the land ecological security of Iron Mine mining environment, mine resource and the social-economic background. On the base of Pressure-State-Response model, the paper set up a matter element evaluation model of land ecological security, and applies it in Chengchao iron mine. The evaluation result proves to be effective in land ecological evaluation.

## 1. Introduction

Land ecological security is vital to human's daily life and development and is the core of sustainable use of land resource. Mine area is the typical ecological fragile zone, whose ecological environment and land ecological security problem has attracted the attention of researchers. The evaluation of land ecological quality of mine[1-2],ecosystem stability [3],vulnerability[4] and health assessment[5],mine ecological civilization [6] and ecological risk assessment [7] made a series of research results. However, the existing research objects are mostly aimed at coal mines, lacking evaluation of underground iron mines. Due to different types of mine resources, mining technology, technical characteristics and social economic background are different. The main ecological environmental problems varies [8]. It demands distinctions between different types of mine to carry out targeted research. Chengchao Iron Mine was founded in 1958, it is a typical integrated underground mines. The mine has the most abundant mineral resources and owns largest production in South Central China. It has made a great contribution to the rapid development of local economy. But at the same time, long-term iron ore mining also brings geological disasters, environmental pollution, ecological destruction and a series of land ecological environment problems, so it is necessary to carry out land ecological security research.

## 2. Basic Understanding and Evaluation Index System of Land Ecological Security in Chengchao Iron Mine

### *2.1. Basic Understanding of Land Ecological Security in Chengchao Iron Mine*

The source power of the land ecological security of ChengChao iron mine is the mining production activities, including mining, row rock, ore dressing and tailings processing four main production procedures. The mining exploration, mining quasi, cutting, ventilation in the mining process, and the transportation, waste rock stacking in rock dumping, and the crushing, grinding, separation in mineral processing, these links will have an impact on the ecological environment. They will act directly or indirectly on the atmosphere, soil, hydrology, rocks, landforms, vegetation and other elements of land ecological environment. As time goes on, these factors will eventually led to the secondary geological disaster, destruction of water resources, landscape and ecological destruction, soil occupation and destruction and other ecological environment effects. In this process, mine resources and environment system, social economic system play an extremely important role. What's more, the main form and degree of land destruction of the ecological environment have an extremely close relationship with the occurrence of ore bodies, mining scale, mining method and process, geological environment conditions and their sensitivity, social economic background conditions. On the other hand, mining production activities, the background of mine resources and environment, socioeconomic background of the three together determine the land ecological security situation and trend of Chengchao Iron mine.

### *2.2. The Evaluation Index System of Land Ecological Safety of Chengchao Iron Mine Based on PSR*

We uses PSR model to construct the evaluation index system of land ecological security in Chengchao Iron Mine. The contents are shown in figure 1.

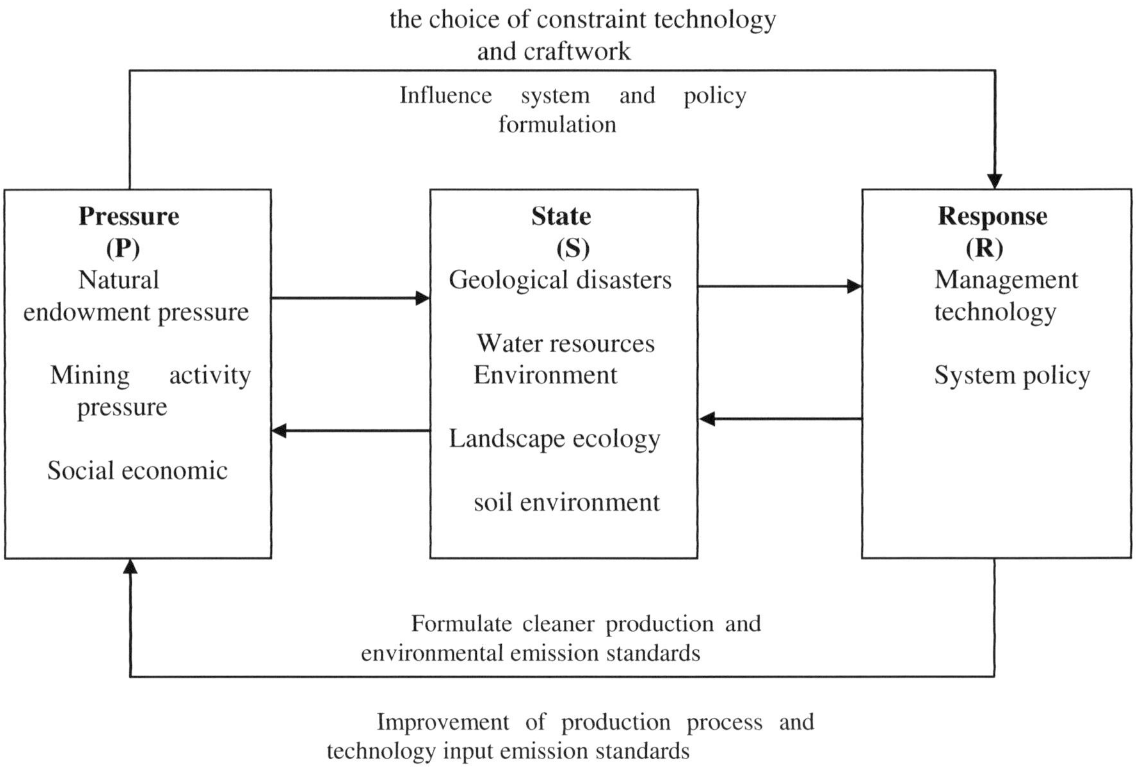


**Figure.1** The associated patterns between P-S-R of Chenchao Iron Mine

When the specific index is selected, aiming at the actual situation of the main problems in Chengchao Iron Mine and land ecological security. The index weight is obtained by expert consultation and analytic hierarchy process (AHP). The index weight of project layer and index layer is in Table 1, and the index weight of Index layer is in Table 3.

**Table 1.**The evaluation index system of the land ecological security

| project | Factor | Index |
|---|---|---|
| Pressure (0.429) | Natural endowment pressure(0.184) | Geomorphic conditions $P_{11}$, Climatic conditions $P_{12}$, Ore grade $P_{13}$, Occurrence of ore bodies $P_{14}$ |
| | Mining activity pressure(0.184) | Mining intensity[1] $P_{21}$, Degree of technological modernization $P_{22}$, Equipment advanced degree $P_{23}$ |
| | Social economic(0.061) | per-capita output value $P_{31}$, Return on net assets $P_{32}$, The cost of mining tons of ore $P_{33}$, Income per worker $P_{34}$ |
| state (0.429) | Geological disasters(0.223) | Disaster area ratio $S_{11}$, Risk of tailings Banks $S_{12}$, Comprehensive risk of goaf $S_{13}$, Waste rock field comprehensive risk $S_{14}$ |
| | Water resources environment | Surface water quality [2]$S_{21}$, Groundwater environment quality [3]$S_{22}$, The influence of groundwater level $S_{23}$ |
| | Landscape ecology(0.034) | Soil erosion area ratio $S_{31}$, Biodiversity index [4]$S_{32}$, Landscape fragmentation index [5]$S_{33}$ |
| | soil environment(0.086) | green coverage ratio $S_{41}$, Land reclamation rate $S_{42}$, Soil environmental quality index [6]$S_{43}$ |
| response (0.143) | Management technology(0.071) | "Three wastes" emission index rate $R_{11}$, Waste rock comprehensive utilization ratio $R_{12}$, Implementation of "three simultaneous" $R_{13}$ |
| | System policy(0.071) | The proportion of management funds to sales revenue $R_{21}$, Science and technology investment accounts for profit ratio $R_{22}$ |

Note: 1. The mining intensity is expressed in tons of ore mined each year; 2. Index of surface water environmental quality $I_W=(\sum_{i=1}^{n} f(x_i))/n$, referring to the literature [9] and [10] for details; 3. Groundwater environmental quality index $F=\sqrt{((\sum_{i=1}^{n} F_i)/n)^2+F_{max}^2)/2}$, referring to the literature [8] for details; 4. Shanon-Wiener index formula is used to calculate; 5. Landscape fragmentation index Ft=(Nt-1)/Nc. Nt is total number of landscape patches, Nc is the ratio of the total area to the minimum patch area was evaluated; 6. The soil environmental quality index calculated by Nemerow index method, referring to the literature [10] for details; 7. Discharge standard rate of "three wastes" $P=\sqrt[3]{C1*C2*C3}$,C1,C2,C3 are the discharge rate of wastewater, waste gas and waste residue are standard respectively.

## 3. The Evaluation Model of Land Ecological Safety in Chengchao Iron Mine

The evaluation of mine land ecological security involves many indexes, and the results of single index evaluation are incompatibility. Matter-element model combine objects, features and their eigenvalues to study the possibility and development law of things in a formalized model. This can be used to solve complex problems that are incompatible. Therefore, this research establishes the matter-element model of the land ecological safety in chengchao iron mine, and the evaluation process is as follows.

### *3.1. Construct the Land Ecological Security Matter-Element*

According to the evaluation index system which is determined by Table 1, the evaluation object M(Land ecological security), feature vector c(the evaluation index) and characteristics of the value x(Parameter values) can build the land ecological security matter-element R=(M, c, x). The land ecological security element of chengchao iron mine is expressed as:

$$R=(M,c,x)=\begin{bmatrix} M & c_1 & x_1 \\ & c_2 & x_2 \\ & \vdots & \vdots \\ & c_n & x_n \end{bmatrix} \quad (1)$$

### *3.2. Determine the Classical Domain and the Matrix of Joint Domain*

The classical domain matrix can be expressed as:

$$R_{oj}=(M_{oj},C_{oji},X_{oji})=\begin{bmatrix} M_{oj} & c_{oj1} & x_{oj1} \\ & c_{oj2} & x_{oj2} \\ & \vdots & \vdots \\ & c_{ojn} & x_{ojn} \end{bmatrix}=\begin{bmatrix} M_{oj} & c_{oj1} & (a_{oj1},b_{ojl}) \\ & c_{oj2} & (a_{oj2},b_{oj2}) \\ & \vdots & \vdots \\ & c_{ojn} & (a_{ojn},b_{ojn}) \end{bmatrix} \quad (2)$$

In the formula: $R_{oj}$ is --- Classical domain element; $M_{oj}$ is ---The evaluation level of $j^{th}$ rank. This evaluation ranks four grades. They are relative safe, not so safe, unsafe, dangerous; $c_{oji}$ is ---The $i^{th}$ index of land ecological security at the j level. In this study $i \in (1,2,\cdots 29)$; $x_{oji}=(a_{oji},b_{oji})$ ---means evaluation index $c_{oij}$ is the range of magnitude in the rank of j which is the classical domain matrix.

Node matter element matrix can be expressed as:

$$R_p=(M_p,C_{pi},X_{pi})=\begin{bmatrix} M_p & c_{pl} & x_{pl} \\ & c_{p2} & x_{p2} \\ & \vdots & \vdots \\ & c_{pn} & x_{ojn} \end{bmatrix}=\begin{bmatrix} M_p & c_{pl} & (a_{pl},b_{pl}) \\ & c_{p2} & (a_{p2},b_{p2}) \\ & \vdots & \vdots \\ & c_{pn} & (a_{pn},b_{pn}) \end{bmatrix} \quad (3)$$

In the formula: $R_p$ is ---node matter element; $M_p$ is ---land ecological security of the object to be evaluated; $c_{pi}$ is the $i^{th}$ evaluation index of Land ecological security; $x_{pi}=(a_{pi},b_{pi})$ is evaluation range of rank $c_{pi}$ on all levels which is the corresponding value range of land ecological security evaluation index.

### *3.3. Determine the Actual Value of Each Index of Chengchao Iron Mine and Establish an Evaluation Element*

Obtaining the actual value of each evaluation index of Chengchao iron mine and establishing an evaluation element $R_o$:

$$R_o=(M_o,C_{oi},X_{oi})=\begin{bmatrix} M_o & c_{o1} & x_{o1} \\ & c_{o2} & x_{o2} \\ & \vdots & \vdots \\ & c_{on} & x_{on} \end{bmatrix} \quad (4)$$

For quantitative indicators, the main data is from the geological topographic map of Chengchao iron mine, the geological report of Chengchao iron mine, the statistics report of Chengchao iron mine, the remote sensing image and field survey data in mining area.For the indicators that cannot be quantified, we can establish index classification table(Table 2) from the index that can reflect the main features of the land ecological security status or have the leading role of land ecological safety aspects. We can also establish index classification table(Table 2) from the index which have a relationship with land ecological safety of underground iron mine. From the index classification table(Table 2),we can set up the description language of relative safe(75-100 points),not so safe(50-75 points),unsafe(25-50 points) and dangerous(0-25 points),these four levels for expert scoring reference, and take the arithmetic mean of experts scoring as the index value.

**Table 2.** The classification basis of the qualitative indexes quantitative indexes

| Index | Classification basis |
|---|---|
| Geological and geomorphic conditions | Degree of tectogenesis; Fracture joint development degree; Hydrogeological condition; Geomorphic types; Elevation and cutting conditions |
| Climatic conditions | Annual rainfall; Storm intensity; Matching condition of light and warm water |
| Occurrence of ore bodies | Buried depth of ore body; Thickness and dip angle; Continuity of ore body; Roof rock types and formation conditions |
| Degree of technological modernization | Scientific development and utilization plan; Green situation of mining metods; Environmental protection degree of mineral processing technology; Completeness of six systems |
| Equipment advanced degree | Advanced level of production equipment; Popularization of environmental protection equipment |
| Tailings reservoir risk | Land occupation scale; Engineering geological condition; Dam condition; Ecological loss size of dam break |
| Comprehensive risk of mined out area | Degree and scale of aggregation; Geological conditions of surrounding rock; The stability of its structure and support; Accident ecological loss size |
| Comprehensive risk of waste rock yard | floor area; Topography and stratigraphic characteristics; Physical properties of waste rock; Influence object and scope |
| Influence of groundwater level | Groundwater development and burial depth; Mining water intake and roadway water inflow; Effects of land disasters on groundwater recharge channels |

### *3.4. Calculation of Correlation Degree and Land Ecological Security Grade Determination*

*3.4.1. Determine the Correlation Function of Index Layer.* Provided $\mathrm{k_j}(x_{oi})$ is the correlation degree of rank j in index layer i, we define $\mathrm{k_j}(x_{oi})$ as follows:

$$\mathrm{k_j}(x_{oi})=\begin{cases}\dfrac{\rho(X_{oi},X_{oji})}{\rho(X_{oi},X_{pi})-\rho(X_{oi},X_{oji})} & ,if\rho(X_{oi},X_{pi})-\rho(X_{oi},X_{oji})\neq 0;\\ -\rho(X_{oi},X_{oji})-1 & ,if\rho(X_{oi},X_{pi})-\rho(X_{oi},X_{oji})=0;\end{cases} \tag{5}$$

In the formula:

$$\rho(X_{oi},\mathrm{X_{oji}})=\left|x_{oi}-\frac{1}{2}(\mathrm{a_{oji}}+b_{oji})\right|-\frac{1}{2}(\mathrm{b_{oji}}-\mathrm{a_{0ji}}) \tag{6}$$

$$\rho(X_{oi},\mathrm{X_{pi}})=\left|x_{oi}-\frac{1}{2}(\mathrm{a_{pi}}+b_{pi})\right|-\frac{1}{2}(\mathrm{b_{pi}}-\mathrm{a_{pi}}) \tag{7}$$

Among them, $\rho(X_{oi},\mathrm{X_{oji}})$ and $\rho(X_{oi},\mathrm{X_{pi}})$ are respectively the distance from index $\mathrm{x_{oi}}$ to the classical field area $\mathrm{x_{oji}}=(\mathrm{a_{oji}},\mathrm{b_{oji}})$ and Joint domain range $\mathrm{x_{pi}}=(\mathrm{a_{pi}},\mathrm{b_{pi}})$.

*3.4.2. Computing Comprehensive Association Degree Layer by Layer.* According to the hierarchy of evaluation indicators determined by Table 1, the comprehensive correlation degree of each level index is calculated layer by layer and start from the bottom one.

Calculation formula:

$$K_j(X_{ol}) = \sum_{i=1}^{m} \{\omega_i * k_j(x_{oli})\}, (j = 1,2,3,4) \tag{8}$$

In the formula: $K_j(X_{ol})$ is comprehensive correlation of index $X_{ol}$ of rank; $k_j(x_{oli})$ is single correlation degree of rank j in every lower index among $X_{ol}$; m is the index numbers of $X_{ol}$; $\omega_i$ is weight of index $x_{oli}$.

We calculate layer by layer, until get the comprehensive correlation grade $K_j(M_o)$ of $M_o$ (land ecological security ) in rank j.

*3.4.3. Grade Assessment and Development Trend Judgment.* Provided $K_j(X_{ol}) = \max\{K_1(X_{ol}), K_2(X_{ol}), K_3(X_{ol}), K_4(X_{ol})\}$, the index $X_{ol}$ belongs to the standard level of land ecological security, referred as j. Provided $K_j(M_o) = \max\{K_1(M_o), K_2(M_o), K_3(M_o), K_4(M_o)\}$ ,then object to be evaluated here in $M_o$ of standard level in land ecological security is J. When $K_j(x) \geq 1.0$, it indicates the evaluated object $M_o$ exceeds the limit, the bigger the number is , the greater potential it has. When $0 \leq K_j(x) < 1$, it indicates the evaluated object is in accordance with J standard, and the bigger the value is, the better the match degree is , which is more close to the upper limit of ranks; When $-1 \leq K_j(x) < 0$ it indicates the evaluated object is not in rank J, but it chance to be transformed into rank j ,and the bigger the number is ,the more chance it has. When $K_j(x) < -1.0$ it indicates the the evaluated object is not in rank J, and it has no chance to be transformed into rank j .

## 4. Evaluation Results and Analysis of the Land Ecological Safety of Chengchao Iron Mine

### *4.1. Data Sources*

Matter-element model of land ecological security evaluation whose classical domain and joint field are refer to "Cleaner production standard(Iron ore mining)","Surface water environmental quality standard"," Groundwater environmental quality standard"," Soil environmental quality standard"," Evaluation conditions of state level green mine pilot units by Ministry of land and resources"," Standards for ecological protection and restoration of mining(Draft for Soliciting Opinions)" and other national standards and norms, and "China Mining Yearbook (2013)" and other statistical data and related research literature. The actual value of Chengchao Iron Mine evaluation index is determined by the above methods described in this paper(Table 3).

**Table 3.** The evaluation index's actual value and weight of Chengchao Iron Mine

| **Index** | $P_{11}$ | $P_{12}$ | $P_{13}$ | $P_{14}$ | $P_{21}$ | $P_{22}$ | $P_{23}$ | $P_{31}$ | $P_{32}$ | $P_{33}$ |
|---|---|---|---|---|---|---|---|---|---|---|
| **Index** | 48 | 78 | 42 | 62 | 63 | 62 | 266 | 27 | 12 | 146 |
| **Weight** | 0.095 | 0.014 | 0.037 | 0.037 | 0.019 | 0.116 | 0.048 | 0.003 | 0.007 | 0.034 |
| **Index** | $P_{34}$ | $S_{11}$ | $S_{12}$ | $S_{13}$ | $S_{14}$ | $S_{21}$ | $S_{22}$ | $S_{23}$ | $S_{31}$ | $S_{32}$ |
| **Index** | 3.4 | 16 | 56 | 38 | 63 | 0.68 | 1.94 | 63 | 12 | 1.1 |
| **Weight** | 0.016 | 0.087 | 0.034 | 0.087 | 0.015 | 0.052 | 0.017 | 0.017 | 0.021 | 0.008 |
| **Index** | $S_{33}$ | $S_{41}$ | $S_{42}$ | $S_{43}$ | $R_{11}$ | $R_{12}$ | $R_{13}$ | $R_{21}$ | $R_{22}$ | |
| **Index** | 0.33 | 72 | 23 | 2.2 | 85 | 37 | 88 | 1.6 | 12 | |
| **Weight** | 0.005 | 0.008 | 0.021 | 0.058 | 0.043 | 0.014 | 0.014 | 0.018 | 0.054 | |

### *4.2. Evaluation Results and Analysis*

According to formula 5-8, the comprehensive correlation degree of each level index is calculated(Table 4). Evaluation results show as follows:

**Table 4.**The results of comprehensive evaluation on ecological security of land in Chengchao Iron Mine

| All levels of indicators | | Comprehensive correlation | | | | Security level |
|---|---|---|---|---|---|---|
| | | K1 | K2 | K3 | K4 | |
| pressure | Natural endowment pressure | -0.045 | 0.016 | -0.021 | -0.074 | not so safe |
| | Mining activity pressure | -0.038 | 0.043 | -0.058 | -0.079 | not so safe |
| | Social economic pressure | 0.057 | 0.014 | 0.030 | -3.599 | relative safe |
| | Pressure synthesis | -0.012 | 0.012 | -0.013 | -0.249 | not so safe |
| state | Geological disaster state | 0.041 | 0.033 | 0.050 | -3.074 | unsafe |
| | State of water resources | -0.008 | 0.030 | -0.002 | -0.145 | not so safe |
| | Landscape ecological status | 0.002 | 0.003 | 0.012 | -1.036 | not so safe |
| | Soil environmental state | -0.184 | 0.109 | 0.068 | 0.025 | not so safe |
| | State synthesis | -0.007 | 0.020 | 0.017 | -0.730 | not so safe |
| response | Management technology response | -0.009 | 0.022 | -0.021 | -0.043 | not so safe |
| | Institutional policy response | -0.015 | 0.012 | -0.009 | -0.025 | not so safe |
| | Response synthesis | -0.002 | 0.002 | -0.002 | -0.005 | not so safe |
| Ecological safety of mine land | | -0.009 | 0.014 | 0.002 | -0.420 | not so safe |

The general situation of land ecological security in chengchao iron mine is(-0.009,0.014,0.002,-0.420), Belongs to "not so safe", but have "unsafe" features, and also shows the greater potential turn to "relative safe".

The general situation of the land ecological safety pressure(P) in Chengchao Iron Mine is (-0.012,0.012,-0.013,-0.249),belongs to "relative safe". The geological conditions in Chengchao iron mine is very complicated, the degree of tectonic movement here is stronger than any other iron ore. What's more, many factories use caving method to mining. These factors put great pressure on the ecological safety of mine land. But on the other hand, the mining equipments in these factories are very advanced. Ore mining costs are low. The economic benefits of these factories are good, and the workers' income in these factories are good. All these factors to some extent alleviate the pressure of mine land ecological security. However, with the depth of iron ore mining increases, and iron ore mining costs increase, especially the lower price of iron ore market. The pressure of mine land ecological security is increasing.

The general situation of the land ecological security state(S) of chengchao iron mine is (-0.007, 0.020, 0.017,-0.730), belongs to "not so safe", but have "unsafe" features, and also shows the greater potential turn to "unsafe". Chengchao iron mine has been mining for years. Ecological environment pollution and landscape ecological destruction have not been timely managed, especially a large number of goaf which is made by caving method to mining that used by the factories here. Tailings, waste rock and other hidden dangers make mine geologic hazard happen frequently, so the Geologic hazard state here belongs to "unsafe".

(4)The general situation of the land ecological security response(R) of chengchao iron mine is (-0.002, 0.002, -0.002, -0.005).belongs to "not so safe". It has relationship with the mine economic benefits, governance funds, science and technology capital investment increased and gradually improve the orresponding management system in recent years. However, the relevant policy measures need to be further implemented, a large number of comprehensive utilization of waste rock remains to be resolved.

## 5. Conclusion

Because of the different types of mine resource endowment, mining technology, technical characteristics and socio-economic background, the ecological environment problems which we faced have much difference. This paper focuses on the potential causal chain of the land ecological security and mining activities, mine resource environment and socioeconomic background of chengchao iron mine, by establishing the evaluation model of ecological security of mine land which is based on PSR, and good application results have been obtained.

We have reviewed a large number of relevant specifications, standards, statistics and previous research results in this research to determine the threshold value of the classical domain of matter element model. However, by the relevant information is incomplete, there are still some threshold value with the subjective, and need to be further improved and perfected.

## 6. Reference

[1] MA Cong-an, WANG Qi-rui. ECOLOGICAL Evaluation Model for Large Opencast Mine and Its Application[J], Journal of Mining and Safety Engineering. 2006,23(4):446-451.

[2] XU Jia-xing, LI Gang, CHEN Guo-liang. DYNAMIC Changes and Evaluation of Land Ecological Quality in Coal Mining Area[J], Journal of China Coal Society. 2013,38(S1):180-185.

[3] WANG Guang-cheng YAN Xu-qian. APPLICATION of Fractal Theory in the Ecosystem Stability Assessment in Mining area[J], Journal of China CoalL Society. 2008,33(4):427-430.

[4] QUAN Zhan-jun, LI Yuan, LI Jun-sheng. ECOLOGICAL Vulnerability of Coal Mining Area: A Case Study of Shengli Coalfield in Xilinguole of Inner Mongolia, China[J], Chinese Journal of Applied Ecology. 2013,24(6):1729-1738.

[5] YAN Xu-qian, Zhang Shun-tang, YUAN Huai-yu. THEORY and Method of Ecosystem Health Assessment in a Mining Area[J], Journal of University of Science and Technology BeiJing. 2005,27(3):351-355.

[6] CHENG Jin-hua, CHEN Jun, YI Xing-hua. RESEARCH on Evaluation Indicators System of Ecological Civilization in Mining Area[J], China Population Resources and Environment.2013,23(2):1-10..

[7] CHANG Qing, QIU Yao, XIE Miao-miao. THEORY and Method of Ecological Risk Assessment for Mining Areas Based on the Land Destruction[J], Acta Ecologica Sinica. 2012,32(16):5164-5174.

[8] WU Qiang. STUDTY of Classification of Geologic Environmental Problems in Mines in China[J], Hydrogeology and Engineering Geology. 2003,(5):107-112.

[9] WEI Bin, YANG Xiao-sheng, WU Ming. RESEARCH Review on Assessment Methodology of Ecological Security[J], Journal of Hunan Agricultural University (Natiural Sciencee). 2009,35(5):572-579.

[10] LI Zhong-cai, LIU Linde, SUN Yu-feng. ASSESSMENT Method of Regional Ecological Security Based on Pressure-State-Response Model[J], Acta Ecologica Sinica. 2010,30(23):6495-6503.

# Development of a Drilling Fluid Drive Downhole Tractor in Oil Field

**Delei Fang, Jianzhong Shang*, Yiying Liu, Wei Wu and Zirong Luo**

Department of Mechanical Engineering, National University of Defense Technology, Changsha 410000, China
E-mail: jz_shang_nudt@163.com

**Abstract**. This paper proposes a drilling fluid drive downhole tractor, which has the advantages of compact structure, large traction, fast speed and high reliability. The overall mechanical structure of the tractor is introduced, the concrete structures including supporting structure and cushion mechanism are designed. And its all-hydraulic drive continuous propulsion principle is analyzed. Finally the simulation analysis of the tractor operation is carried out to prove that the traction motion scheme is feasible.

## 1. Introduction

In the development of oil field, the technology of continuous tubing is widely used, and the operations underground of continuous tubing must be assisted by downhole tractor. Downhole tractor is the underground tool that can provide traction at the bottom of the well, which operates in the complex and narrow environment with high temperature and high pressure. Therefore, downhole tractor must have a compact structure to adapt to the complex environment and provide enough traction to meet larger load requirements [1-2].

There are two kinds of downhole tractors according to its structure and working principle. One is the wheel tractor, which can realize the motion by driving wheel which contacts with oil pipe wall. The main research companies are Welltec and Sondex[3-4]. The other type is the telescopic tractor, which can realize the motion by the two centralizers alternately supporting to the wall. One holding arms can be held on the pipe wall, while the other moves in opposite direction to realize motion of the tractor. Products of this type in SmarTract and Omega are famous[5].

However, the existing tractors have the following disadvantages. In the aspect of structure, the wheel tractor is relatively mature. But the driving force is small and the tractor is easier to skid in the inner wall of the oil pipe. And when it comes to the irregular inner wall, the tractor stability is poor. In the drive mode, the current tractors are most electric drive mode, which driven by the motor drive screw or the gear. Although the design is simplified, the structure of the tractor is complicated and the mechanical efficiency is low. Most hydraulic drive tractors are controlled by electric. All electrical

IOP Conf. Series: Earth and Environmental Science **111** (2018) 012007 doi:10.1088/1755-1315/111/1/012007

components and hydraulic components are difficult to work in high temperature and high pressure for a long time, greatly reducing the tractor reliability in the underground[6-7].

Therefore, in order to solve the shortage of existing tractor, this paper proposes a drilling fluid drive downhole tractor, which is driven from drilling fluid, not electrically driven. Firstly the paper introduces the overall mechanical structure, supporting structure and cushion mechanism. Then its all-hydraulic drive continuous propulsion mechanism is analyzed. Finally the simulation analysis of the tractor operation is carried out to prove that the traction motion scheme is feasible.

## 2. Structure Design of Downhole Tractor

### *2.1. Overall Mechanical Structure*

The overall structure of downhole tractor is shown in figure 1, which consists of three parts: the front supporting arm, hydraulic control unit and the rear supporting arm.

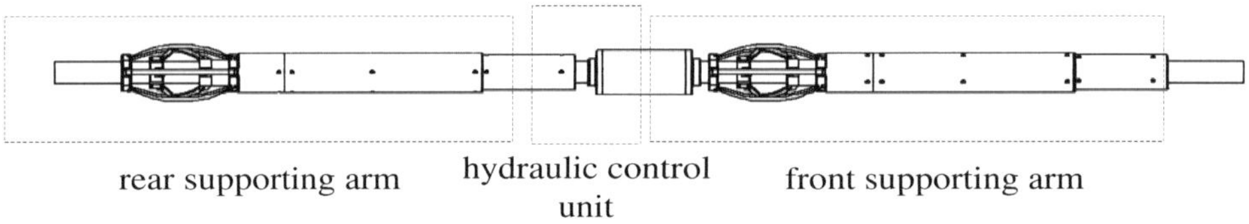


**Figure 1**. Overall structure model of downhole tractor

The specific structure of the supporting arm is shown in figure 2. The main composition contains flexible supporting unit, supporting cylinder, walking cylinder and cushion cylinder.

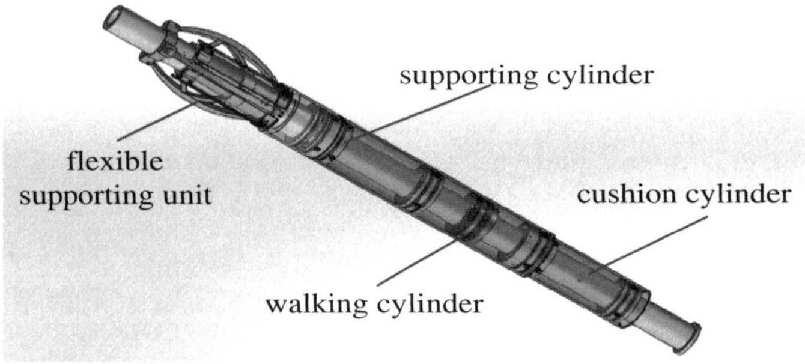


**Figure 2.** Structure model of supporting arm

### *2.2. Design of Supporting Mechanism*

The supporting arm is the key module to ensure that the tractor can advance normally. The supporting arm enables the supporting claw to hold the well wall, ensuring that the part of the mechanism is fixed, and the other sections of tractor move relatively.

Figure 3 shows the three-dimensional model diagram of supporting arm. Where, the major components include, supporting cylinder 12, walking cylinder 13, cushion cylinder 14, center pipe 4, supporting piston121, walking piston 131, cage 111, sliding block112, flexible supporting bar1131, fixed block 1132, scroll wheel 1133, supporting bar A1135, supporting bar B 1135.

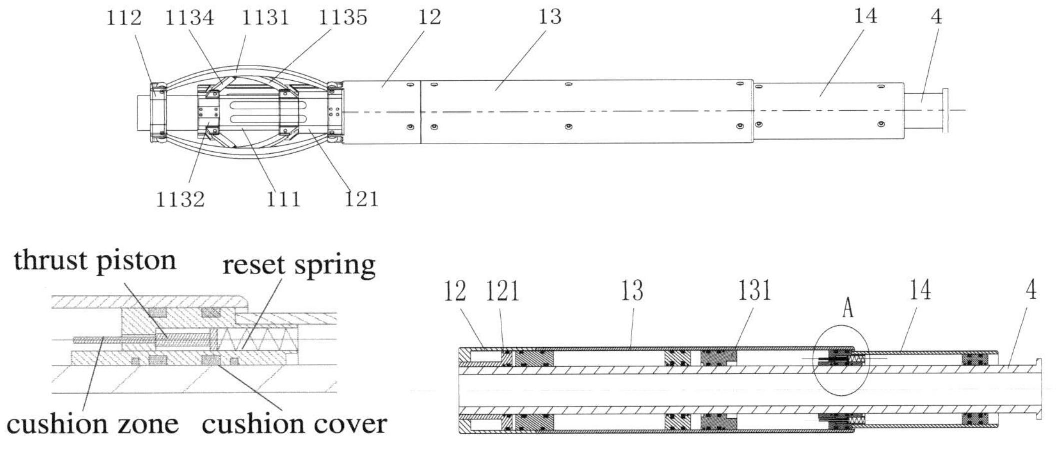


**Figure 3** Three-dimensional model diagram of supporting mechanism.

The cushion mechanism is the key module to ensure that the tractor can realize changing direction. As shown at point A, the left side of the cushion mechanism is a walking piston with a slow punch. In the middle, the cylinder head is cushioned, and the thrust piston and compression spring are arranged, as well as the relevant sealing elements.

### *2.3. Mechanical Analysis*

The force diagram of supporting mechanism is shown in figure 4.

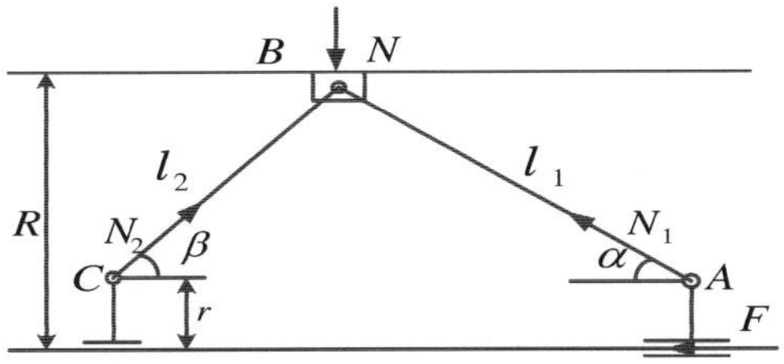


**Figure 4.** force diagram of the supporting mechanism

*F* is the traction of the tractor, *N* is the pressure on the supporting arm forced by the wall. The coordinate system is set at point *C*, according to the theory of virtual work,

$$F \cdot \delta x_A = N \cdot \delta y_B \tag{1}$$

Based on the coordinate relation,

$$\begin{cases} x_A = l_1 \cdot \cos\alpha + l_2 \cdot \cos\beta \\ y_B = l_2 \cdot \sin\beta = l_1 \cdot \sin\alpha \end{cases} \tag{2}$$

The relationship 2 is differentiable,

$$\begin{cases} \delta x_A = -l_1 \cdot \sin\alpha \cdot \delta\alpha - l_2 \cdot \sin\beta \cdot \delta\beta \\ \delta y_B = l_2 \cdot \cos\beta \cdot \delta\beta = l_1 \cdot \cos\alpha \cdot \delta\alpha \end{cases} \tag{3}$$

Also,

$$\delta\alpha \cdot \sqrt{l_1^2 - l_2^2 \cdot \sin^2\beta} = \delta\beta \cdot l_2 \cdot \cos\beta \tag{4}$$

The formula (4) is substituted into equation (5),

$$\delta x_A = -\frac{l_2 \cdot l_2 \cdot \cos\beta \sin\beta\delta\beta}{\sqrt{l_1^2 - l_2^2 \cdot \sin^2\beta}} - l_2 \cdot \sin\beta \cdot \delta\beta \tag{5}$$

Finally, the relationship between $F$ and $N$ is

$$F = -\frac{-N \cdot \cos\beta}{\left(1 + \frac{l_2 \cdot \cos\beta}{\sqrt{l_1^2 - l_2^2 \cdot \sin^2\beta}}\right) \cdot \sin\beta} \tag{6}$$

The relationship between traction and positive pressure can be studied by relational expression (6), which can lay a foundation for the establishment of downhole traction performance index.

## 3. Motion Principle of Hydraulic Drive Tractor

### *3.1. Motion Principle*

The schematic diagram of drilling fluid drive telescopic tractor is shown in figure 5. The basic structure of the supporting arm and the two working conditions (supporting arm extension and supporting arm contraction) .

In this paper, the driver of the tractor is aided by drilling fluid. When the high pressure filtered fluid flows into the supporting circuit, the piston drives the supporting grip to lock the well wall. The high pressure fluid flows into left cavity of walking cylinder, pushing the drilling centre pipe forward. Low pressure fluid returns from right cavity of walking cylinder, and the fluid crossing the hydraulic control unit returns to the outer layer.

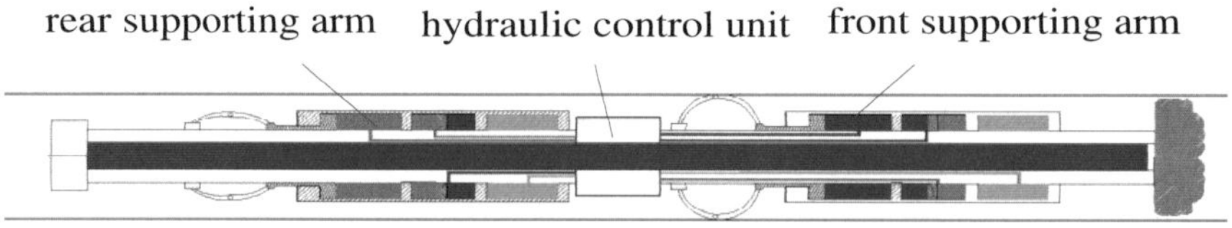


**Figure 5.** Schematic diagram of telescopic tractor

When the supporting cylinder fills with low pressure fluid, the supporting grip retreats. The fluid of walking cylinder left cavity is low pressure fluid, and flowing across hydraulic control unit returns to the outer layer. The right cavity of the walking cylinder is filled the oil, pushing the support arm move forward relative to the drilling centre pipe.

### *3.2. Operation Process*

The tractor can move forward continuously through the contraction and expansion of the front and the rear supporting arms, as shown in figure 6 (the tractor moves to the right).

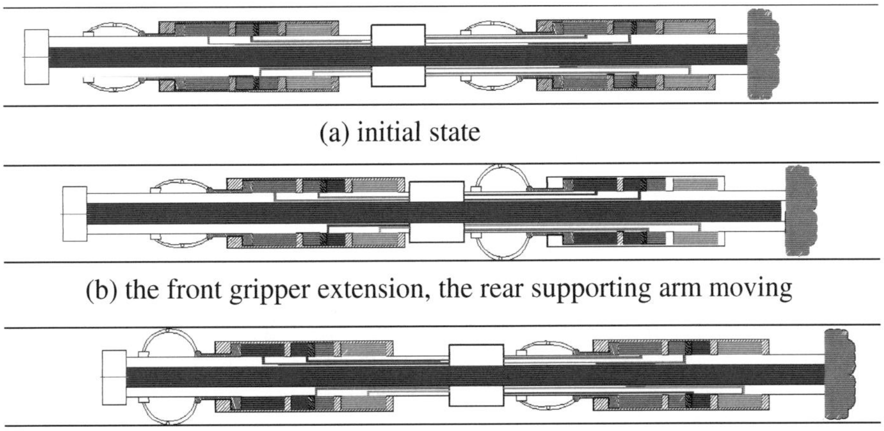

(a) initial state

(b) the front gripper extension, the rear supporting arm moving

(c) the rear gripper extension, the front supporting arm moving

**Figure 6.** Schematic diagram of continuous propulsion

Figure 6 (a) is the initial state of tractor. Each hydraulic cylinder chamber has no pressure fluid and each supporting arm is in a contraction state.

When the high pressure filtered fluid flows into the supporting circuit. Operation 1 is shown in Figure 6 (b), after the tractor is started, the front supporting cylinder is pressed into high pressure fluid (red liquid path), and the front supporting clam holds the inner wall. The left cavity of the front walking cylinder is pressured high pressure fluid (red liquid path), and the drilling centre tube is moving forward. The rear supports cylinder relieves, and the rear supporting cylinder press the low pressure fluid (blue liquid path). The right cavity of the rear walking cylinder is pressured high pressure fluid (red liquid path), the rear support arm is moving forward relative to the drill pipe. The current moving cylinder piston moves to the right side, touching the cushion tank (yellow liquid path), which leads to movement of reversing valve, making fluid circuit change.

Operation 2 is shown in Figure 6 (c). The action of the reversing valve causes the fluid circuit to change. The rear supporting cylinder is pressed into high pressure fluid (red liquid path), and the rear supporting clam holds the inner wall. The left cavity of the rear walking cylinder is pressured high pressure fluid (red liquid path), and the drilling centre tube is moving forward. The front supports cylinder relieves, and the front supporting cylinder press the low pressure fluid (blue liquid path). The right cavity of the front walking cylinder is pressured high pressure fluid (red liquid path), the front support arm is moving forward relative to the drill pipe. The current rear moving cylinder piston moves to the right side, touching the cushion tank (yellow liquid path), which leads to movement of reversing valve, making fluid circuit change. Such periodic cycle operation can realize the continuous advance of downhole tractor

## 4. Simulation of Operation Process

Based on the motion principle, the tractor operation process is simulated by software. The simulation parameters are set as follows: the work pressure is 14 *MPa*, and the input flow is 45 L/min. because of the modular design of the walking cylinder and the supporting cylinder, they have the same size and structure, and the cylinder stroke is 150 mm. The time of simulation is set 20s, and some curves are shown in below figures.

Because the front and the rear cylinders are the same, the rear walking cylinder is chosen as the analysis object. As shown in figure 7, the dynamic characteristic in the rear cylinder is got.

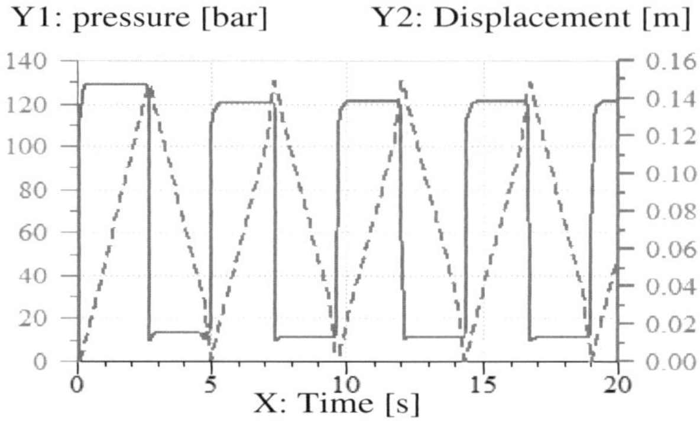


**Figure 7.** Displacements of walking cylinders

In the figure 7, the red curve represents the piston displacement of the cylinder. The green line represents the pressure change of the left cavity in walking cylinder. It can be seen that the changes of displacement and pressure are both periodic. In the initial stage, the pressure fluctuates, and after one cycle the pressure tends to be stable. It can be seen from the slope of displacement curve, the walking speed is not stable in the whole process, which may be influenced by the buffering action.

As shown in figure 8, the horizontal coordinate is the time, and the vertical coordinate represents the displacement curves of the walking cylinders piston. In the figure, you can see that both the movements of the cylinder pistons are periodic, and the motion cycle is about 4.67 s. because of the hydraulic cylinder stroke is 0.15 m, so the walking speed is calculated indirectly of 0.03 m/s. The two piston displacements characteristics are the same quantity and opposite direction, which is due to motion principle of the downhole tractor.

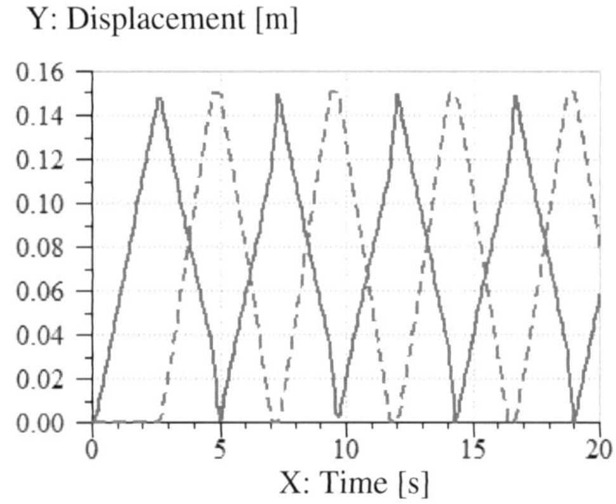


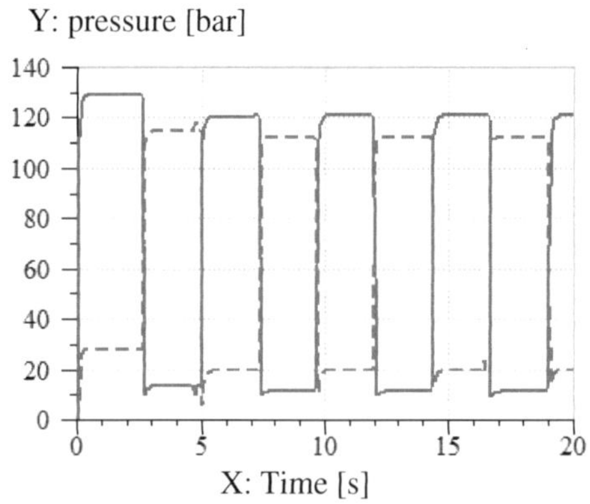


**Figure 8.** Displacements of walking cylinders **Figure 9.** Pressures of walking cylinders

Figure 9 shows the left cavity pressure curves of two walking hydraulic cylinders. As shown in the figure, the left cavity pressures of the two cylinder changes alternately. In the initial state, pressure in the rear cylinder fluctuates, and then operates smoothly later. We can see that the pressure in the rear walking cylinder is a little larger than the pressure in the front walking cylinder, which may be caused by the different resistance in thrust and drag.

## 5. Conclusion

In order to solve the shortage of the existing tractor, this paper proposes a drilling fluid drive downhole tractor. The main advantage of the tractor is that the whole process is driven by drilling fluid, which increases the reliability of the complex environment. The overall mechanical structure was introduced firstly in the paper, and the concrete structure was designed, including supporting structure and cushion mechanism. Subsequently its all-hydraulic drive continuous propulsion operation was analyzed. Finally the simulation analysis of the tractor operation was carried out, which can prove that the traction motion scheme is feasible.

## 6. Acknowledgments

This paper was supported by the National Natural Science Foundation of China (51675527).

## 7. References

[1] Stolte C, Wu C, Carroll D, et al. Coiled tubing and well test experiences on sichuan horizontal shale gas wells[C]// International Petroleum Technology Conference, Beijing, China, March 26–28, 2013: 1–5.

[2] Siddiqui D, Iqbal F, Naseem M, et al. Successful integration of innovative technologies enabled operator to drill and complete first horizontal well in depleted shale inter-bedded limestone reservoir in Pakistan-a milestone achieved[C]// Abu Dhabi International Petroleum Exhibition and Conference, Society of Petroleum Engineers, 2014: 1–10.

[3] S. Roh, H. R. Choi. Differential-Drive In-pipe Robot for Moving Inside Urban Gas Pipelines [J]. IEEE Transactions On Robotics, 2005, 21(1): 1~17.

[4] Billingham M, Etoukhy A M, Hashem M K, et al. Conveyance—down and out in the oil field[J]. Oilfield Review, 2011, 23(2): 18–31.

[5] Tractor[EB/OL]. Aberdeen. Omega Completion Technology Company, 2012[2016-05-08]. http://www.versa-line.com/files/tractor.pdf

[6] Liu Qingyou, Ren Tao, Chen Yonghua. Characteristic analysis of a novel in-pipe driving robot[J]. Mechatronics, 2013, 23(4): 419 - 428.

[7] Logging while Tractoring[EB/OL]. Houston, GE Oil and Gas Company, 2016[2016-05-08]. https://www.geoilandgas.com/oilfield/wireline-technology/logging-while-tractoring.

IOP Conf. Series: Earth and Environmental Science **111** (2018) 012008 doi:10.1088/1755-1315/111/1/012008

# Experimental Research of a New Wave Energy Conversion Device

**Zhongyue Lu[1], Jianzhong Shang[1], Zirong Luo[1], Chongfei Sun[1] and Gewei Chen[1]**

1 National University of Defense Technology, Changsha, China
E-mail: jz_shang_nudt@163.com

**Abstract.** With the increasing tension of contemporary social energy, the development and utilization of renewable energy has become an important development direction. As an important part of renewable energy, wave energy has the characteristics of green environmental protection and abundant reserves, attracting more investment and research. For small marine equipment energy supply problem, this paper puts forward a micro wave energy conversion device as the basic of heaving motion of waves in the ocean. This paper designed a new type of power output device can solve the micro wave energy conversion problem.

## 1. Introduction

With the deepening of the marine exploration, functionality and performance of marine equipment put forward higher requirements[1, 2], such as detection of more powerful function, more intelligent control level, more reliable, more accurate navigation algorithm working state and longer working hours. However, these functions and performance can't be achieved without a premise: good battery life[3]. As an important part of marine energy, wave energy has the characteristics of large reserves, green environment and renewable energy. It is estimated that the annual amount of waves in the world's oceans can be in the range of 0.1 x 1010 to 1.010kW, with an average of 10~50kW per meter[4, 5]. If these renewable sources of energy absorption, transformation and storage, and provide energy for marine equipment, this will greatly improve the battery life and independent survival ability, realize long-endurance of offshore platform and a wide range of work[6]. This is of great significance to improve human's exploration and exploitation of marine resources and environment[7]. As the energy conversion method owns its pros and cons. To be specific, the pneumatic has high reliability but low conversion efficiency[8]; the hydrodynamic device has stable energy conversion efficiency, but the wave size will influence the absorbing performance; and the direct electrical device got the high conversion efficiency though, the complex mechanical transmission could be moderator variable[9].In this paper, a new type of micro-wave energy conversion device is studied by model description, analysis, and experimental research is also carried out.

## 2. Model Analysis

The energy absorber is mainly composed of rotating blade group, stabilizing wing, center shaft and blade baffle, as shown in figure 1. The blade group is mounted on the outer center wheel, and the upper and lower blade groups are connected through the central shaft. As the device moves up and down with the floating float, the water will impact the blades, as shown in figure 2. The upper and lower blades are installed in the opposite direction, when the impact is hitting, the upper and lower blade groups are subjected to the opposite lateral thrust. It generates a relative rotation and drives the connected generator to spin the electricity. The blade baffle is used to adjust the maximum angle of the blade group, so as to observe the effect of the angle of the blade on the absorption effect of the wave. Four stabilizer center symmetrically installed around the center shaft, the relative rotation with the upper and lower blade groups is also possible. When the device is in the water, the upper and lower blade relative rotating, the stabilizer remain relatively static in the water due to resistance function of the environment, so that can make the device not serious deviation, so as to improve the efficiency of energy conversion.

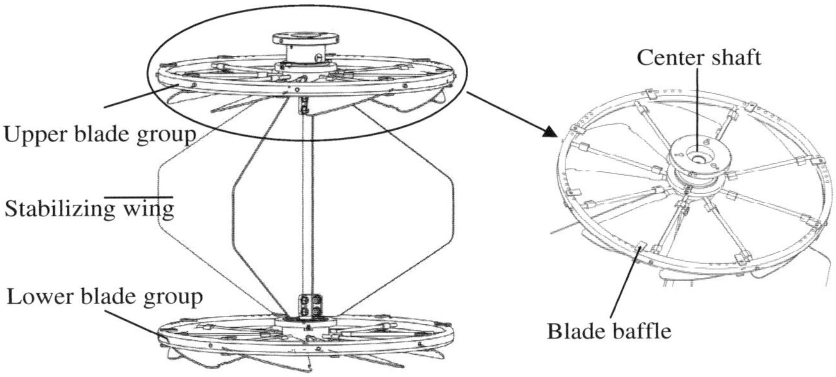


**Figure 1.** Structure of the wave energy convertor

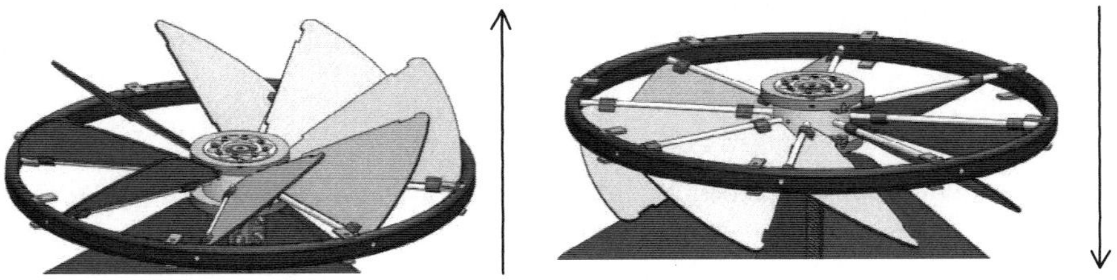

**Figure 2.** The impact of flow on blades

The model adopts the following assumptions: no friction, homogeneous, incompressible, steady state flow. And to calculate the maximum power of the device, assume wave-structure interaction of absorber (A) is equal to absorber (B), so the thrust on absorber (A) is also equal to absorber (B). Absorber (A) and absorber (B) have same design parameters. The parameters on absorber (A) are shown in figure 10, so the formula can be described as:

$$\begin{aligned} \omega_1 &= \omega_2 = \omega \\ J_1 &= J_2 = J \end{aligned} \tag{1}$$

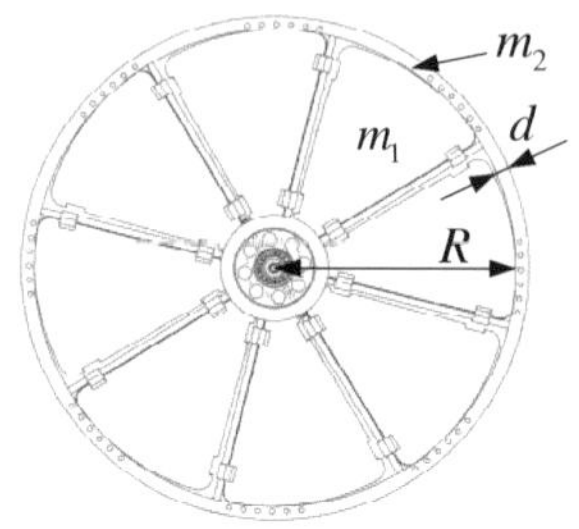


**Figure 3.** The parameters on one blade group

As shown in figure 11(a), T is the thrust on one blade, F is the thrust in circumferential direction.

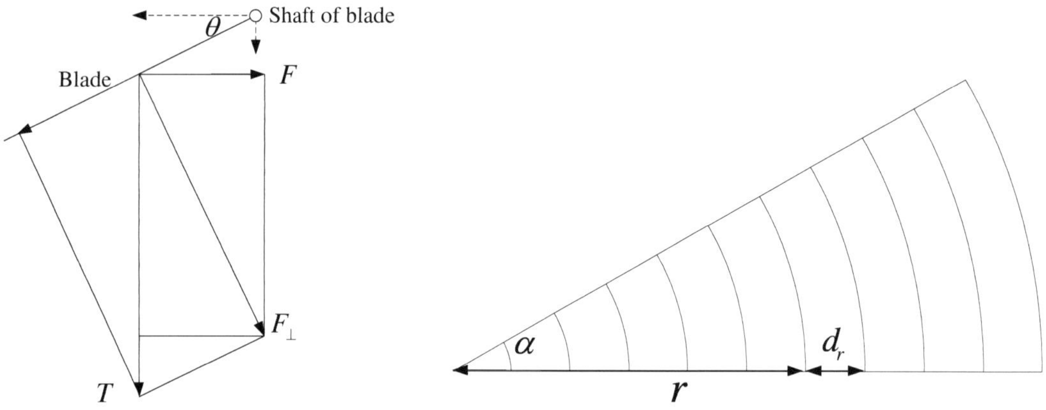


(a) Force analysis from the side of any blade (b) Finite element analysis of any blade

**Figure 4.** Analysis of any blade

$$\begin{aligned} F_{\perp} &= T\cos\theta \\ F &= F_{\perp}\sin\theta \end{aligned} \quad (2)$$

Then the circumferential thrust on unit area is calculated as:

$$f = \frac{F}{\pi R^2 Z\cos\theta} \quad (3)$$

As the analysis shown in figure 11(b), the circumferential thrust on any blade is calculated as:

$$dT_F = \frac{F}{\pi R^2 Z\cos\theta}\alpha \cdot rdr \quad (4)$$

Then the circumferential torque is calculated as:

$$dM_1 = \frac{F}{\pi R^2 Z\cos\theta}\alpha g r^2 dr \quad (5)$$

$$M = \frac{F}{\pi R^2 \cos\theta}\alpha\int_0^R r^2 dr = \frac{2\alpha R \cdot T \cdot \sin\theta}{3\pi} = 0.823N \cdot m \quad (6)$$

## 3. Results and Discussion

Analysis based on relevant hydraulic theory indicates that, dominant factors that influence efficiency of power generation include blade angle, shape and size.

**Figure 5.** The structure of the platform

Build experiment platform under the laboratory environment as shown in figure 5. To verify the influence of maximum blade angle to output voltage value, set the amplitude is 150mm, the period is 2s, and set the electrical load is $10\Omega$. The test chose maximum blade angle ($\theta$, as shown in figure ) to be $\pm 25°$, $\pm 20°$, $\pm 15°$ and $\pm 10°$ respectively. The results are shown in figure 6.

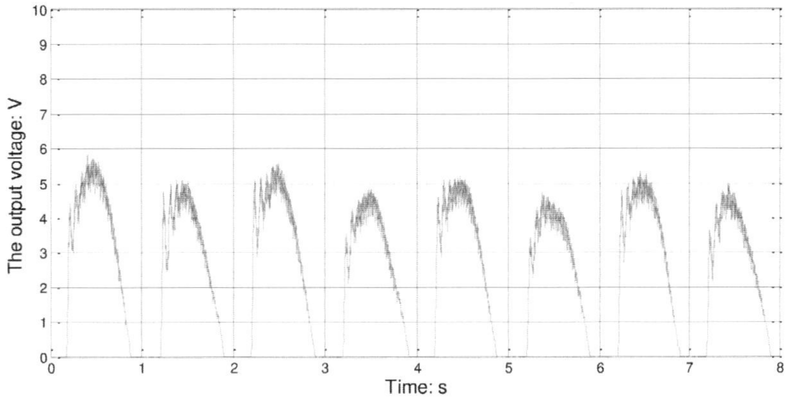


(a) The maximum blade angle is $\pm 25°$

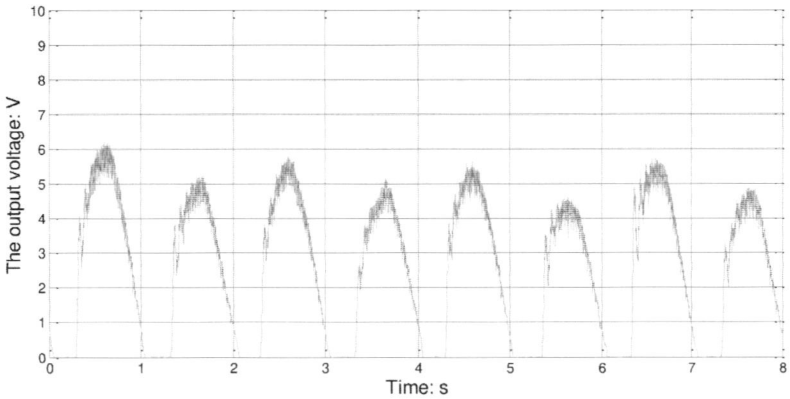


(b) The maximum blade angle is $\pm 20°$

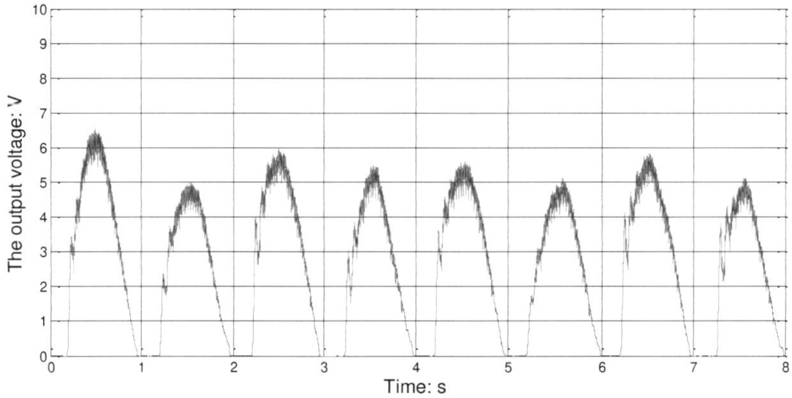


(c) The maximum blade angle is ±15°

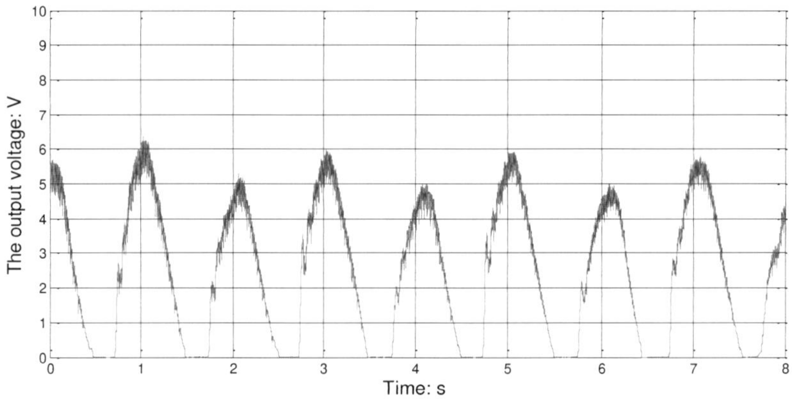


(d) The maximum blade angle is ±10°

**Figure 6.** The output voltage corresponding to different the different maximum blade angle

It can be seen that, while the maximum blade angle is ±25°, ±20°, ±15° and ±10°, the maximum output voltage is about 5V, 5.5V, 5.7V and 5.9V respectively while the platform running stable. It also can be see that, as the maximum blade angle decreases, the output voltage increases, but when reduced to a certain value, the output voltage no longer increase. While the maximum blade angle is 0, the blade cannot rotate, so the absorber cannot obtain power. It is because that, while the maximum blade angle decrease, the circumferential component of thrust on the absorber will decrease, lead to the output voltage decrease. Even the circumferential torque cannot meet the starting torque of generator when the maximum blade angle decreases to 0. This experiment based on the experimental conditions described above indicates that, the output voltage will reach the maximum value when the maximum blade angle within the scope of ±15°~±10°.

## 4. Conclusion

A new kind of micro wave energy converter was designed in this paper. The innovation of the device is that it can converts wave energy into rotating mechanical energy, then into electrical energy. Based on the actual characteristics of power generation device and the power output of this wave power generation device, this paper builds experimental platform and verifies the experiment. The actual output characteristic curve and power generation efficiency of the generator are obtained, and the efficiency is less than 50%. Get the device output voltage under the different maximum blade angle.

When the blade angle is within the scope of $\pm 15° \sim \pm 10°$, the average output voltage is the largest, close to 6V. Future work such as simulation analysis and experiment in actual sea condition need to be done to carry on optimizing the design of the absorber mechanism, and to pick up the optimal parameters group more precisely and improve the energy conversion efficiency.

## 5. Acknowledgments

This work was financially supported by NSFC(Natural Science Foundation of China) E050701.

## 6. References

[1] Scientific applications research associates. A power take-off solution to convert the forces from ocean wave motion to usable electricity[EB/OL]. http://www.sara.com.2013-08-25/201-09-21.

[2] Taylor G W, Burns J R, Kammann S M, et al. The energy harvesting eel: a small subsurface ocean river power generator[J]. J. of Oceanic Engineering, 2001, 26(4): 539~547.

[3] Murray R, Rastegar J. Novel two-stage piezoelectric-based ocean wave energy harvesters for moored or unmoored buoys [C]. Proceedings of SPIE-Active and passive smart structure and integrated systems, 2009.

[4] Gardner FE. Learning experience of AWS pilot plant tests offshore Portugal[C]. Proceedings of 6th European Wave Energy Conference, 2005, 149~154.

[5] Gardner FE. Learning experience of AWS pilot plant tests offshore Portugal[C]. Proceedings of 6th European Wave Energy Conference, 2005, 149~154.

[6] Aurelien Babarit, Michel Guglielmi, Alain H.Clement. Declutching control of a wave energy converter [J]. Ocean engineering, 2009, 36(12):1015~1024.

[7] Whittaker T, Collier D, Folley M, Osterried M, Henry A, Crowley M. The development of Oyster-a shallow water surging wave energy converter[C]. Proceedings of 7th European Wave Tidal Energy Conference, Porto, Portugal, 2007, 15~23.

[8] cuotto M, Falca˜o AF de O. Wells and impulse turbines in an OWC wave power plant: a comparison. In: Proceedings of 6th European Wave Tidal Energy Conference; 2005. p. 463–9.

[9] Danielsson O, Thorburn K, Leijon M. Direct drive linear generators. In: Cruz J, editor. Ocean wave energy. Berlin: Springer; 2008. p. 220–4.

IOP Conf. Series: Earth and Environmental Science **111** (2018) 012009 doi:10.1088/1755-1315/111/1/012009

# Study on Waste Heat Utilization Device of High-Temperature Freshwater in the Modern Marine Diesel Engine

**Shuaijun Wang[1, a], Chentao Liu[2] and Yao Zhou[3]**

[1,2,3]Marine College of Zhejiang Ocean University, Zhoushan 316000, P. R. China
Email: [a]wshj2694@126.com

**Abstract.** Based on using the waste heat recycling from high temperature freshwater in marine diesel engine to heat fuel oil tank, lubrication oil tank and settling tank and so on to achieve energy saving, improve fuel efficiency as the goal, study on waste heat utilization device of high-temperature freshwater in the modern marine diesel engine to make the combustion chamber effectively cooled by high-temperature freshwater and the inner liner freshwater temperature heat is effectively utilized and so on to improve the overall efficiency of the power plant of the ship and the diesel optimum working condition.

## 1. Introduction

With the development of shipping industry, energy shortage, higher fuel cost, sustainable growth of the demand of fuel and International Maritime Organization (IMO)'s attention to energy saving make energy saving be viewed as an extreme significant problem by all shipping companies[1] [2]. To achieve win-win benefits of economy and environmental protection, we should study on energy-saving technology and applying it into practice. Therefore, to recycle and utilize ship waste heat is an important and effective way.

Waste heat recovery mainly consists of recycling waste heat from exhaust gas and high temperature chilled freshwater. In recent years, experts and scholars devote themselves into the research of recovering energy of ship from exhaust gas. So far, they have obtained the certain achievement. But it's far from adequate that only recycle waste heat from exhaust gas. It is of great significance to ship energy saving that make full use of the waste heat recycled from high temperature freshwater.

In the ship which takes diesel engine as the main propulsion, most of heat energy produced by burning fuel is dissipated through cooling water, exhaust and so on, only 40-50% can be converted to mechanical energy[3].The general temperature of the high temperature freshwater in the inner liner of diesel engine is 70-85°C. Obviously, the freshwater contains a large amount of energy, but most of them will be wasted as the freshwater flows out of the ship. According to Heat Balance of the Diesel Engine,25%-35% of total heat is dissipated by liner cooling water. Presently, reusing ship waste heat has become the inevitable trend of energy conservation and emission reduction[4][5].

## 2. The Status of High Temperature Freshwater Cooling System in Modern Marine Diesel Engine

### *2.1. High temperature Freshwater Cooling System*

During the operation of marine diesel engine, a great deal of heat produced by fuel in the air cylinder makes the surface temperature of the combustion chamber increase which may affect the normal work of the unit. to ensure that the temperature is in a normal working range, some heat should be taken away by smoke and the rest part needs to be processed by cooling medium.

IOP Conf. Series: Earth and Environmental Science **111** (2018) 012009 doi:10.1088/1755-1315/111/1/012009

As shown in Figure 1,the high temperature freshwater enters the cooling water chamber of cylinder liner through the cooling water main pipe K.After cooling the cylinder liner, it enters the cooling jacket of the cylinder head by two cooling water joints. When the lower part of the cylinder head is cooled, it goes upward through water holes into cylinder head and exhaust valve seat to form the confined space to cool the exhaust valve seat. After that, a part of water cools the upper part of the exhaust valve, the rest of the water flows out of cylinder head and gathers at the outlet main pipe. Finally the first part will flow out into the main pipe.

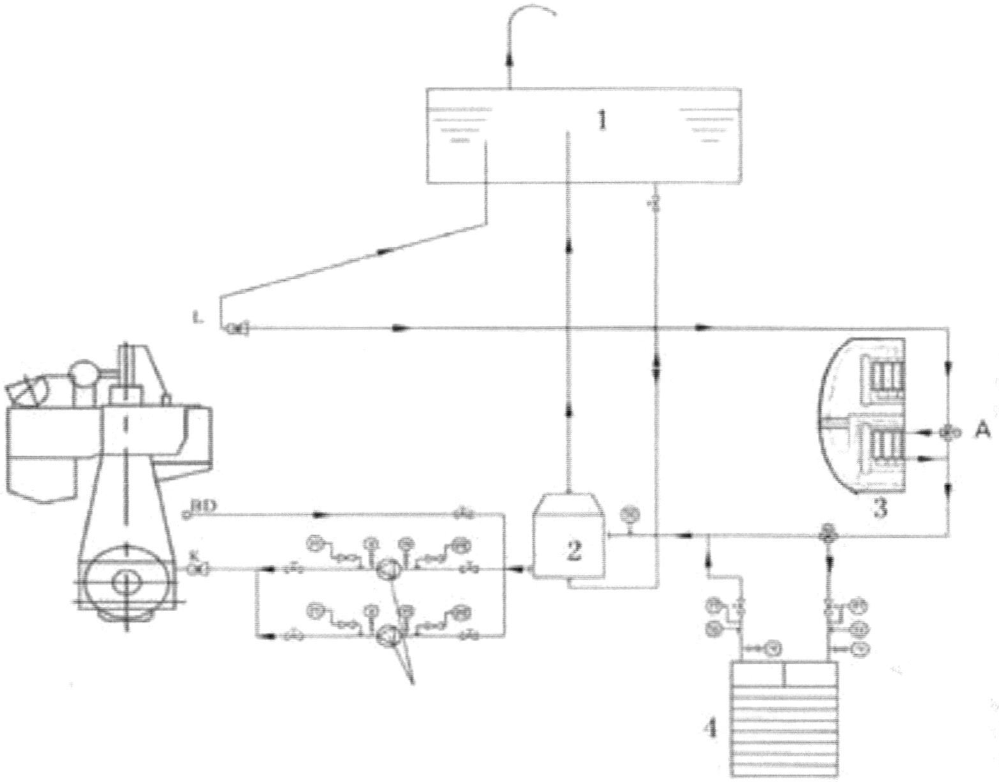


**Figure.1** High temperature freshwater cooling system
1-expansion tank; 2-dearation tank; 3-evaporator
4-high temperature freshwater cooler

When cooling water flows out from the outlet main pipe L, it will pass through the first temperature control triple valve A which is a bypass-valve of the evaporator, the temperature is generally set to about 70°C. The temperature of water which flows into the jacket water cooler can be regulated by the by-pass valve. When the evaporator is running, the triple valve makes a part of high temperature freshwater pass through the evaporator, but the freshwater will all pass by if it stops working.

After passing through the triple valve of evaporator, the cooling water will flow into the second triple temperature control valve and its temperature is set to about 80°C.Then the water will be divided into two parts, one bypasses the cooler while the other goes into the high temperature freshwater cooler. At last, they will flow into dearation tank together. In the system, the temperature of the water that flows out of the main engine can be detected by the sensor installed on the water outlet main pipe of cylinder liner. Afterwards, the temperature can be maintained between 80 and 85°C by controlling the flow of by-pass water and the water flowing into cooler with the help of regulator. When the temperature of the water that flows into cylinder liner is below normal value, the opening degree of valve will be regulated to reduce the flow of the water cooler. If the temperature is higher than normal value, it will do the opposite. Once air gets into the system, it will enter into the expansion tank through the upper pipe of the dearation tank and be released into air. In addition, there's a stream of water from expansion tank flowing directly into dearation tank to play the role of replenishing water and increasing the pressure head of water pump and flowing into two parallel running cylinder pump finally.

### *2.2. Defects and Influences of Existing Systems*

When the marine diesel engine is running, heat loss in the engine is one of the most important aspects among numerous defects. In the central cooling system, the temperature of sea water will gradually rise because of interchanging heat with low temperature water. However, the heat can't be recycled for that most of sea water is expelled from the ship, which may reduce the thermal efficiency of the marine diesel engine.

Once the seawater pump is plugged, it cannot reach the cooling effect. What's more, seawater is so corrosive that the cooler and pipeline may be corroded ,which may arouse freshwater pollution, increase the maintenance cost of central cooling system of marine diesel engine and shorten the lifespan of equipment and pipelines[6].

## 3. Waste Heat Utilization Device of High-Temperature Freshwater in the Modern Marine Diesel Engine

### *3.1. Working Principle*

A large amount of heat is wasted in the central cooling water system. To solve this problem effectively, based on heat transfer theory and the character that water's temperature is higher than fuel oil tank, lubrication oil tank and settling tank, we come up with an idea to improve the cooling water system of the main engine and design a waste heat utilization device of high-temperature freshwater which can replace the traditional low temperature freshwater cooling system and sea water cooling system. It will cool the combustion chamber while applying the waste heat into heating fuel oil tank, lubrication oil tank and settling tank, etc.

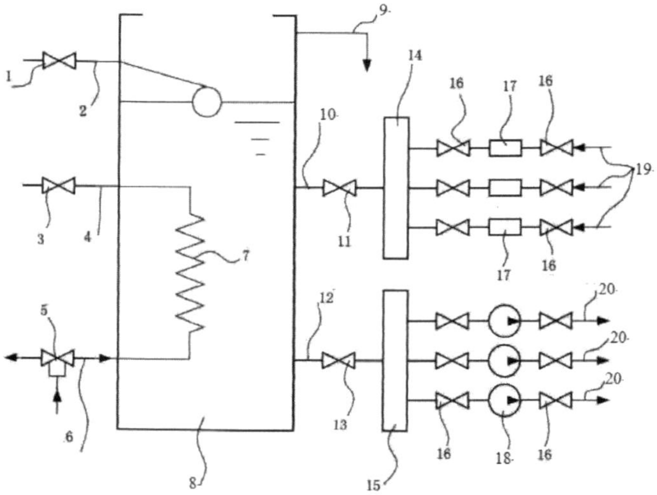


**Figure.2** Block diagram of high temperature freshwater cooling water system for marine main engine
1-water replenishing valve; 2-water replenishing pipe; 3-outlet valve; 4-outlet pipe(high temperature freshwater inlet pipe of main engine); 5-triple temperature control valve; 6-inlet pipe(high temperature freshwater outlet pipe of main engine); 7-heat exchange tube; 8-heat exchange tank; 9-overflow pipe; 10-return pipe; 11-return valve; 12-outlet pipe; 13-outlet valve; 14-return junction box; 15- hot water distributor; 16-non-return valve; 17-inspection glasses; 18-centrifugal pump; 19-inlet pipe; 20-outlet pipe;

As shown in Figure 2, the system consists of high temperature freshwater system, heat exchange tank, residual heat utilization terminal cooler and heat exchange tube. The temperature of freshwater rises for taking away the excess heat of the main engine. To achieve the double effect of cooling high temperature freshwater and heating the water in the tank, bring the outlet water into the heat exchange tank when it flows to the triple valve which can adjust the opening degree automatically maintain the temperature between 60-70°C. The hot water flows out from the water outlet pipe and then flows to heating terminal tube through the main valve, hot water distributor, non-return valve and so on. That means the hot water will heat fuel oil tank, lubrication oil tank and settling tank, etc. At last,it flows back to the tank and cycles again and again.

### *3.2. The Function of Components*

Waste heat utilization device of high-temperature freshwater in the marine main engine is composed of many independent components. The functions of these components are as follow:

*3.2.1. Heat Exchange Tank.* The open-designed tank is used to store the water. The opening at the top of the tank is beneficial to the expansion of the hot water and convenient for the marine engineer to observe. The lower part is equipped with an outlet pipe to export the heated water and the upper part is equipped with a replenish pipe with a water replenishing valve. Once the water is not enough, the sensor installed at the lowest water-level sends out water shortage signal to replenish water. To avoid the overflow, the overflow pipe is installed above the water replenishing valve. There are return pipe and valve letting the water flow back to tank.

*3.2.2. Heat Exchange Tube.* Heat exchange tube plays the role of supplying the freshwater into the tank and exchanging heat with inner water. The inlet pipe is equipped with a triple temperature control valve and connected to outlet of the main engine. The outlet of heat exchange tube is connected with the high temperature water pump by outlet pipe. The triple valve can regulate the flow according to the change of temperature in the tank. The triple temperature control valve is respectively connected with main engine cylinder liner, heat exchange tank and high temperature freshwater cooler in the main engine cylinder liner. There is an outlet valve installed between the outlet of heat exchange tube and the inlet of high temperature freshwater pump.

*3.2.3. Hot Water Distributor.* The main function of hot water distributor is to distribute the hot water to different water pipes. The inlet of hot water distributor is connected with outlet pipe of the tank and the outlet is connected with outlet pipe. Each outlet pipe has a centrifugal pump, of which non-return valve is installed on both sides, and is connected to the heating terminal tube to heat different equipment.

*3.2.4. Heated Terminal.* The heated terminal includes fuel oil tank, lubrication oil tank and settling tank. At the bottom of the fuel oil tank, heating coils is installed to maintain its liquidity and make it easy to be transferred. To accelerate the separation of precipitation in oil, engineers can regulate the flow and heat the fuel to the proper temperature. Lubricating oil which is pumped from the lubricating oil circulating tank and settling tank flows to lubrication oil purifier after heated by heating coils. Afterwards, sludge will be sent to the sludge tank. In the end, the lubricating oil return to the lubricating oil circulating tank.

Settling tank is used for storing the oil from fuel oil tank and heating the fuel oil to make the sludge settle into bottom and remove most of the water. The treated fuel oil will be transferred to daily service tank.

## 4. Maintenance of the Utilization Device

### *4.1. Adjustment of Fresh Water Temperature*

It is forbidden to make the temperature of freshwater outlet too low or too high, the instructions should be followed strictly. For medium-speed diesel engine, the outlet temperature can be controlled at

70-80°C and 60-70°C for the low speed engine. In addition, the temperature between inlet and outlet is controlled within 12°C [7]. Once found that the individual cylinder temperature is abnormal, the marine engineer must evaluate whether thermal load is abnormal. If fuel injection quantity and fuel delivery is normal, engineer can adjust the cylinder outlet valve to make the water temperature return to normal. Engineer should remove some part of load and make the cooling water temperature gradually decreased if the diesel engine were overheated. It's forbidden to stop diesel engine immediately or replace the high temperature water with cooling water.

### *4.2. Inspection of Cooling Water Flow in Each Cylinder*

The high temperature freshwater should be cooled to make sure the water is in the normal range when the main engine is working. So an automatic temperature control device at the outlet of the system is needed. If the temperature reaches the limit value, the sea water cooling system can start automatically to keep the freshwater cooling system from high temperature. The more heated terminals are, the higher the heat exchange efficiency is, but the temperature should range from 60°C to 70°C. To utilize the waste heat effectively and protect high temperature freshwater cooling system from damage, the quantities of fuel oil tanks, lubrication oil tanks and settling tanks should be appropriate.

### *4.3. Inspection of Components and Parameters of Waste Heat Utilization Device of High-Temperature Freshwater*

Need-to-check components include the triple valve, heat exchange tank, heat exchange tube, water supply device, overflow device, outlet pipe, hot water distributor, centrifugal pump, return main pipe, observation glasses and each valve. The parameters of tank level, liquid level of hot water distributor and return manifold should be checked and estimated whether meet the requirements. In the process of operation, tank level, temperature and pipeline circulation should be checked regularly. In addition, engineers should observe if each devices is working properly, replace water and clean the tank regularly according to the quality of water.

## 5. Conclusions

In this paper, based on the performance of heat-exchange device and water, the marine diesel engine cooling system will be replaced by waste heat utilization device of high-temperature freshwater. Some conclusions are as follow:

(1)The device can cool the combustion chamber effectively, for which the main engine can work properly.

(2)The device can recycle a large amount of heat, improve the efficiency of marine diesel engine and the power of propulsion plant.

(3) It can reduce maintenance cost in circumstance of protecting flange and valve from the damage caused by high temperature steam that heat fuel oil tank, lubrication oil tank and settling tank with the heat from high temperature freshwater. Meanwhile, it can improve the utilization of energy.

## 6. References

[1] Zeng Qingdong. On Requirement of International Emission Law for Improvement of Ship Power train Technology and Countermeasures [J].Journal of Wuhan University of Technology,2015,39(2):350-353
[2] Wang Qiang.Optimization of power reduction and emission reduction technology for marine main engine[D].Dalian: Dalian Maritime University,2013.
[3] Huang Busong.The Implementation of EEDI Impact of Marine Diesel Engines[J].Energy Conservation &Environmental Protection in Transportation, 2013(5):50-53.
[4] Li Kangkang. Research on Optimization Design and Temperature Intelligent Control of Main Engine Cooling Water System[D].Wuhan:Wuhan University of Technology,2012.
[5] Mei Lei.Study on Novel Recovery System for Marine Diesel Engine Waste Heat [D].Wuhan: Wuhan University of Technology,2011.
[6] Tu Huan.Modeling and Simulation of ENGARD Marine Central Cooling System[D].Wuhan: Wuhan University of Technology,2013.
[7] Li bin. Marine diesel engine[M].Dalian Maritime University Press,2014.

# Monitoring Dust Concentrations over Desert and Cropland during Dust Storms Using Visibility Sensors

**Mingyuan Du[1,a], Seiichiro Yonemura[1,b], Yanbo Shen[2,c], Wanfu Wang[3,d]**

[1]Institute for Agro-Environmental Sciences, National Agriculture and Food Research Organization, Tsukuba, Japan
[2]National Climate Centre (NCC), China Meteorological Administration (CMA), Beijing, China
[3]The Conservation Institute, Dunhuang Academy, Dunhuang, China
E-mail:[a]dumy@affrc.go.jp, [b]yone@affrc.go.jp, [c]shenyb@cma.gov.cn, [d]wwanfu@hotmail.com

**Abstract.** Directly measuring the dust concentration during a dust storm is very important for studying dust emission and wind erosion processes. The Mira visibility sensors were introduced for directly measuring dust concentrations over desert and cropland during dust storm events. The output of the sensor was compared with that of optical particle counter. We found that the Mira visibility sensors can directly measure the particle numbers concentrations. Characteristics of dust concentrations at the two ground surfaces during dust storms were analyzed. By using the Mira visibility sensors and optical particle counters and other meteorological elements in Dunhuang, China, dust emission processes and characteristics over desert and oasis during dust storms was discussed.

## 1. Introduction

Dust and dust storms have a considerable effect on visibility [1], air quality [2], human health [3] and the climate system [4]. Even though these dust storm effects had been broadly studied, the processes at the sources such as the frequency, duration, intensity or aerosol number-size distribution had not been adequately described. In order to clarify the process of dust outbreak and the relationship between ground surface conditions, such as surface vegetation cover, soil water content, temperature, etc., and dust outbreak and to obtain basic data for applying it to dust outbreak model in the course of parameterization and verification, as a part of the Japan-China Cooperation Project on aeolian dust experiment on climate impact (ADEC), continuous field observation of dust concentration and meteorological elements on different surfaces such as sandy desert and agricultural field were induced [5]. Two automatic observation stations have settled in Dunhuang, China. One of the stations was settled on a Gobi Desert in March 2001 and another has settled on a cropland inside the Dunhuang oasis in September 2001. Directly measuring the dust concentration during a dust storm is very important for studying dust emission and wind erosion processes. However, it is very difficult to measure. Unfortunately, the size distribution of emission fluxes has seldom been measured in natural source areas and, for lack of sufficient field data, it has often been assumed in the past to be independent of meteorological conditions prevailing during emission [6, 7]. A visibility sensor was introduced for directly measuring dust concentration at the two sites. The output of the sensor was compared with that of optical particle counter. Characteristics of dust concentrations at the two ground surfaces during dust storms were analyzed.

The purpose of this study is to test if the visibility sensor can be used for directly measuring dust concentrations during dust storms and give out the observational results of dust concentration variations at ground surface during dust storms in Dunhuang, China and to clarify the process of dust concentration change during the dust storm.

## 2. Study Area and Data

Figure.2-1shows the locations and topographies of the observation sites. Station 1 has settled in a flat moveable sandy desert or Gobi Desert with moveable sand area near the famous Mogao Grottoes 20km apart from Dunhuang Oasis. The land surface of the Gobi Desert site was covered by gravel (about 2-10 mm in diameter) and removable sand and dust particles. Station 2 has settled in an agriculture field in the central part of Dunhuang Oasis. This site is a cotton field during May to November and the other time is bare soil and under vary dry condition after the ploughing work, including harrowing, leveling, plowing, fertilizing and seeding. Both sites located in very flat area as shown in Fig. 3. The flat area has a diameter of over 2km. However, the oasis site is inside of windbreak network. The grid of the network is about 200-500m*200-500m. The station is settled in the center of a grid about 400m*400m.

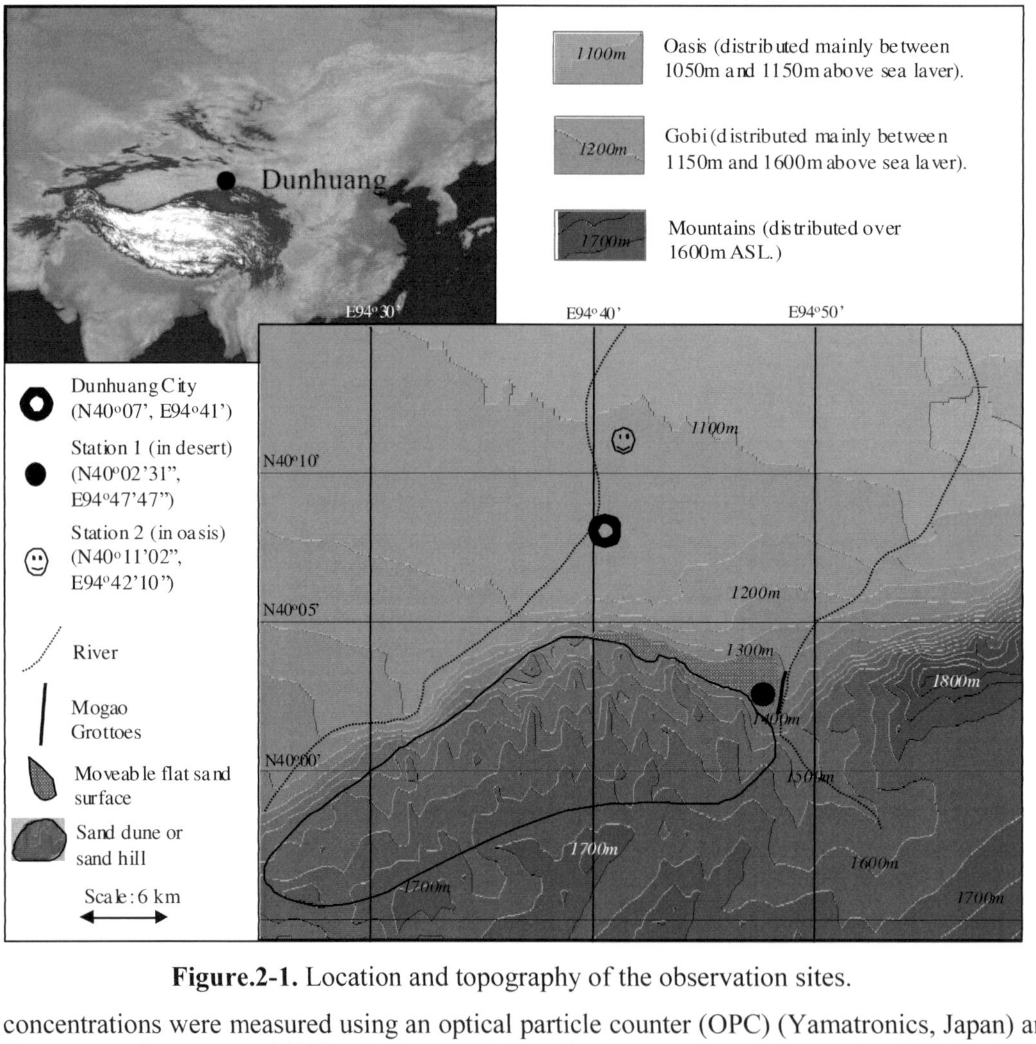


**Figure.2-1.** Location and topography of the observation sites.

Dust concentrations were measured using an optical particle counter (OPC) (Yamatronics, Japan) and a visibility sensor, the MIRA Visibility Sensor 3544 (Aanderaa Instruments, Norway), which is designed to fulfill the demand for a small, low power (DC) unit to be operated with Aanderaa Measuring Stations. The sensor consists of an aluminum body containing all necessary solid-state electronics and two vertical legs. It is furnished with a standard Aanderaa meteorological sensor foot. It is a rugged, watertight, corrosion free and solid-state sensor with minimum maintenance requirements. In one leg of the body, an infrared (IR) light emitting diode is installed at an angle of 25 degrees. The opposite leg contains an IR photo detector as shown in Fig.2-2. The diode and the photo-detector are protected against clogging by a cover.

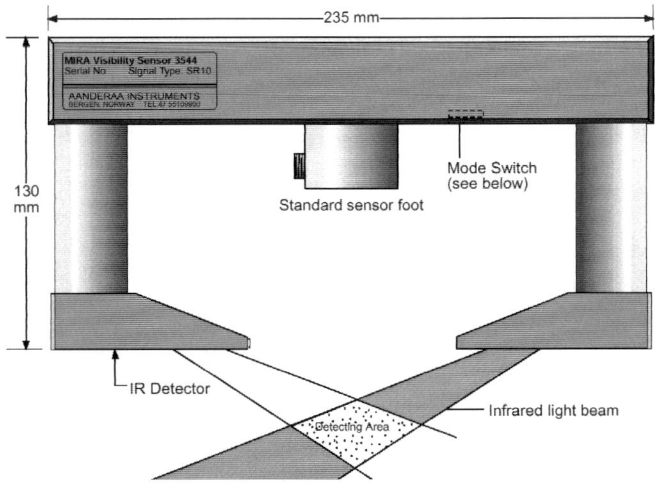


**Figure.2-2.** The Mira visibility sensor.

CR10X Measurement and Control Systems (Campbell Scientific, Inc.) are used for all meteorological elements (wind, air temperature, humidity, pressure, radiation, soil temperature and water content, heat flux etc.), except visibility. The energy balance between the incoming spectral short-wave and long-wave radiation, from 0.3 to 50 μm, relative to the surface reflected short-wave and outgoing long-wave radiation were measured by The CNR 1 Net Radiometer (Kipp & Zonen INC.). The station in desert is a 10m tower with 4 levels wind and air temperature and humidity observations at 1m, 2m, 4m and 10m and 2 levels (3 levels from September of 2001) of visibility observation. The station in oasis is a 4m tower with 3 levels as shown in Fig.2-3.

**Figure.2-3.** View of the experimental sites.

Dust concentrations were measured using an optical particle counter (OPC) (Yamatronics, Japan) and a visibility sensor, the Mira Visibility Sensor 3544 (Aanderaa Instruments, Norway), which is designed to fulfil the demand for a small, low power (DC) unit to be operated with Aanderaa Measuring Stations. Due to the visibility sensor is designed to detect fog and haze, we define a Dust Concentration Index (DCI) for use it to detect dust as

$$DCI = (a - N) / a \quad (1)$$

Where the coefficients a = 1022, the maximum raw data reading and N is the raw data reading. DCI will be between 0 (no heavy dust event) and 1 (the heaviest dust event as visibility = 0). Fast sampling mode of the visibility sensor was used. In this mode, the light beam is transmitted every 6 seconds and average data for 10 minutes was recorded.

All the elements (wind, temperature, humidity, radiation and pressure) above the ground is sampled every 10 second. Soil temperature and moisture (TDR) are sampled every 10 minutes. One minutes sampling data are averaged and recorded for 10 minutes intervals during ordinary observation period and sampling data are recorded during IOP period (Intensive Observation Period). Maximum and minimum of wind and their standard deviation within the recording 10 minutes intervals were also recorded. The details of the observation stations and observation system has introduced by Du et al. [8]

## 3. Results

### *3.1. Use of visibility sensor for measuring dust concentration*

Figure.3-1 shows a comparison of DCI and particle numbers concentration for a dust event on April 8-9, 2002. It is obviously that DCI has a very good relationship to the dust numbers concentration. For the total particles numbers lager than 0.5μm, DCI changes with the particle numbers concentration exponentially and it changes proportionally with the particle numbers concentration larger than 5.0μm as shown in Fig.3-2 (observation period was from April to September 2002). That is,

$$\text{Particles numbers lager than } 0.5\mu m = 2934.1e^{3.2452DCI} \quad (2)$$

$$\text{Particle numbers larger than } 5.0\mu m = 197370DCI + 44489 \quad (3)$$

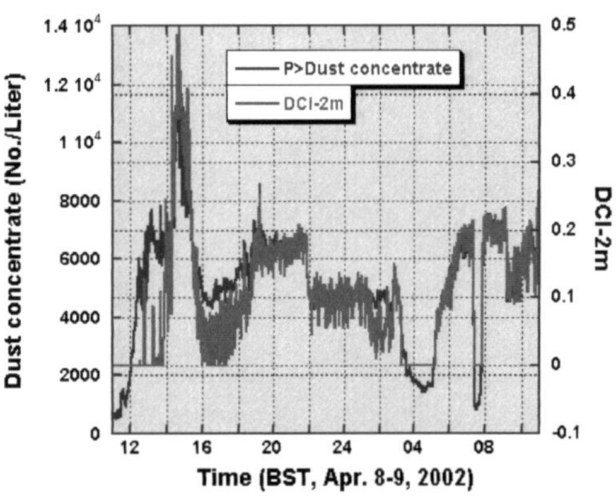


**Figure.3-1.** Comparison of DCI and dust particle numbers on Apr. 8-9, 2003 over desert in Dunhuang, China

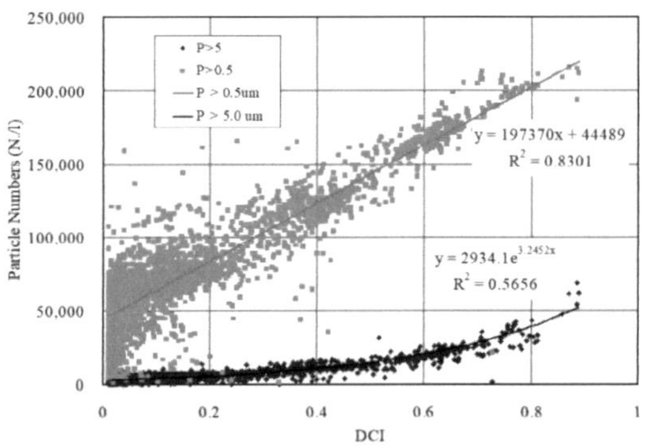


**Figure.3-2.** Relationship between DCI and dust concentration over desert in Dunhuang, China (data from Apr. to Sep. 2002)

### 3.2. Dust storm process

Most dust storms in spring in Dunhuang are probably generated by cold frontal systems with dry squall lines. As shown in Fig. 6, air pressure, wind speed and relative humidity were the lowest of the day before the dust storm passing the station on April 13, 2002. Cloud appeared (radiation decreased compared with that on April 12) and wind direction changed several hours before dust storm. When dust storm passing the station, strong wind speed appeared and DCI at 9m became bigger than that at 2m. Air Pressure increased and temperature decreased with relative humidity increasing with time after the dust front passing.

The process of a dust storm can be divided into 4 stages in Dunhuang, China as follows.

*3.2.1. Pre-emission stage.* Pre-erosion stage lasts several hours before the dust storm comes. In this stage, wind speed becomes weak and weak (usually lower than 4m/s (sometimes below 1.0m/s)) and wind direction changes (usually clockwise). Air pressure decreases to lowest value and there will be some cloud appears.

*3.2.2. Dust outbreak stage.* This stage only lasts about one hour occurred when cold front passing. As wind speed suddenly increasing, DCI increase very quickly. The most important features of this stage are the variation of roughness length and friction velocity and the dust concentration (DCI) at 9m is larger than that at 3m. Dust concentration in the air (DCI) increases exponentially with wind speed or friction velocity when strong wind blown (wind speed at 10m is over about 6m/s or friction velocity is over about 0.2m/s). Due to saltation occurs in the sandy desert surface, wind profile and roughness length change with wind speed.

*3.2.3. Dust passing stage.* Dust passing stage can last from one hour to over 10 hours. During this stage, air pressure, air temperature and soil temperature decrease and humidity increase with strong wind blows. Sometime wind speed may decrease or increase for several meters as shown in figure 6. DCI at 9m is lower than that at 3m.

*3.2.4. Calm down stage.* This stage lasts several hours too. DCI decrease as wind speed decrease. Sometime DCI decreased very quickly due to rainfall occurred. Usually air temperature and surface soil temperature will be about several degrees lower that that before the dust event as shown in Fig.3-3

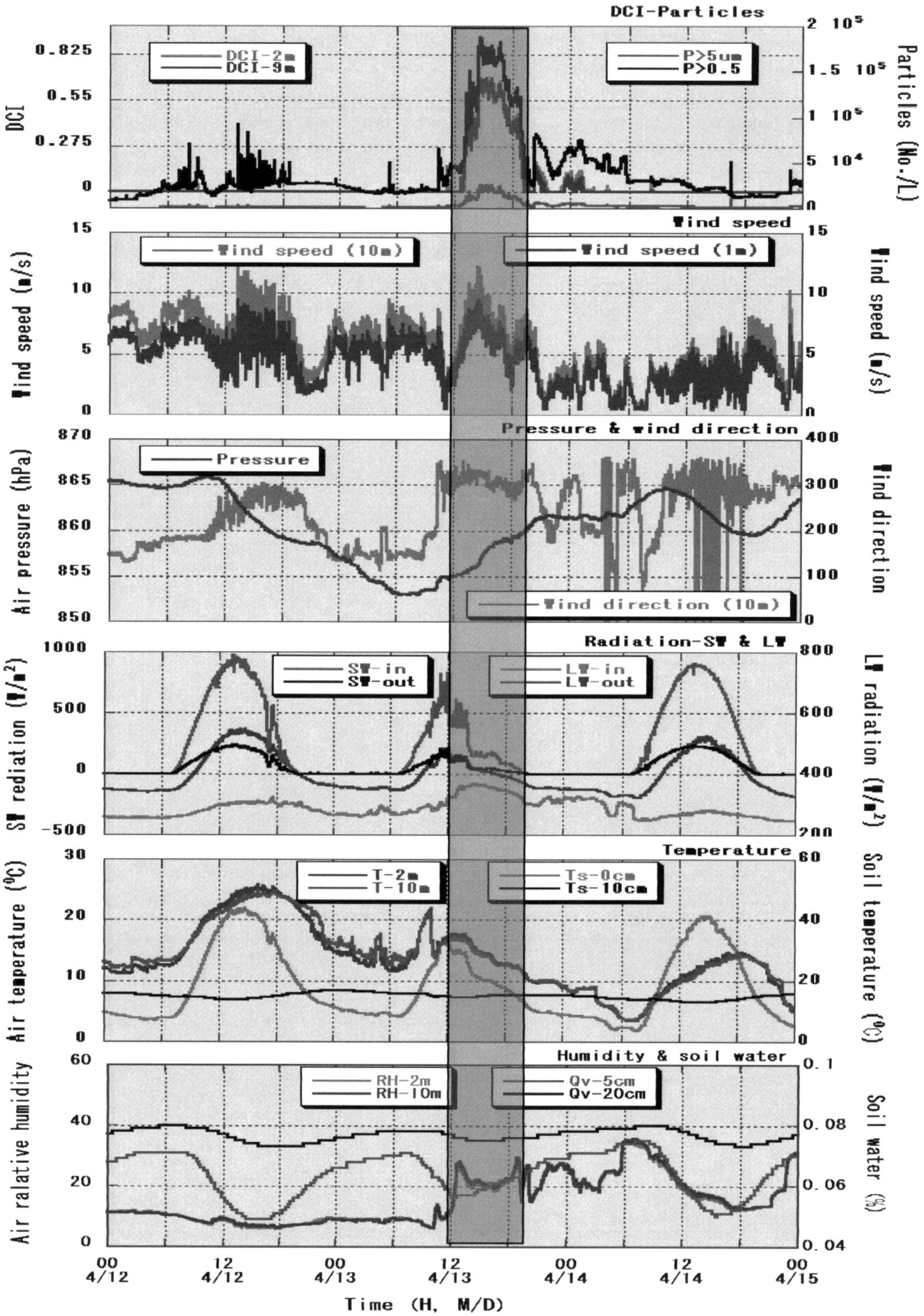


**Figure.3-3.** Variations of Several Elements during a Typical Case of Dust Storm from April 12 to15, 2002 in Dunhuang China, including DCI, Solar Radiation, Pressure, Wind Speed and Direction, Air Temperature and Humidity.

### *3.3. Difference of dust concentration between desert and bare cropland*

Figure.3-4. shows a comparison of dust concentrations both the particle numbers concentration and the Dust Concentration Index (DCI) between desert and bared cropland. It shows that although the DCI over the cropland was higher than that over the desert during the dust storm, dust concentration observed by the optical particle counter did not show much difference between the two sites. However, during the calm down stage of the dust storm, numbers of small particles such as 0.5μm to 1.0μm over the cropland was much more than that over the desert as DCI shows.

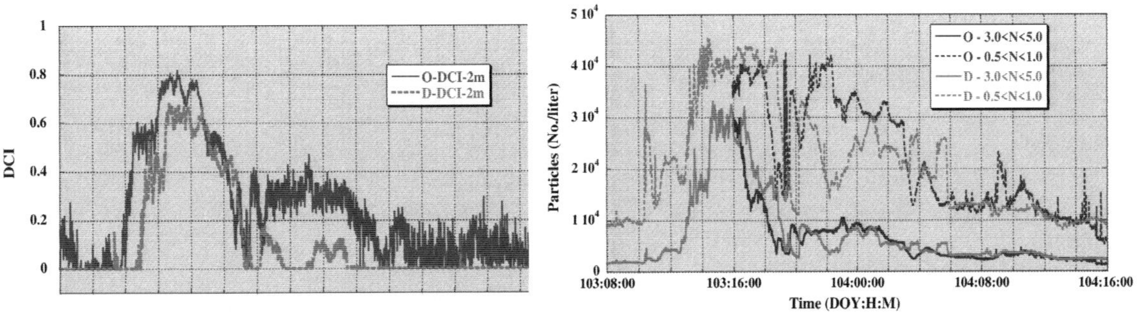


**Figure.3-4** Comparison of Dust Concentrations (up: DCI; down: particle numbers) between Desert and Cropland on Apr. 13, 2002 in Dunhuang, China

Optical particle counter observation also shows that higher particle numbers concentration over cropland when there is no dust emission. This is because there are many fine dusts on the surface of the cropland, especially on the field road in the oasis in Dunhuang. These fine particles contribute the higher dust concentration over the cropland.

## 4.Conclusions

The Mira visibility sensor can be used for directly measuring dust numbers concentration during dust event after approving by other instruments such as optical particle counter. It has very high response and can measure dust concentration directly with an interval of less than one minute. By changing its output to dust concentration index (DCI), it will proportional increase with particle numbers concentration larger than 0.5μm and exponential increase with the particle numbers concentration larger than 5.0μm.

By using the Mira visibility sensor and optical particle counter with other meteorological elements in Dunhuang, China, dust emission processes by a dust storm can be divided into 4 stages as 1) pre-emission; 2) outbreak; 3) passing and 4) calming down stages.

During a dust storm, dust concentration over cropland is almost the same as that over desert. However, concentration of fine dust (smaller than 1.0μm) over cropland for the other period is higher than that over desert in Dunhuang, China.

## 5. References

[1] D.A. Gillette, "Dust production by wind erosion: necessary conditions and estimates of vertical fluxes of dust and visibility reduction by dust," in Physics of Desertification, F. El-Baz, M.H.A. Hassan, Eds., Martinus Nijhoff Publishers, Dordrecht, 1986, pp. 361-371.

[2] Y.C. Chan, G. McTainsh, J. Leys, H. McGowan, K. Tews, "Influence of the 23 October 2002 dust storm on the air quality of four Australian cities," Water Air and Soil Pollution, 164, 2005, pp. 329-348.

[3] Z.Q. Meng, Q.X. Zhang, "Oxidative damage of dust storm fine particles instillation on lungs, hearts and livers of rats," Environmental Toxicology and Pharmacology, 22, 2006, pp. 277-282.

[4] R. Arimoto, "Eolian dust and climate: relationships to sources, tropospheric chemistry, transport and deposition," Earth-Science Reviews, 54, 2001, pp. 29-42.

[5] Mikami M, Shi G.Y, Uno I, Yabuki S, Iwasaka Y, Yasui M, Aoki T, Tanaka T.Y, Kurosaki Y, Masuda K, Uchiyama A, Matsuki A, Sakai T, Takemi T, Nakawo M, Seino N, Ishizuka M,

Satake S, Fujita K, Hara Y, Kai K, Kanayama S, Hayashi M, Du M, Kanai Y, Yamada Y, Zhang X.Y, Shen Z, Zhou H, Abe O, Nagai T, Tsutsumi Y, Chiba M, Suzuki J., "Aeolian dust experiment on climate impact: An overview of Japan-China joint project ADEC," Global and Planetary Change, 52, 2006, pp.142-172.
[6] T. Allen, Particle size measurement, 3rd ed., Springer, 1981.
[7] Hinds, W.C., Aerosol technology: properties, behavior, and measurement of airborne particles, 2nd ed., John Wiley & Sons, 1982.
[8] Du, M., Yonemura, S., Maki, T., Shen, Z., Shen, Y. and Wang, W., "Observational study on the relationship between ground surface condition and aeolian dust outbreak in Dunhuang, China," Arid Land Geograpgy, 26, 2003, pp.25-31.

# Effects of Comprehensive Technologies on the Improvement of Acidified Vineyard Soils

**Hongguo Jiang[1,2], Qiunan Wang[3], Feng Xu[4], Jun Jin[2], Guoyu Wang[1], Jianguo Liu[1,a]**

[1]School of Environmental & Safety Engineering, Changzhou University, Changzhou 213164, China
[2]Changzhou Station for Horticultural Technology Spread, Changzhou 213001, China
[3]Wujin Station for Horticultural Technology Guidance, Changzhou 213161, China
[4]Changzhou Center for Agricultural Technology Spread, Changzhou 213001, China
E-mail: [a]liujianguo@cczu.edu.cn

**Abstract.** Soil acidification is an important factor that restricts the yield and quality of fruits. In this study, the comprehensive improving technologies were applied on the vineyards in which the soil pH is below 5.5. The technologies include application of soil conditioner, organic fertilizer and bacterial manure, and growth of green manure and natural grass. The results show that the comprehensive improving technologies can raise the pH of 0-15 cm soil layer by 0.5-0.8 unit and the pH of 15-30 cm soil layer by 0.3-0.6 unit. The soil bulk densities are decreased by 0.77-10.42%. The contents of organic matter, total N, available P and K in the soils are all increased. Therefore, the soil fertilities are improved. The yields of grape fruits are increased by 12.77-14.94%, and the contents of soluble solid in the grapes are raised by 7.01-9.55%, by the comprehensive measure of seaweed liquid silicon plus sheep manure plus growth of green manure. The comprehensive measure of soil conditioner Naduoli No. 1 plus bacterial manure plus natural grass increases the yields of grape by 7.67%, raises the content of soluble solid in the grape by 8.6%. But the effect of the comprehensive measure of unslaked lime plus sheep manure plus growth of green manure is not clear.

## 1. Introduction

According to some surveys, soil acidification in vineyard is common because of the application of compound fertilizer of nitrogen and phosphorus [1]. The soil pH is below 5.5 in many vineyards, and even under 4.5 in some vineyards [2]. The soil acidification will lead to soil hardening, decrease of the ratio C/N, and wash away of many mineral nutrients (Ca, Mg, P, etc.) [3]. Soil acidification can also increase release of some toxic metals (Cd, Cr, Pb, Mn, Al, etc.). In acid condition, the soil fixed Al will become soluble $Al^{3+}$ and $Al(OH)^{2+}$ that are toxic to plant root. As a result, the abilities of plant root in absorbing water and nutrients are restricted, and the growth and development of plant is inhibited [4].
Soil acidification will also cause unbalance of nutrients in the soils. Excessive supply of nitrogen and phosphorus fertilizers will results in the accumulation of $H^+$ in the soils. Alkaline ion, such as $Ca^{2+}$, $Mg^{2+}$, $K^+$, etc., adsorbed by soil colloidal particles will be displaced by $H^+$ [5]. The accumulation of $H^+$ in soil will accelerate soil acidification, which can damage the plasmalemma of root cell, and cause the runoff of nutrients from root cell and decomposing of plant root [6]. Therefore, soil acidification in vineyard will result in decrease of grape yield, and the decline of grape qualities, such as single fruit weight, content of soluble solid, hardness, color and luster, and taste [7]. In some places of China, Soil acidification has become an important threat to the growth, yield and quality of grape, and the sustainable development of grape industry.

IOP Conf. Series: Earth and Environmental Science **111** (2018) 012011 doi:10.1088/1755-1315/111/1/012011

In this study, the experiments were designed according to the relevant standard of China (NY/T2271-2012), and locational experiments were applied in some vineyard in Changzhou, China. The effects of different comprehensive technologies on improving acidified vineyard soils, and on the yield and qualities of grape fruit, were studied. The technologies include application of different soil conditioner, organic fertilizer and bacterial manure, and growth of green manure and natural grass. The results will provide scientific basis to the improvement of acidified vineyard soils, and to the elevation of grape yield and quality, in the middle and lower reaches of Changjiang River in China.

## 2. Materials and methods

### *2.1. Experimental location and period*
The experiments were carried out in two vineyards, e.g. the vineyard of Chaoqun Fruit Co. Ltd. and the vineyard of Ruiyuanyuan Grape Specialized Cooperative. The experimental period was from August 2016 to September 2017.

### *2.2. Grape varieties for experiment*
Grape varieties for this experiment are two-year-old Sunshine Rose Grape in Chaoqun Fruit Co. Ltd., and fourteen-year old Summer Black Grape in Ruiyuanyuan Grape Specialized Cooperative.

### *2.3. Soil additives*
The soil additives are: (1) seaweed liquid silicon (pH 9.5-11.0, containing Si, K, humic acid and alginic acid), (2) soil conditioner Naduoli No. 1 (containing glutamic acid), (3) unslaked lime, (4) sheep manure, (5) bacterial manure, (6) growth of green manure, broad bean or hairy vetch.

### *2.4. Experimental design*
The experiment treatments are shown in Table.2-1.

**Table.2-1.** The Comprehensive Measures on Improving Acidified Vineyard Soils

| Location | Code Name of Measure | Treatments |
|---|---|---|
| Chaoqun Fruit Co. Ltd. | A | Seaweed liquid silicon 30 L/hm$^2$ + sheep manure 15 ton/ hm$^2$ + growth of hairy vetch |
| | B | Seaweed liquid silicon 30 L/hm$^2$ + sheep manure 15 ton/ hm$^2$ + growth of broad bean |
| | C | Unslaked lime 750 kg/hm$^2$ + sheep manure 15ton/ hm$^2$ + growth of hairy vetch |
| | D | Unslaked lime 750 kg/hm$^2$ + sheep manure 15ton/ hm$^2$ + growth of broad bean |
| | Control 1 | Sheep manure 15ton/ hm$^2$ |
| Ruiyuanyuan Grape Specialized Cooperative | E | Soil conditioner Naduoli No. 1 30 L/hm$^2$ + bacterial manure 1050 kg/hm$^2$ + natural grass |
| | Control 2 | Natural grass |

### *2.5. Determination of soils and fruits*
The soils were sampled at 0-15 cm and 15-30 cm before and after the treatments, and ten soil samples were taken for each treatment at each sampling time. The following properties were tested: pH, bulk density, and contents of organic matter, total N, available P and available K.

Grape fruits were harvested at maturity. Fruit yields were weighted, and contents of soluble solid of the fruits were tested with Digital Refractometer (PR-101α, Japan) [8].

## 3. Results and discussion

### 3.1. *Effects of different comprehensive measures on physical and chemical properties of acidified vineyard soils*

*3.1.1. Soil pH.*It can be seen from Table.3-1 that the comprehensive measures are effective in improving soil pH of the acidified vineyards, and the soil pH is raised by 0.3-0.8 unit. But the effects vary with the measures and soil layers.

**Table.3-1.** Effects of Comprehensive Measures on the pH of Acidified Vineyard Soils

| Comprehensive Measure | **0-15 cm Soil Layer** | | | **15-30 cm Soil Layer** | | |
|---|---|---|---|---|---|---|
| | ***Before Treatment*** | ***After Treatment*** | ***Change*** | ***Before Treatment*** | ***After Treatment*** | ***Change*** |
| A | 5.5 | 6.0 | 0.5 | 5.6 | 5.9 | 0.3 |
| B | 5.6 | 6.1 | 0.5 | 5.7 | 6.0 | 0.3 |
| C | 5.4 | 6.1 | 0.7 | 5.3 | 5.9 | 0.6 |
| D | 5.4 | 6.2 | 0.8 | 5.4 | 6.1 | 0.7 |
| Control 1 | 5.4 | 5.5 | 0.1 | 5.6 | 5.7 | 0.1 |
| E | 5.3 | 6.1 | 0.8 | 5.9 | 6.5 | 0.6 |
| Control 2 | 5.3 | 5.4 | 0.1 | 5.6 | 5.6 | 0 |

The soil pH is increased by 0.6-0.8 unit by the measure C (Unslaked lime + sheep manure + growth of hairy vetch), measure D (Unslaked lime + sheep manure + growth of broad bean) and measure E (Soil conditioner Naduoli No. 1 + bacterial manure + natural grass), but it is only raised by 0.3-0.5 unit by the measure A (Seaweed liquid silicon + sheep manure + growth of hairy vetch) and measure B (Seaweed liquid silicon + sheep manure + growth of broad bean). The effects are higher in 0-15 cm soil layer than in 15-30 cm soil layer. The soil pH of the controls changes little.

*3.1.2. Organic matter contents.*Table.3-2 shows that the contents of organic matter in the vineyard soils are increased by the comprehensive measures, but the effects differ with the measures and soil layers.

In 0-15 cm soil layer, the contents of organic matter are raised by 5.70-10.55%, and the effects of different measures are in the order: measure B > measure E > measure A > measure D > measure C. In 15-30 cm soil layer, the effects are smaller than in 0-15 cm soil layer, and vary little among the comprehensive measures (from 4.78% to 5.01%). For the controls, organic matter contents of the vineyard soils change small during the experiments.

The results also show that natural grass can increase organic matter contents of vineyard soil largely.

**Table .3-2.** Effects of Comprehensive Measures on Organic Matter Contents of the Vineyard Soils (g/kg)

| Comprehensive Measure | **0-15 cm Soil Layer** | | | **15-30 cm Soil Layer** | | |
|---|---|---|---|---|---|---|
| | ***Before Treatment*** | ***After Treatment*** | ***± (%)*** | ***Before Treatment*** | ***After Treatment*** | ***± (%)*** |
| A | 14.88 | 16.36 | 9.95 | 13.57 | 14.22 | 4.79 |
| B | 17.73 | 19.60 | 10.55 | 16.02 | 16.79 | 4.81 |
| C | 20.17 | 21.32 | 5.70 | 19.38 | 20.17 | 4.08 |
| D | 19.38 | 20.55 | 6.04 | 16.97 | 17.82 | 5.01 |
| Control 1 | 15.68 | 16.37 | 4.42 | 14.65 | 14.69 | 0.27 |
| E | 22.17 | 24.36 | 9.88 | 17.59 | 18.43 | 4.78 |
| Control 2 | 21.97 | 21.98 | 0.05 | 17.48 | 17.50 | 0.11 |

*3.1.3. Total N contents.*The comprehensive measures are effective in increasing total N contents of the soil (Table.3-3). The effects differ with the measures and soil layers. In 0-15 cm soil layer, total N contents of the soils are raised by 8.27-16.39%, and the effects of different measures are in the order: measure B > measure A > measure D > measure C > measure E. In 15-30 cm soil layer, the effects are also smaller than in 0-15 cm soil layer, and the increasing rates range from 7.07% to 11.00%. The effects of different measures are in the order: measure B > measure D > measure C > measure E > measure A.

**Table.3-3.** Effects of Comprehensive Measures on Total N Contents of the Vineyard Soils (g/kg)

| Comprehensive Measure | **0-15 cm Soil Layer** | | | **15-30 cm Soil Layer** | | |
|---|---|---|---|---|---|---|
| | ***Before Treatment*** | ***After Treatment*** | *± (%)* | ***Before Treatment*** | ***After Treatment*** | *± (%)* |
| A | 1.14 | 1.27 | 11.40 | 0.99 | 1.06 | 7.07 |
| B | 1.22 | 1.42 | 16.39 | 1.00 | 1.11 | 11.00 |
| C | 1.32 | 1.45 | 9.85 | 1.15 | 1.24 | 7.83 |
| D | 1.06 | 1.18 | 11.32 | 0.94 | 1.04 | 10.64 |
| Control 1 | 0.99 | 1.05 | 6.06 | 0.93 | 0.95 | 2.15 |
| E | 1.33 | 1.44 | 8.27 | 0.97 | 1.04 | 7.22 |
| Control 2 | 1.33 | 1.35 | 1.50 | 0.97 | 0.98 | 1.03 |

*3.1.4. Available P contents.*Among the comprehensive measures, most measures (measure A, B, C and D) have no obvious effect on soil available P contents, and the increasing rates are below 5% (Table.3-4). But the effect of measure E is high, and increasing rate is 22.68% in 0-15 cm soil layer, and 16.79% in 15-30 cm soil layer.

*3.1.5. Available K contents.*It can be seen from Table.3-5 that the measures are generally effective on available K contents of the soils, but the effects vary with the measures and soil layers. The effects of different measures are in the order: measure B > measure A > measure E > measure C > measure D, irrespective of 0-15 cm or 15-30 cm soil layer.

**Table.3-4.** Effects of Comprehensive Measures on Available P Contents of the Vineyard Soils (mg/kg)

| Comprehensive Measure | **0-15 cm Soil Layer** | | | **15-30 cm Soil Layer** | | |
|---|---|---|---|---|---|---|
| | ***Before Treatment*** | ***After Treatment*** | *± (%)* | ***Before Treatment*** | ***After Treatment*** | *± (%)* |
| A | 49.98 | 52.08 | 4.20 | 36.80 | 38.12 | 3.59 |
| B | 50.14 | 52.22 | 4.15 | 43.56 | 44.99 | 3.28 |
| C | 65.46 | 67.67 | 3.38 | 49.57 | 50.58 | 2.04 |
| D | 60.53 | 62.79 | 3.73 | 40.26 | 41.84 | 3.92 |
| Control 1 | 38.56 | 40.33 | 4.59 | 28.80 | 29.72 | 3.19 |
| E | 36.28 | 44.51 | 22.68 | 24.90 | 29.08 | 16.79 |
| Control 2 | 36.28 | 37.45 | 3.22 | 24.90 | 25.72 | 3.29 |

**Table.3-5.** Effects of Comprehensive Measures on Available K Contents of the Vineyard Soils (mg/kg)

| Comprehensive Measure | **0-15 cm Soil Layer** | | | **15-30 cm Soil Layer** | | |
|---|---|---|---|---|---|---|
| | ***Before Treatment*** | ***After Treatment*** | *± (%)* | ***Before Treatment*** | ***After Treatment*** | *± (%)* |
| A | 210.60 | 238.45 | 13.22 | 234.20 | 256.25 | 9.42 |
| B | 166.90 | 191.32 | 14.63 | 156.00 | 175.26 | 12.35 |
| C | 203.10 | 221.07 | 8.85 | 190.10 | 205.15 | 7.92 |
| D | 210.50 | 221.36 | 5.16 | 204.90 | 213.92 | 4.40 |
| Control 1 | 167.30 | 175.32 | 4.79 | 154.50 | 159.45 | 3.20 |
| E | 176.63 | 194.40 | 10.06 | 124.74 | 135.12 | 8.32 |
| Control 2 | 176.63 | 188.66 | 6.81 | 124.74 | 128.94 | 3.37 |

*3.1.6. Bulk densities.*The effects of different comprehensive measures on bulk densities of the vineyard soils differ largely among the measures, and the decreasing rates on soil bulk densities range from -10.42% to -0.77% (Table.3-6). The effects of different measures are in the order: measure E > measure A > measure B > measure D > measure C. For the controls, the soil bulk densities change small during the experiments.

**Table.3-6.** Effects of comprehensive measures on bulk densities of the vineyard soils (g/cm$^3$)

| | Comprehensive Measures | | | | | | |
|---|---|---|---|---|---|---|---|
| | A | B | C | D | *Control 1* | *E* | *Control 2* |
| Before treatment | 1.43 | 1.39 | 1.3 | 1.38 | 1.45 | 1.44 | 1.46 |
| After treatment | 1.34 | 1.32 | 1.29 | 1.33 | 1.44 | 1.29 | 1.47 |
| ± (%) | -6.29 | -5.04 | -0.77 | -3.62 | -0.69 | -10.42 | 0.68 |

The results indicate that the soils become loose and have more pores after the application of the comprehensive measures, which will benefit the exchange of water, fertilizer, gas and heat, the activities of microorganism, the transfer of nutrients from soil to plant, and the growth of plants [9]. Some researches also presented that soil bulk densities decreased, soil total porosities increased, and gas permeability of the soils was improved after application of soil conditioners [10-12].

## *4. Effects of different comprehensive measures on the yields and soluble solid contents of grape fruits*

### *4.1. Yield*

Compared to the controls, the comprehensive measures can increase the yields of grape fruits, but the increasing rates differ largely among the measures (Figur.4-1). The yields of grape fruits are raised by 3.30-14.94% by the measures, and the increasing effects are in the order: measure B > measure A > measure E > measure D > measure C.

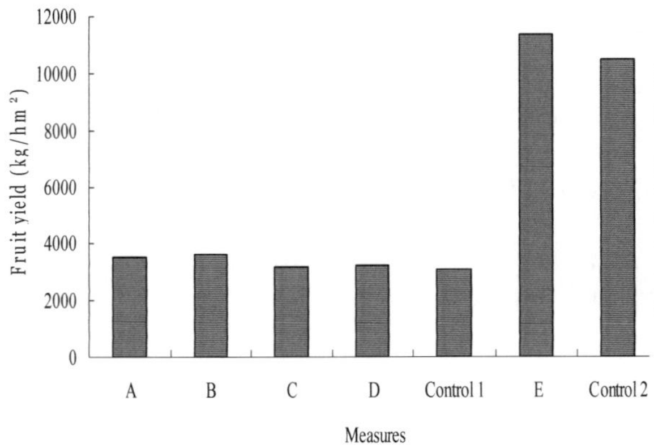


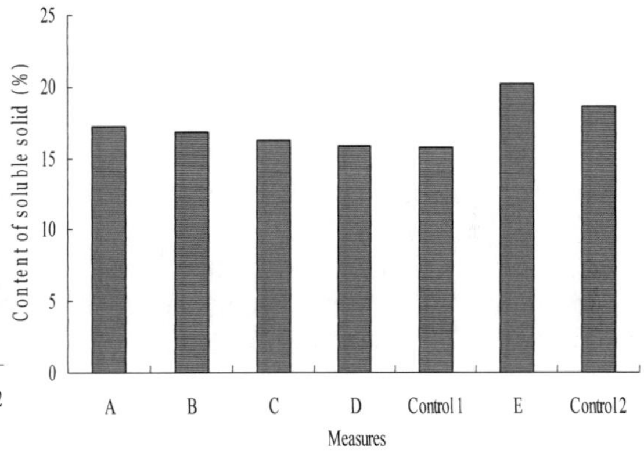


**Figure.4-1.** Effects of different comprehensive measures on the yields of grape fruit

**Figure.4-2.** Effects of different comprehensive measures on contents of soluble solid in grape fruit

### *4.2. Contents of soluble solid.*

Effects of the comprehensive measures on the contents of soluble solid in grape fruits also vary among the measures, and the increasing rates range from 0.64% to 9.55% (Figure.4-2). The effects are in the order: measure A > measure E > measure B > measure C > measure D.

## 5. Conclusions

(1) The comprehensive improving measures can effectively raise the pH of acidified vineyard soils. The effects of measure C (Unslaked lime + sheep manure + growth of hairy vetch), measure D (Unslaked lime + sheep manure + growth of broad bean) and measure E (Soil conditioner Naduoli No. 1 + bacterial manure + natural grass) are better than that of measure A (Seaweed liquid silicon + sheep

manure + growth of hairy vetch) and measure B (Seaweed liquid silicon + sheep manure + growth of broad bean).
(2) The comprehensive measures can obviously decrease soil bulk density, and the effects are in the order: measure E > measure A > measure B > measure D > measure C.
(3) The comprehensive measures can largely increase the organic matter contents of 0-15 cm soil layer. The effects of measures B, E and A are better than that of measures D and C.
(4) The comprehensive measures can effectively increase total N and available K contents in the soils. The total N contents are raised by 7.07-16.39%, and available K contents by 4.40-14.63%, by the measures. For increasing available P contents, only measure E is highly effective.
(5) As a result of soil quality improvement, the yields and quality of grape fruits are largely increased by three comprehensive measures (measure A, B and E). The yields are raised by 7.67-14.94%, and the soluble solid contents by 7.01-9.55%, by the three measures.

## 6. Acknowledgment
This research was financially supported by the Agricultural "Three New Projects" of Jiangsu Province (SXGC [2016]076)

## 7. References
[1] J. Y. Wang, Q. H. Liu, J. L. Li, S. A. Jin, and Y. B. Yuan, "Analysis on the Characteristic and Cause of Orchard Soil Acidification in the Area of Shandong Peninsula," Chinese Agricultural Science Bulletin, vol. 26, Oct. 2010, pp. 164-169.
[2] R. Q. Zhang, J. C. Yang, X. Sun, L. L. Zhao, Y. X. Sun, W. H. Xu, and Z. W. Jiang, "Research on Melioration Effect of Two Kinds of Biochars on Orchard Acidified Soil," Shandong Agricultural Sciences, vol. 48, Feb. 2016, pp. 74-79, doi: 10.14083/j.issn.1001-4942.
[3] Y. P. Li, H. Y. Hu, Z. J. Li, Q. B. Kong, C. S. Dai, and Y. Q. Zhang, "Effects of soil conditioner on red soil pH and yield and quality of water spinach," Soils and Fertilizers Sciences in China, vol. 6, Dec. 2014, pp. 21-26, doi: 10.11838/sfsc.20140604.
[4] J. X. Yi, L. X. Lü, and G. D. Liu, "Research on soil acidification and acidic soil's melioration," Journal of South China University of Tropical Agriculture, vol. 12, Jan. 2006, pp. 23-27, doi: 10.3969/j.issn.1674-7054.
[5] H. L. Chen, "Reason and harm of soil acidification in solar greenhouse and its comprehensive prevention technology," Tianjin Agriculture and Forestry Technology, vol. 4, Apr. 2009, pp. 37, doi: 10.16013/j.cnki.1002-0659.
[6] K. Z. Xie, P. Z. Xu, C. Yan, F. B. Zhang, J. S. Chen, S. H. Tang, X. Huang, and W. J. Gu, "Study the effects of soil improvement on acid soil in the south of China," Chinese Agricultural Science Bulletin, vol. 25, Nov. 2009, pp. 160-165.
[7] S. C. Lu, Quality status and improvement of orchard soil, Beijing: Chemical Industry Press, 2013.
[8] L. Xu, J. Yang, Y. X. Wang, J. T. Ye, B. X. Ma, and C. Lu, "Detection of Soluble Solids Content of Postharvest Grape Based on Hyperspectral Imaging. Journal of Henan Agricultural Sciences, vol. 46, Mar. 2017, pp. 143-147, doi: 10.15933/j.cnki.1004-3268.
[9] H. W. Zhang, M. J. Long, and F. S. Zeng, "Study of amelioration to lateritic red soil by copolymer of humic acids grafted vinyl monomer," Research of Soil and Water Conservation, vol. 2, Feb. 2001, pp. 115-118, doi: 10.3969/j.issn.1005-3409.
[10] D. G. Fan, Z. H. Yuan, and Z. Q. Song, "Primary report on the application of soil conditioner in dryland orchard," Jiangxi Horticulture, vol. 5, May. 2004, pp. 12-13, doi: 10.3969/j.issn.1006-4958.
[11] D. L. Li, C. L. He, and X. M. Men, "Effect of different soil conditioners applied on Shajiang black soil," Soil, vol. 4, Apr. 2000, pp. 210-214, doi: 10.3321/j.issn:0253-9829.
[12] X. B. Wang, and D. X. Cai, "Research and application of soil conditioner in agriculture," Plant Nutrition and Fertilizer Science, vol. 6, Apr. 2000, pp. 457-463, doi: 10.11674/zwyf.

IOP Conf. Series: Earth and Environmental Science **111** (2018) 012012 doi:10.1088/1755-1315/111/1/012012

# Removal of Microbial Contamination from Surface by Plasma

**Xinxin Feng[1], Hongxia Liu[1,a]*, Zhenxing Shen[1], Taobo Wang[1]**

[1]Department of Environmental Science and Engineering, Xi'an Jiaotong University, Shaanxi, P.R. China

E-mail: [a]hxliu72@xjtu.edu.cn

**Abstract.** Microbial contamination is closely associated with human and environmental health, they can be tested on food surfaces, medical devices, packing material and so on. In this paper the removal of the microbial contamination from surface using plasma treatment is investigated. The Escherichia coli (E. coli) has been chosen as a bio-indicator enabling to evaluate the effect of plasma assisted microbial inactivation. Oxygen gas was as the working gas. The plasma RF power, plasma exposition time, gas flow and the concentration of organic pollutant were varied in order to see the effect of the plasma treatment on the Gram-negative germ removal. After the treatment, the microbial abatement was evaluated by the standard plate count method. This proved a positive effect of the plasma treatment on Gram-negative germ removal. The kinetics and mathematical model of removal were studied after plasma treatment, and then the removing course of E. coli was analyzed. This work is meaningful for deepening our understanding of the fundamental scientific principles regarding microbial contamination from surface by plasma.

## 1. Introduction

Contamination of surfaces by micro-organisms is a major source of problems in health care [1]. For example, surgical instruments were microbially contaminated, which would lead to increased surgical infection incidence [2, 3]. Furthermore, microbiological contamination of food surfaces causes some foodborne illnesses. In 2015, FoodNet identified 20107 confirmed cases of infections, 4531 hospitalizations and 77 deaths caused by nine pathogens transmitted through food at 10 sites, which encompassed 15% of the US population [4]. To avoid risks to the safety of patients and personnel, therefore, controlling microbial contamination has become an important issue and a hot research topic in the worldwide.

In general, many different methods have been used to inactivate microorganism, which widely used two methods including autoclaving and exposure to gases (ethylene) oxide. Both methods, although effective, suffer from drawbacks such as exposure to extremely high temperatures (> 100 ◦C) in the case of autoclaves and toxic exposure in the case of ethylene oxide. Another concern with the latter is long aeration process (up to 24 h) and, importantly, creates a serious threat for both personnel and the environment [5]. Especially, due to some thermo sensitive materials or food, the decontamination technique of microbial from surfaces of those materials and food, without using toxic materials or high temperatures, is arrestive more and more in the medical field and food industry [4,6,7]. For these reasons, development of new cold techniques of microbial elimination is extremely important. Plasma-based inactivation technique is one of the most serious current alternatives to gaseous sterilization. As a chemically active medium, plasma is formed by excitation, dissociation, and ionization of any gaseous or vaporous substance, including nontoxic substances and even inert gases. Such active particles exist only while the discharge glows, and they disappear almost immediately after turning off the discharge. Furthermore, adequate processes do affect only slightly the bulk material, do not produce toxic by-products, are fast and cost effective. These circumstances completely solve the problems of safety,

ecology and economy. Therefore, elimination of microbial contamination by low temperature plasma has already been regarded as one of the most promising inactivation techniques [1,4,8-12].
For a fair estimation of the efficiency and application range of the plasma removal technique, it is necessary to first investigate the main factors engaged in the gas-discharge plasma. In accordance with these reasons, the paper firstly discusses the influence of the treatment conditions and organic pollutant on the removing efficiency of bio-indicator, *Escherichia coli* (*E. coli*), in an oxygen discharge generator. Then, the removing kinetics is discussed on the basis of above obtained data.

## 2. Experimental

### *2.1. Materials*
E. coli 8099 slant lawn incubated at 37℃ for 24 h was oscillated and eluted by phosphate buffer solution (PBS), and then the eluent was diluted to form certain concentration suspensions of bacilli. This suspension was taken 0.01 mL to spread uniformly on 25×25 mm glass sheet and dried naturally at room temperature [13].

### *2.2. Decontamination testing*
The experimental arrangement used is shown schematically in Fig.2-1. RF generator is the type of SY-500W, whose frequency is 13.56 MHz and output power can be adjusted continuously in order to match SP-II matcher. Reflection power can be close to zero by using and adjusting inductive coupled discharge. The reaction chamber is Pyrex glass tube (length 120 mm, diameter 45 mm), where inductance-coupling discharge is applied.

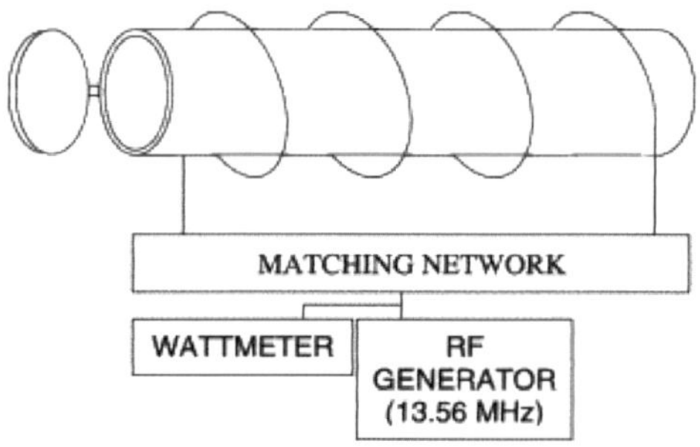


**Figure.2-1.** Schematic view of experimental arrangement .

During the experiment, sheets contaminated by bacilli were placed on the carrier and inserted into the reactor in order to carry out the treatment. The purity of oxygen was more than 99.99%. The conditions of treatment were at a discharge power of 20-100 W, exposition time of 20-120 s, and oxygen flow of 20-100 $cm^3$/min.
After treatments, several process steps must be applied to the test sheets: first the bacilli have been removed from the sheets into phosphate buffer solution (PBS), then they were transferred to Petri dish which contained nutritional agar and incubated at $37^0$C for 48 h prior to determining the resulting number of colony forming units. Germicidal effect (GE) was determined by the following equation,

$$GE = \log N_0 - \log N_t \tag{1}$$

Where $N_0$ and $N_t$ are the number of colony forming units of control and exposed, respectively [13,14].

### *2.3. Influence of organic pollutant measurement*
Serum of moggy was as the organic pollutant and mixed pro-rata with suspensions of bacilli. The concentration of serum was set by 0, 20%, 40%, 60%, respectively. The mixed solution was used to

make samples and then treated them in above mentioned reactor. The method and calculation were in the way described in section 2.1 and 2.2.

## 3. Results and Discussion

### *3.1. Germicidal Effect at Various Treatment Conditions*

For studying the germicidal efficiency of oxygen plasma, the samples were positioned at the center of the induction coil in the Pyrex glass tube and the GE was as a function of the plasma RF power, the exposition time and the oxygen flow. The GE showed a strong dependence on above mentioned treatment conditions, which are plotted in Figs.3-1.

The GE changes undulately in a narrow range with increasing the RF power up to 80 W, as shown in Fig.3-2, and subsequently, a large increase in GE shows between 80 and 100 W, the peak GE value reaches 3.42, i.e. the germicidal efficiency reaches 99.96%. The reason is that above 80 W of RF power, the ionization degree of oxygen gas and average energy of reactive species are augmented rapidly, consequently the probability of action on bacilli is enhanced.

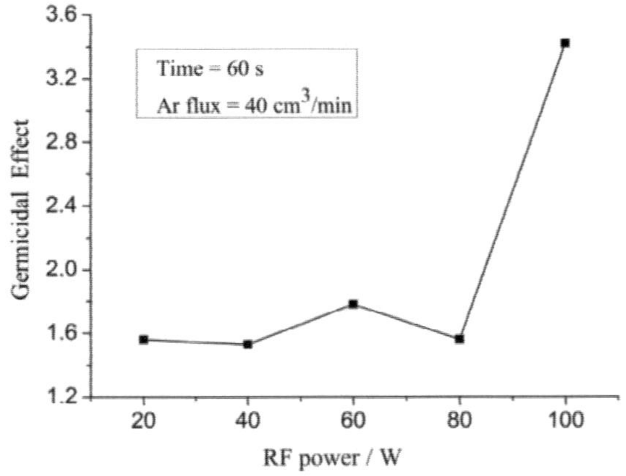


**Figure.3-1.** Effect of rf power on Germicidal Effect of *E. coli*

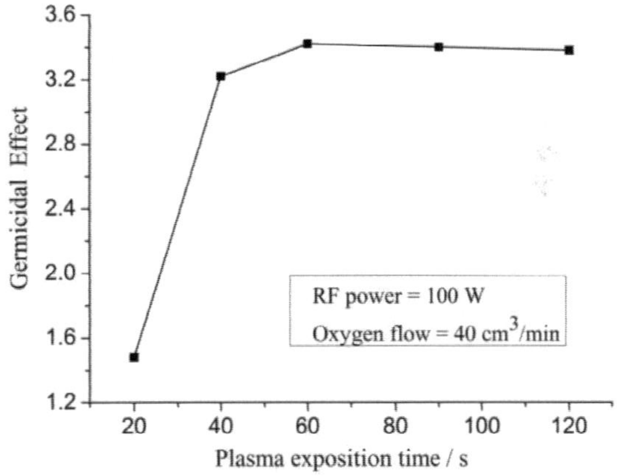


**Figure.3-2.** Effect of exposition time on Germicidal Effect of *E. coli*

From Fig.3-2, the GE increases with increasing the plasma exposition time up to 50 s, then it tends to stable beyond 50 s. Clearly, the longer exposition time cannot help improving GE. The interaction between bacilli and reaction species has almost completed at 60 s. We consider that a longer treatment time than 60 s with a higher oxygen plasma RF power than 100 W causes heavy degradation and damages on the substrate, especially for the medical polymer [15-17]. Therefore we find that the optimum plasma treatment conditions are 100 W power and 60 s exposition time to obtain the maximum GE. For the subsequent investigation of the effect of oxygen flow on GE, we fix the two conditions for sample decontamination as the optimum plasma treatment conditions.

Fig.3-3. shows the effect of oxygen flow on GE. The GE increases firstly but decreases later with increasing oxygen flow. When discharge power and treatment time are fixed, the energy of system received is a constant. At low oxygen flow, though the frequency of reaction species colliding with the bacilli is less than that at high flow, the average energy of them is higher than that with high flow. Thus, under 40 $cm^3$/min, with increasing the gas flow, the efficiency of bacterial decontamination is enhanced and GE increases. Beyond 40 $cm^3$/min, the number of reaction species is large but the average energy and residence time is lower, so the action on bacilli is comparatively small. Consequently, when the oxygen flow increases, while other two conditions are fixed, the GE plays down. In a word, action on bacilli is the combining effect of the energy and the amount of reaction species and residence time. All these values in Figs.3-4 indicate that the optimum oxygen plasma treatment conditions are of 100 W, 60 s and 40 $cm^3$/min for the maximum GE from surface.

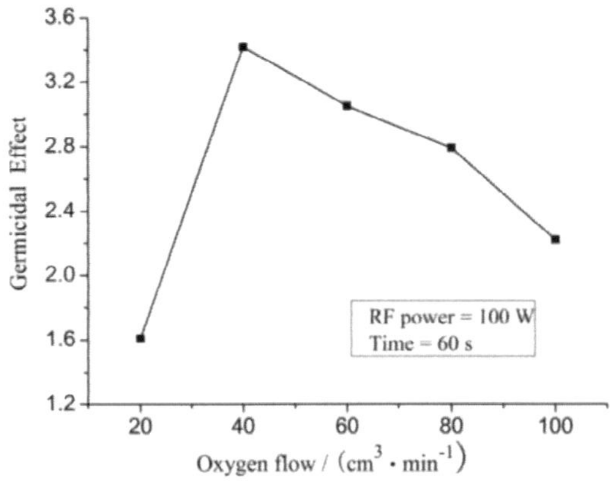


**Figure.3-3.** Effect of oxygen flow on Germicidal Effect of *E. coli*

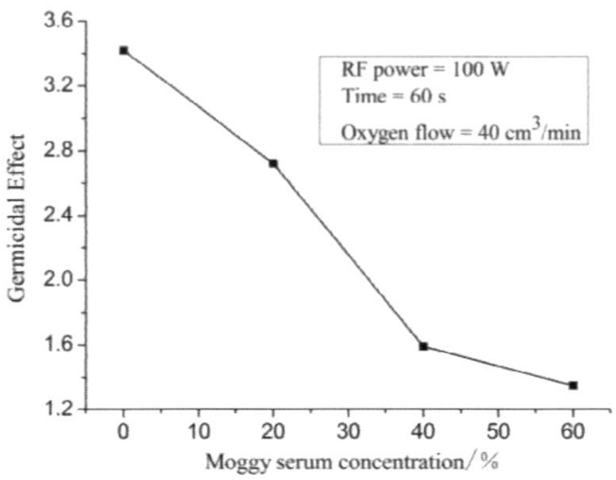


**Figure.3-4.** Germicidal Effect as function of the concentration of organic pollutant.

*3.2. Influence of Organic Pollutant on Germicidal Effect*

Fig. 5 shows the typical influence of different serum concentration of moggy on germicidal effect. From Fig. 3-4, germicidal effect plays down sharply with increasing the serum concentration of moggy. This conclusion indicates that elimination efficiency of bacilli by oxygen plasma is very unsatisfactory when organic pollutant, for example, the serum of moggy, is present. The reason could be as follows [18].

(1) Organic pollutant can form a protection course on microbial surface which can hold back the interaction between activated particles such as electron, ions and radicals and microorganisms so as to microorganisms produce adaptability to these activated particles. (2) The interaction between organic pollutant and activated particles can produce compound with lower solubility which has mechanical protecting action to microorganisms. (3) Some activated particles react to organic pollutant such as neutralization so as to their concentration acted on microorganisms reduces.

Germicidal kinetics from surface by oxygen. Using the Origin software to fit the experimental data, the kinetic curve of removing bacteria by oxygen plasma was obtained, as shown in Fig.3-5. After comparing with typical curves of inactivation kinetics of microorganism in classical report (Fig.3-6) [18], we found that the curve of germicidal kinetics of E. coli from surface by oxygen plasma belongs to typical concave curve. It indicates that *E. coil* was killed by oxygen plasma undergoing the steps as follows: susceptive stirps were changed to resistive ones before they were killed [13,19]. Based on the reported choices within the fitting functions of inactivation curves, the mathematical model of oxygen plasma decontamination was obtained, which is shown by equation (2).

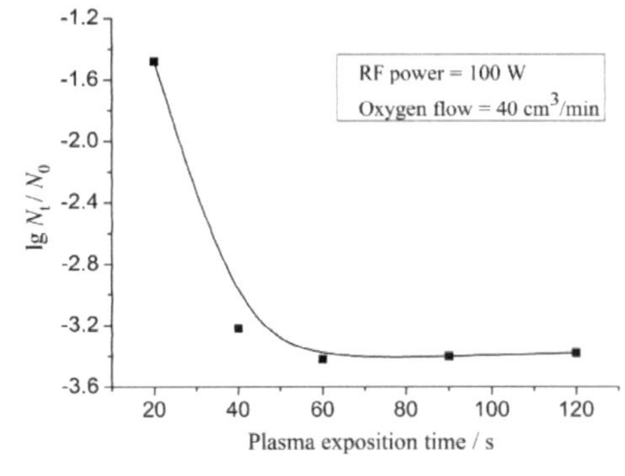


**Figure.3-5.**Germicidal kinetics curve of *E. coli* by oxygen plasma.

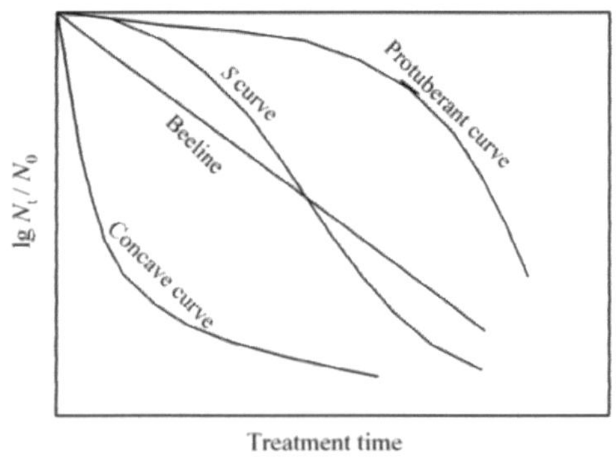


**Figure.3-6.**Representative germicidal kinetics curves of microorganism.

### *3.3. n plasma*

The template is used to format your paper and style the text. All margins, column widths, line spaces, and text fonts are prescribed; please do not alter them. You may note peculiarities. For example, the head margin in this template measures proportionately more than is customary. This measurement and others are deliberate, using specifications that anticipate your paper as one part of the entire proceedings, and not as an independent document. Please do not revise any of the current designations.

$$\log \frac{N_t}{N_0} = A_2 + (A_1 - A_2) \Big/ (1 + e^{(t-t_0)/B}) \quad (R^2 = 0.999) \tag{2}$$

Where $A_1$= -1.51535, $A_2$= -3.34452, B=3.41965.

Transforming above equation after filling the values of every parameters in it, the final function of inactivation kinetics of E. coil from surface by oxygen plasma is obtained as follows.

$$\frac{N_t}{N_0} = \frac{K_1}{1 + K_2 e^{\lambda_1 t}} - \lambda_2 \tag{3}$$

Where $K_1$=1.82917, $K_2$=0.10794, $\lambda_1$=0.29243, $\lambda_2$=3.34452.

## 4. Conclusions

In this paper, the removal of the microbial contamination from surface using plasma treatment is investigated. The main results are summarized as follows:

A) We found the optimum oxygen plasma treatment conditions were 100 W power, 60 s treatment time and 40 $cm^3$/min oxygen flow to obtain the maximum germicidal effect of *E. coli*.

B) Because the influence of organic pollutant on germicidal effect was great, we also known the polluted appliances with blood and salt couldn't be treated by this kind of oxygen plasma. If do, the cleaning procedure was absolutely necessary.

C) Fitting the experimental data using Origin software, the inactivation kinetics curve and mathematical model of oxygen plasma treatment were obtained, and then the obituary course of *E. coli* was analyzed. Equation (3), Constructed with the simple format, can provide the theoretic gist and guidance for forecasting the bacterial removal efficiency from surface, improving the treatment method and designing the plasma equipment.

This work is meaningful for deepening our understanding of the mechanism principles regarding microbial contamination from surface by plasma, providing valuable and feasible references for researchers and designers in microbial contamination control. Controlling the contamination of micro-ogranisms of surfaces by promising technology plasma decreases the safety risks of environment and human.

## 5. Acknowledgment

This work was financially supported by the Natural Science Basic Research Plan in Shaanxi Province o f China (NO.2017JM5067), the Fundamental Research Funds for the Central Universities (NO. xjj2015 132) and the National Key Research and Development Program (NO. 2016YFC0204201).

## 6. References

[1] K. Fricke, I. Koban, H. Tresp, L. Jablonowski, K. Schroder and A. Kramer et al., "Atmospheric Pressure Plasma: A High-Performance Tool for the Efficient Removal of Biofilms", Plos One, vol. 7, August 2012.

[2] S. J. Dancer, M. Stewart, C. Coulombe, A. Gregori, M. Virdi, "Surgical site infections linked to contaminated surgical instruments", J. Hosp Infect, vol. 81, pp. 231-238, August 2012.

[3] Y. Saito, H. Kobayashi, Y. Uetera, H. Yasuhara, T. Kajiura and T. Okubo, "Microbial contamination of surgical instruments used for laparotomy", Am. J. Infect. Control, vol. 42, pp. 43- 47, January 2014.

[4] C. Pignata, D. D'Angelo, E. Fea and G. Gilli, "A review on microbiological decontamination of fresh produce with nonthermal plasma", J. Appl. Microbiol, vol. 122, pp. 1438-1455, June 2017.

[5] S. Ashish, P. Amy, S. Ovidiu, and J. C. George, "Bacterial Inactivation Using an RF-Powered Atmospheric Pressure Plasma", IEEE Trans. Plasma Sci, vol. 34, pp. 1290-1296, August 2006.
[6] J. Ehlbeck, U. Schnabel, M. Polak, J. Winter, Th. von Woedtke and R. Brandenburg et al, "Low temperature atmospheric pressure plasma sources for microbial decontamination", J. Phys. D Appl. Phys, vol. 44, January 2011.
[7] K. D Weltmann and T. von Woedtke, "Campus PlasmaMed-From Basic Research to Clinical Proof", IEEE Trans Plasma Sci, vol. 39, pp. 1015–1025, April 2011.
[8] V. Scholtz, J. Pazlarova, H. Souskova, J. Khun and J. Julak, "Nonthermal plasma-A tool for decontamination and disinfection", Biotechnol. Adv, vol. 33, pp. 1108–1119, November 2015.
[9] D. Bermúdez-Aguirre, E. Wemlinger, P. Pedrow, G. Barbosa-Cánovas and M. Garcia-Perez, "Effect of atmospheric pressure cold plasma (APCP) on the inactivation of Escherichia coli in fresh produce", Food Control, vol. 34, pp. 149-157, November 2013.
[10] C. Pignata, D. D'Angelo, D. Basso, M. C. Cavallero, S. Beneventi and D. Tartaro, "Low-temperature, low-pressure gas plasma application on Aspergillus brasiliensis, Escherichia coli and pistachios", J. Appl. Microbiol, vol. 116, pp. 1137-1148, May 2014.
[11] A. Fernandez, N. Shearer, D. R.Wilson and A. Thompson, "Effect of microbial loading on the efficiency of cold atmospheric gas plasma inactivation of Salmonella enterica serovar Typhimurium", J. Food Microbiol, vol. 152, pp. 175-180, January2012.
[12] J. E. Kim, D. U. Lee and S. C. Min, "Microbial decontamination of red pepper powder by cold plasma", J. Food Microbiol, vol. 38, pp. 128-136, April 2014.
[13] K. Y. Lee, B. J. Park, and D. H. Lee, "Sterilization of Escherichia coli and MRSA using microwave-induced argon plasma at atmospheric pressure", Surf. Coat. Technol, vol. 193, pp. 35-38, April 2005.
[14] C. Mok and T. Lee, "Dielectric barrier discharge plasma inactivation of Escherichia coli", Food Eng. Prog, vol. 16, pp. 33–39, 2012.
[15] A. Suganya, G. Shanmugavelayutham and C. S Rodriiguez, "Study on plasma pre-functionalized PVC film grafted with $TiO_2$/PVP to improve blood compatible and antibacterial properties", J. Phys. D: Appl. Phys, vol. 50, March 2017.
[16] H. X. Liu and J. R. Chen, "Influence of remote oxygen plasma sterilization on surface property of substrate material", J. Xi'an Jiaotong University, vol. 43, pp. 82-86, March 2009.
[17] H. X. Liu, H. J. Zhang, and J. R. Chen, "Surface wettability and surface structure of remote $O_2$ plasma treated poly (tetrafluoroethylene)", J. Xi'an Jiaotong University, vol. 44, pp. 120-125, March 2010.
[18] T. C. Penna and C. A. Ferraz, "Cleaning of blood-contaminated reprocessed angiographic catheters and spinal needles", Infect. Cont. Hosp. Ep, vol. 21, pp. 499-504, August 2000.
[19] A. Prokop and A. E.Humphrey, "Mechanism of thermal death of bacterial spores, Electron Microscopic Observations", Folia Microbiol, vol. 17, pp. 435-443, 1972.

IOP Conf. Series: Earth and Environmental Science **111** (2018) 012013 doi:10.1088/1755-1315/111/1/012013

# The Characteristics in the Sensitivity of Microfiber Fabry-Perot Interferometric Transducers

**Xiuxin Wang[1,a*], Zhangyong Li[1], Jinzhao Lin[1], Wei Wang[1], Yin Tian[1] and Yu Pang[1]**

[1]Medical Electronics and Information Technology Engineering Research Center, the University Innovation Team Construction Plan of Smart Medical System and Core Technology, the Enhancement Plan of Chongqing Key Laboratory of Photo electronic Information Sensing and Transmitting Technology. CQUPT, Chongqing, China
E-mail:[a]wangxiuxin300@163.com

**Abstract.** We inscribe a Fabry-Perot (FP) resonator in the microfiber utilizing the 193-nm UV exposure and the phase mask technique. Some new characteristics in contrast to the conventional counterparts are measured, which are attributed to the index change in the grating and the dispersion of the effective grating length, respectively. The FP spectral dependencies on external strain, temperature, and refractive index are investigated. Our fabricated structures can have potential of acting as ultrasonic transducers and photo acoustic imaging.

## 1. Introduction

For sensing, the ordinary fiber inducing to the external environment changes need to carry out special processing, which gives birth to the side of polishing type fiber, the corroded fiber and the microfiber. L. M. Tong et al. fabricated an optical fiber with a 50 nm diameter, forming a smooth and uniform optical waveguide as "microfiber" in 2003 [1]. Microfibers with sub wavelength diameters have been caught in great research interests due to their light confinement, surface optical intensity, mechanical strength, compactness and large waveguide dispersions [2]. Microfibers can be realized by heating the ordinary fiber. Due to the lateral direction of the sub wavelength waveguide, the evanescent field effect is the most advantageous characteristic, which allows an obvious interaction between the core fiber and the surrounding environment as transducers [3]. On the other hand, because of Fiber Bragg Grating (FBG) is equipped with intrinsic selectivity of wavelength, Fabry-Perot (FP) interferometric transducer based on FBG has unique properties [4]. The structure of FP interferometric transducer uses FBG instead of traditional high reflectivity mirror, which simplifies the traditional complex structure into simple linear structure and provides convenience for element packaging and array manufacturing [5-7]. The two FBG structure not only simple structure, wavelength division multiplexing capability, convenient manufacture, but also FP resonator can be obtained with high sensitivity and long the cavity compared with the traditional dissipative type short cavity fiber [8-9].

In recent years, compared with the fiber FP interferometric transducer produced by the traditional fusion method, the microfiber FP interferometric transducer can provide a higher mechanical strength by directly engraving the grating in the optical fiber [10]. In addition, the symmetrical structure is considered to be the simplest method to solve the double sensitivities in temperature and stress of single FBG sensing, which avoids the problem both mutual restraint of sensitivity and detection precision in the dynamic stress sensing system of single FBG interferometry transducer[11-12]. Two FBGs FP interferometric transducers are used as an important structure in many fields, such as a filter, a resonant cavity of fiber laser, a hydrophone to detect sound waves and so on [13-14]. Combining the microfiber

Bragg grating with the Fabry-Perot resonator can greatly reduce the volume of the device and realize the temperature, stress, vibration, curvature, acoustic wave, concentration, refractive index and other parameters of the sensing measurements. The microfiber FP interferometric transducer has shown many new physical phenomena and has been extended to biomedical applications [15-16].
Besides the small size, the fabricated structure exhibits a high index contrast in the periodic structure and has potential in the application of quantum dynamic experiments. Even though, the milling procedure is relatively complex and time consuming because of the low efficiency of nanostructure. To improve the mechanical properties of microfiber devices, high manufacture efficiency with no distinct physical damage is required.
In this paper we fabricate a microfiber-based FP resonator by the use of 193-nm UV exposure. The responses of FP spectrum to external strain, temperature, and external refractive index are investigated. The research results are valuable for future applications of the related devices.

## 2. Fabrication Of Microfiber Fabry-Perot Interferometric Transducers

In the experiment, we use 193nm excimer laser to output ultraviolet laser as shown in Figure 2-1. The output pulse power is 3mJ and the repetition rate is 200Hz. Methods using fused tapered multimode fiber made of the microfiber with low loss and no obvious loss peak. The slight change in the diameter of the microfibers can effect on vary of the effective refractive index. The second FBG once engraved with the first FBG is not match; the reflection wavelength will drift a lot, which cannot produce resonance phenomenon. Therefore, it is necessary to use auxiliary light to ensure that Fabry-Perot cavity is located in the uniform area of the whole microfiber.

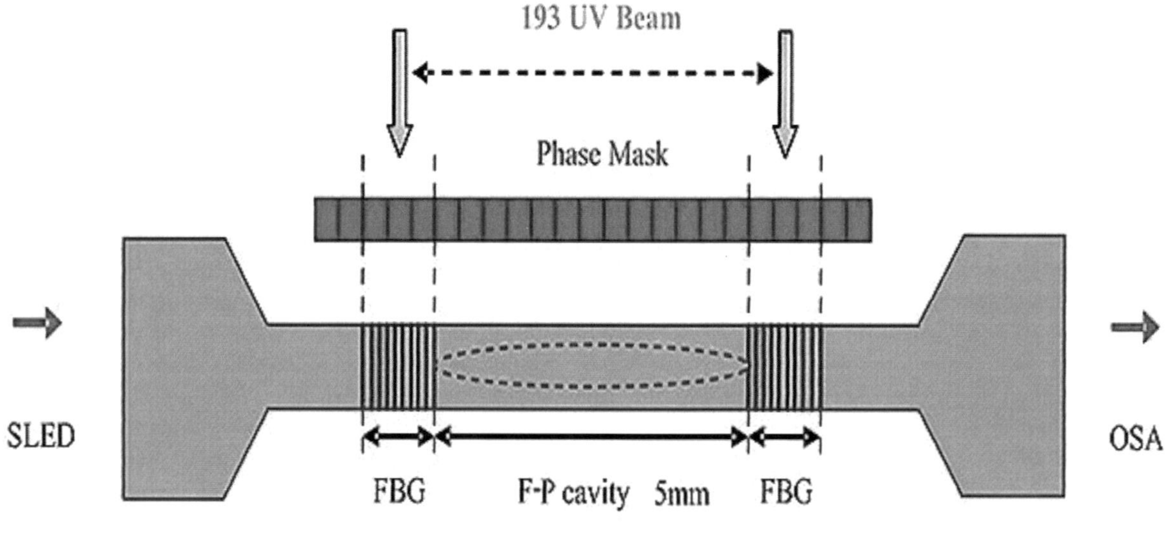


**Figure.2-1.** The fabrication of microfiber grating Fabry-Perot interferometer.

The transmission spectrum of microfiber Fabry-Perot interferometer is shown in Figure 2-2, whose diameter is d=6.1μm and length is $L_0$=5mm. A and B represent the interference fringes of each order mode, respectively. The central wavelengths of the interference peaks in the fundamental and high order modes appear at 1543.101 nm ($HE_{11}$ mode) and 1530.465 nm ($HE_{21}$ mode) respectively. The transmission depths are 24.131dB and 5.816dB, respectively. The bandwidth of Fabry-Perot interferometer is much narrower than that of the M-Z interferometer with microfibers, which is determined by the multi-beam interference principle. The peak spacing and bandwidth of the interference peaks are determined by the effective cavity length and the reflectivity of mFBG. The smaller the effective cavity length is, the higher the reflectivity is. Then the peak spacing of interference fringes is greater with the smaller the bandwidth.

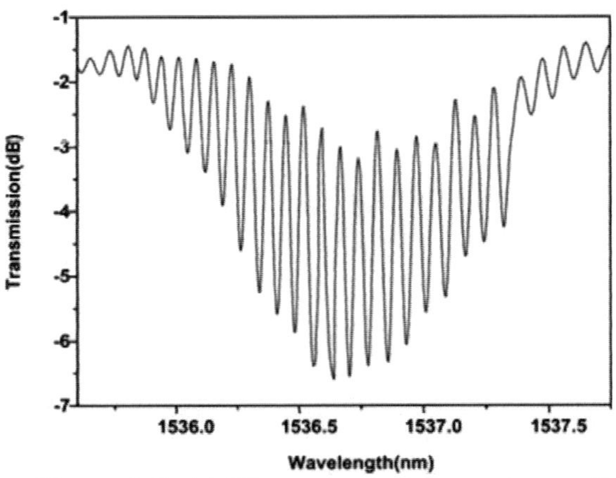


**Figure.2-2.** (a) The transmission spectrum of microfiber grating Fabry-Perot interferometer.

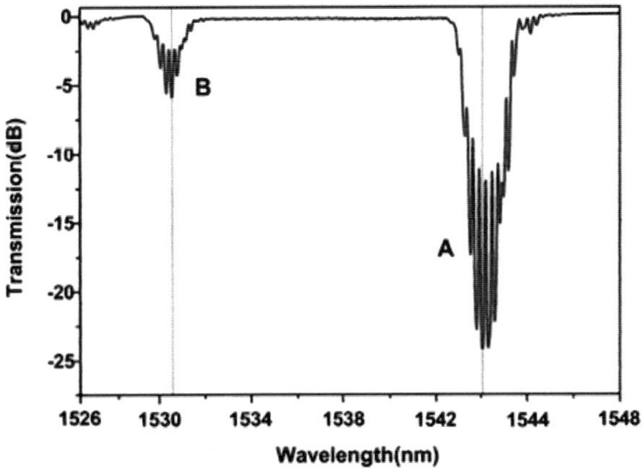


**Figure.2-2.** (b) The transmission spectrum of microfiber Fabry-Perot interferometer.

## 3. Experiments and Results

*3.1. Response to Temperature*

The microfiber Fabry-Perot interferometer is sensitive to temperature. According to the formula of interference peak:

$$2\bar{n}\left(L_0 + 2L_{eff}\right) = m\lambda \tag{1}$$

Through differentiating (1), the temperature response of the microfiber Fabry-Perot interferometer can be expressed as:

$$\begin{aligned}\frac{d\lambda_B}{dT} &= 2\bar{n}\frac{d(L_0 + 2L_{eff})}{dT} + 2(L_0 + 2L_{eff})\frac{d\bar{n}}{dT} \\ &= \lambda_B(\alpha + \beta)\end{aligned} \tag{2}$$

Where, $\alpha = \frac{1}{L_0 + 2L_{eff}} \cdot \frac{d\left(L_0 + 2L_{eff}\right)}{dT}$ indicates the thermal expansion coefficient, $\beta = \frac{1}{\bar{n}} \cdot \frac{d\bar{n}}{dT}$ indicates the thermal light coefficient. Obviously, it can be seen that the temperature can cause changes in the interference peaks. The microfiber Fabry-Perot interferometer is measured using temperature measurement system, as shown in Figure 3-1. In order to improve the accuracy of the measurement, the temperature of electric furnace rises from 40°C to 140°C continuously at room temperature. The wavelength drift is recorded during continuous warming.

Figure 3-2 shows the wavelength drift of the interference peak with different modes, which is caused by temperature. It can be seen from Figure 3-2 that the higher order interference peak wavelength drifts more as the temperature increases. It is due to that the higher order mode has a larger fiber evanescent field distribution. Then it shows obvious changes with temperature rising and perceives temperature changes outside more easily.

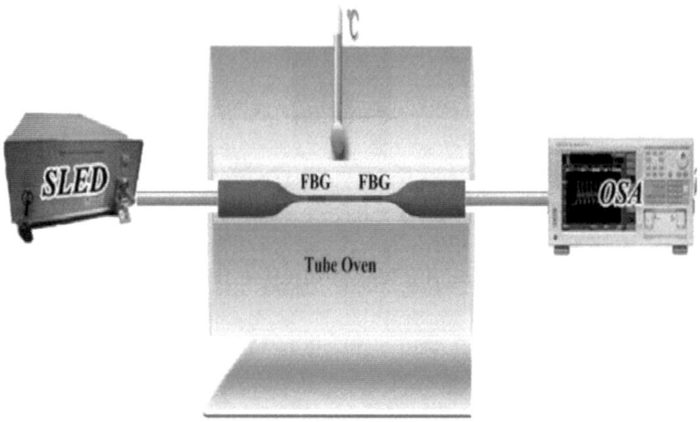


**Figure.3-1.** The schematic diagram of temperature measurement system.

It can be seen from Figure 3-3 that the interference peaks in the fundamental and high order modes show linear increment with the same peak spacing when the temperature gradually increases. Not like the refractive index which has big difference, the temperature sensitivity is 12.6 pm/°C and 12.3 pm/°C, respectively. The temperature sensitivity of an ordinary Bragg grating is 10 pm/°C at 1550nm. It is clear that the temperature sensitivity of microfiber Fabry-Perot interferometer is almost close to that of ordinary grating. It is due to the fact that the $SiO_2$ material has a similar thermal-light coefficient.

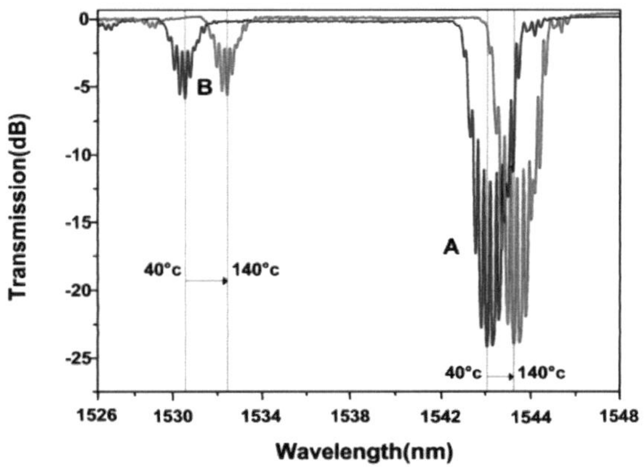


(a)

**Figure.3-2.**The interference peak wavelength with different modes and transmission spectrum.

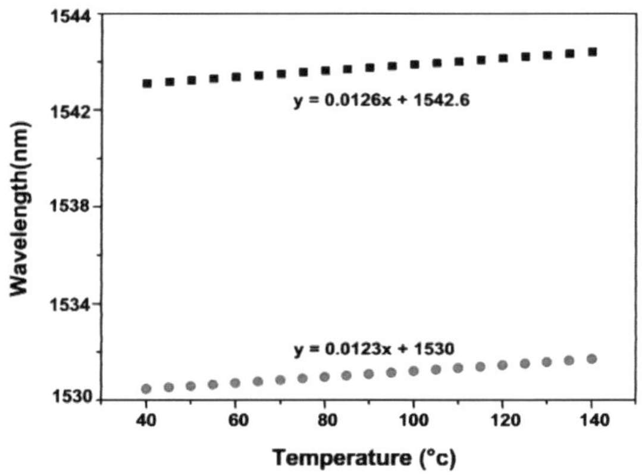


(b)

**Figure.3-3.**The interference peak wavelength with different modes and temperature response.

## 4. Conclusion

We have proposed and demonstrated a necessary condition for the ultrasonic transducers needed based on microfiber grating for photo acoustic imaging [16]. In this paper, the fabrication method, the spectral characteristics and the sensitivity of Fabry-Perot interferometric transducers based on Bragg gratings are studied. First, the theory and the output spectrum of the microfiber grating Fabry-Perot interferometric transducers are discussed. Second, the influence of reflectivity and cavity length on the output spectrum of fiber grating Fabry-Perot interferometric transducers is obtained. Third, through to the microfiber Fabry-Perot interferometric spectral analysis, it found that the different mode interference on peak temperature and refractive index sensing measurement shows different characteristics, and further to the causes of these similarities and differences between the sensitivity of parameters is analyzed.

## 5. Acknowledgment

This work was supported in part by the National Natural Science Foundation of China (Grant No. 61605021, 61571070, 61471075, 61671091); in part by Scientific and Technological Research Program of Chongqing Municipal Education Commission (Grant No. KJ1600441); in part by Natural Science Foundation of Chong Qing CSTC (cstc2017jcyjBX0057), in part by Chongqing Key Laboratory Improvement Plan (Grant No. cstc2014pt-sy40001); in part by University Innovation Team Construction Plan Funding Project of Chongqing (Grant No. CXTDG201602009); in part by Doctoral Foundation of Chongqing University of Posts and Telecommunications (Grant No. A2015-26) and in part by Youth Foundation of Chongqing University of Posts and Telecommunications (Grant No. A2015-58).

## 6. References

[1] L. Tong, R. R. Gattass, J. B. Ashcom, S. He, J. Lou, M. Shen, I. Maxwell, and E. Mazur, "Sub wavelength diameter silica wires for low- loss optical wave guiding," Nature, vol. 426, pp. 816-819, November, 2003.

[2] G. Brambilla, F. Xu, P. Horak, Y. Jung, F. Koizumi, N. P. Sessions, E. Koukharenko, X. Feng, G. S. Murugan, J. S. Wilkinson, and D. J. Richardson, "Optical fiber nanowires and microwires: Fabrication and applications," Adv. Opt. Photon., vol. 1, pp.107-161, January, 2009.

[3] C. E. Lee and H. F. Taylor, "Sensors for smart structures based on the Fabry-Perot interferometer," in Fiber optic smart structures, Eric Udd: New York, 1995, pp. 249–269.

[4] T. Zhu, D. Wu, M. Liu, and D. W. Duan, "In-line fiber optic interferometric sensors in single-mode fibers," Sensors (Basel), vol. 12, pp.10430-10449, August, 2012.

[5] S. Legoubin, M. Douay, P. Bernage, P. Niay, S. Boj, and E. Delevaque, "Free spectral range variations of grating-based Fabry-Perot filters photowritten in optical fibers," J. Opt. Soc. Am. A., vol. 12, pp. 1687-1694, April, 1995.

[6] G. E. Town, K. Sugden, J. A. R. Williams, I. Bennion, and S. B. Poole, "Wide-band Fabry-Perot-like filters in optical fiber," IEEE Photon. Technol. Lett., vol. 7, pp. 78–80, January, 1995.

[7] B. Jacobsson, V. Pasiskevicius, and F. Laurell, "Single-longitudinal-mode Nd-laser with a Bragg-grating Fabry-Perot cavity," Opt. Express, vol.14, pp. 9284-9292, October, 2006.

[8] Y. O. Barmenkov, D. Zalvidea, S. Torres-Peiró, J. L. Cruz, and M. V. Andrés, "Effective length of short Fabry-Perot cavity formed by uniform fiber Bragg gratings," Opt. Express, vol. 14, pp. 6394-6399, July, 2006.

[9] S. C. Kaddu, D. J. Booth, D. D. Garchev, and S. F. Collins, "Intrinsic fiber Fabry-Perot sensors based on collocated Bragg gratings," Opt. Commun., vol. 142, pp.189-192, June, 1997.

[10] S. S. Wang, Z. F. Hu, Y. H. Li, and L. M. Tong, "All-fiber Fabry-Perot resonators based on microfiber Sagnac loop mirrors," Opt. Lett., vol. 34, pp. 253-255, February, 2009.

[11] X. B. Zhang, J. L. Li, Y. Li, W. Y. Wang, F. F. Pang, Y. Q. Liu, and T. Y. Wang, "Sensing properties of intrinsic Fabry-Perot interferometers in fiber tapers," Proc. SPIE 8421, vol. 842189, pp. 8421891-4, November, 2012.

[12] W. Liang, Y. Huang, Y. Xu, R. K. Lee, and A. Yariv, "Highly sensitive fiber Bragg grating refractive index sensors," Appl. Phys. Lett., vol. 86, pp. 151122 , March, 2005.

[13] J. J. Zhang, Q. Z. Sun, R. B. Liang, J. H. Wo, D. M. Liu, and P. Shum, "Microfiber Fabry-Perot interferometer fabricated by taper- drawing technique and its application as a radio frequency interrogated refractive index sensor," Opt. Lett., vol.37, pp.2925-2927, July, 2012.

[14] K. P. Nayak, F. Le Kien, Y. Kawai, K. Hakuta, K. Nakajima, H. T. Miyazaki, and Y. Sugimoto, "Cavity formation on an optical nanofiber using focused ion beam milling technique," Opt. Express vol. 19, pp. 14040-14050, July, 2011.

[15] M. Ding, P. Wang, T. Lee, and G. Brambilla, "A microfiber cavity with minimal-volume confinement," Appl. Phys. Lett., vol. 99, pp. 051105, Octorber, 2011.

[16] X. X. Wang, L. Jin, J. Li, Y. Ran, and B. O. Guan, "Microfiber Interferometric Acoustic Transducers," Opt. Express, vol. 22, pp.8126-8135, April, 2014.

# Scenario Study on PM emission Reduction in Cement Industry

**Qian Tang[1], Xiaojun Chen[1]*, Xin Xia[2], Lijuan Wang[1], Huili Wang[1], Ling Jin[1] and Zhen Yan[1]**

[1]Chinese Academy for Environmental Planning, Beijing, China
[2]North China Electric Power University, Beijing, China
Email: chenxj@caep.org.cn

**Abstract**. Cement industry is one of the high pollution industries in China. Evaluation of the primary particulate matter (PM) emission status and the reduction potential is not only important for our understanding of the effectiveness of current pollution control measures but also vital for future policy design. In this study, PM emitted from cement producing process in 2014 was calculated using an emission factor method. Three PM emission control scenarios were set up considering source control, process management and end-of-pipe treatment, and the PM emission reduction by 2020 under the three scenarios was predicted, respectively. In 2014, the primary PM emission from cement industry was 1.95 million tons. By 2020, the productions of cement and clinker were expected to increase by 12% and 7%, respectively, and the PM emission would increase by about 10%. By implementation of GB4915-2013 and comprehensive control of fugitive PM emission, the PM emission would probably be reduced by 34%. Another 7% decrease would be expected from source control. The second scenario can be considered as an assessment of the effectiveness of the revised emission standard, and this research can be used as a technical support to the environmental management authorities to make relevant policies.

## 1. Introduction

Cement is one of the most important and widely used building materials in the world, and its production and consumption is closely related to the economic growth, industrialization and infrastructure development in a country or area [1,2]. Cement production and consumption in China showed a rapid growth trend and the cement production increased by 133% from 2005 to 2014[3]. China has been world No.1 cement producer and consumer in the last two decade.

Cement industry has been identified as a main source of air pollution. A large amount of air pollutants emitted from cement production, including SO2, NOx, CO, especially primary particulate matter (PM) [4]. PM is discharged in every stages of cement production. About 70% organised PM emission occurred in the kilns and other processes, such as raw materials handling, coal milling, belt transportation, coal preparation, are main sources of fugitive PM emission.

In literature, most researchers studied the application and effectiveness of the dust removal facilities and some papers estimated air pollutants emission status in cement industry years ago[5,6].The National Emission Standard of Air Pollutions for Cement Industry was revised and enacted in 2014, and it was supposed to have a significant impact on pollution control in cement industry. Systematic evaluation of the current PM emission status and estimation of PM emission reduction potential is essential for current air pollution prevention policy assessment and future policy making. In this research, we developed a method to estimate both organised and fugitive PM emission, and predicted PM emission reduction potential under three different scenarios.

## 2. Methodology

In this study, the current PM emission status in cement industry was evaluated using an emission factor method. The PM emission reduction potential in 2020 was estimated under three different scenarios (Table 1). Only direct primary PM emitted from cement production was considered.

**Table 1.** Three PM emission control scenarios in 2020

| Scenario | Clinker and Cement Production | PM Control Measures |
|---|---|---|
| Scenario 1 (BAU) | Predicted according to the economic growth expectation during the 13$^{th}$ Five-year-Plan period and relevant cement industrial policies | Same as 2014 |
| Scenario 2 | Same as Scenario 1 | Strict implementation of GB4915-2013, and take comprehensive measures to control fugitive PM emission |
| Scenario 3 | Elimination of backward productivity and setting cap of clinker and cement production | Same as Scenario 2 |

The primary PM emission from cement industry can be calculated as (1).

$$PM = \sum EF_{1i} \cdot P_{1i} + \sum e_{1i} \cdot P_{1i} + \sum EF_{2i} \cdot P_{2i} + \sum e_{2i} \cdot P_{2i} \quad (1)$$

Where, $EF_1$ and EF2 denote the emission factor of organised PM from clinker production and cement grinding, respectively;

e1 and e2 denote the emission factor of fugitive PM from clinker production and cement grinding, respectively;

P1 and P2 denote the clinker and cement production, respectively;

i denotes the ith production line.

The emission factors in the base year were obtained by studying the PM generated from cement production process, the efficiencies of different dust removal facilities[5-10] and online monitoring data from clinker kilns of different scales. The emission factors in scenario 2 and 3 can be predicted mainly according to the quantities of flue gas[11] and emission standards (GB4915-2013) for each cement production process. Fugitive emission factors were estimated from PM2.5 emission accounting technical specification for cement industry[12] and the percentage of PM2.5 in TSP emitted from each cement production process[6,13]. The data were shown in Table 2.

**Table 2.** Emission factors of fugitive PM from clinker production and cement grinding in Scenario 1

| Product | Process | Scale | E(kg/t product) |
|---|---|---|---|
| Clinker | New dry process rotary kilns | ≥4000t clinker/day | 0.1 |
| | | <4000t clinker/day | 0.2 |
| | | <2000t clinker/day | 0.4 |
| | Shaft kilns | - | 0.7 |
| | Others | - | 0.6 |
| Cement | Cement mills | ≥0.6 million tonns cement/year | 0.2 |
| | | <0.6 million tonns cement/year | 0.3 |

## 3. Results and Discussion

### *3.1. PM Emissions from Cement Industry in 2014*

In 2014, the productions of cement and clinker in China were 2.5 billion tons and 1.4 billion tons, respectively. The total amount of PM emitted from cement industry was 1.95 million tons. The fugitive PM emission was comparable to the organised PM emission. The organised PM emitted from clinker production was the major contributor, accounted for 37% of the total emission, followed by the fugitive emission from cement grinding, the proportion was 32% (shown in Figure 1).

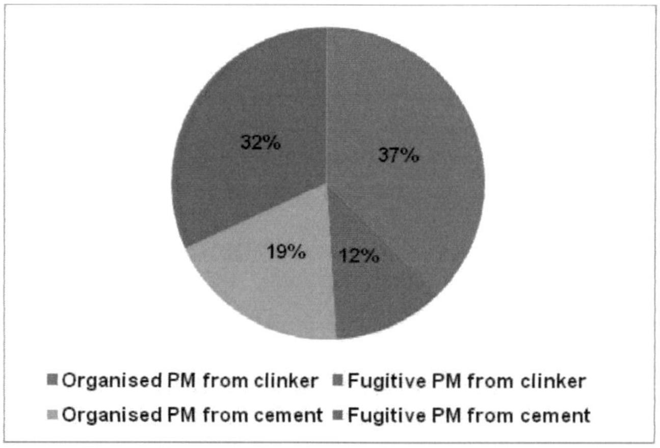


**Figure 1.** PM emission from each cement production process.

The quantities of PM emissions from cement industry in different regions vary in a relatively wide range (Figure 2). PM emissions in Guizhou province and Sichuang province exceeded 120,000 tons, which was over 20 times of that in Beijing and Tianjin.

The total amount of PM emissions is not only positively associated with the clinker and cement production, but also depends on the level of pollution control management. The effectiveness of PM emission control during clinker/cement production can be described as PM emission intensity, which was defined as the amount of PM emission per unit clinker/cement production. The average PM emission intensity during clinker production in mainland China was 0.68kg/t. The highest emission intensities were observed in Guizhou province, Heilongjiang province and Xinjiang autonomous region, which were above 1kg/t (shown in Figure 3). It was probably due to the relatively insufficient emission control management.

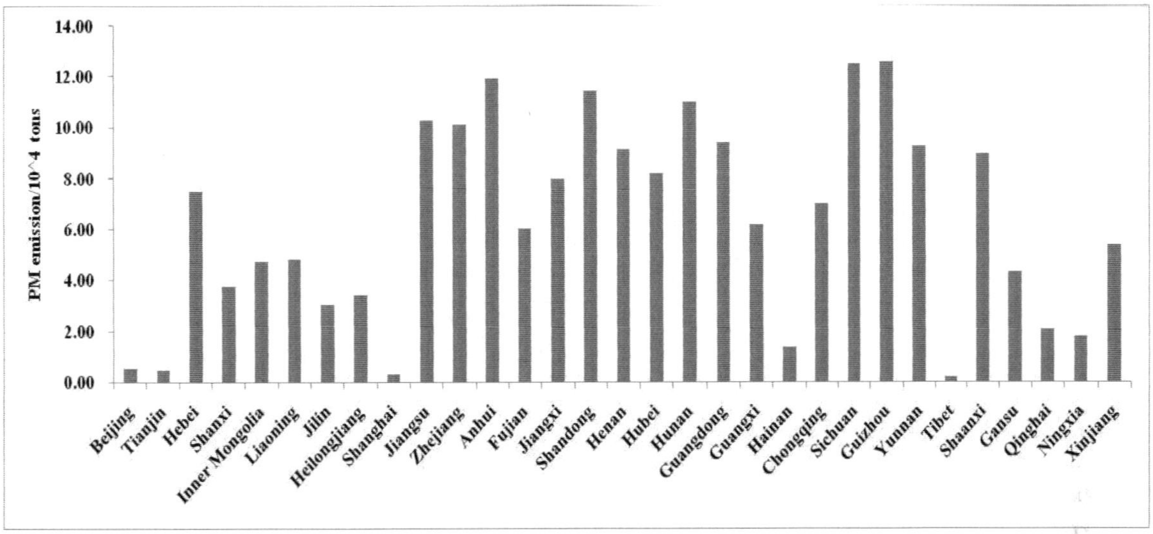


**Figure 2.** PM emission from cement industry in different provincial administrative regions in 2014.

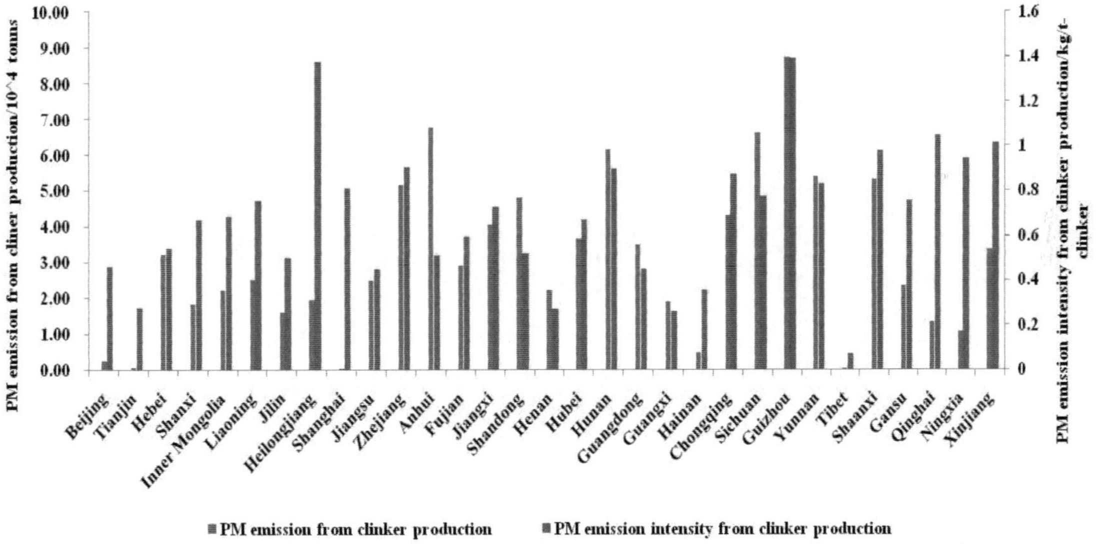


**Figure 3.** The quantities and intensities of PM emission from clinker production in different provincial administrative regions in China in 2014.

### *3.2. Scenario Study of PM Emissions from Cement Industry in 2020*

The production of clinker and cement in 2020 is closely associated with the economic development, industrialization and urbanization. It also depends on the relevant industrial policies. The annual average growth rate of GDP in China is expected to be 6.5% during the 13th Five-year Plan period [14]. The scale and quality of urbanization will be promoted, and the infrastructure construction will be further strengthened. There is likely to be strong demand for cement. At the other side, the issue of production overcapacity is perplexing the cement industry and enterprises within the industry. The average operation rate of the rotary kilns in 2015 was about 67%, and in certain areas, the operation rate was even lower than 50%. It is vital to resolve the serious problem of excess production capacity. Considering comprehensively all of the above-mentioned factors, the annual average growth rate of cement and clinker production in China during the 13th Five-year Plan was predicted to be 3% in this study, and the productions of clinker and cement in each of the provinces, autonomous regions and municipalities were estimated considering their current production status and the relevant policies in the 13th Five-year Plan period.

In 2020, the productions of cement and clinker would be 2.78 billion tons and 1.51 billion tons, an increase of 12% and 7%, respectively, compared with 2014. Jiangsu province and Henan province would still be the two largest cement production provinces, and their annual production would exceed 200 million tons. Under the pressure of elimination of excess capacity and air quality improvement in the Beijing-Tianjin-Hebei region, a decrease of the cement production in the two municipalities and a slight increase in Hebei province would be expected. The details are shown in Figure 4. In the BAU scenario, the PM emission control measures were supposed to be the same as 2014, and the total amount of PM emitted from cement industry in 2020 would be 2.15 million tons, an increase of 10% compared with 2014.

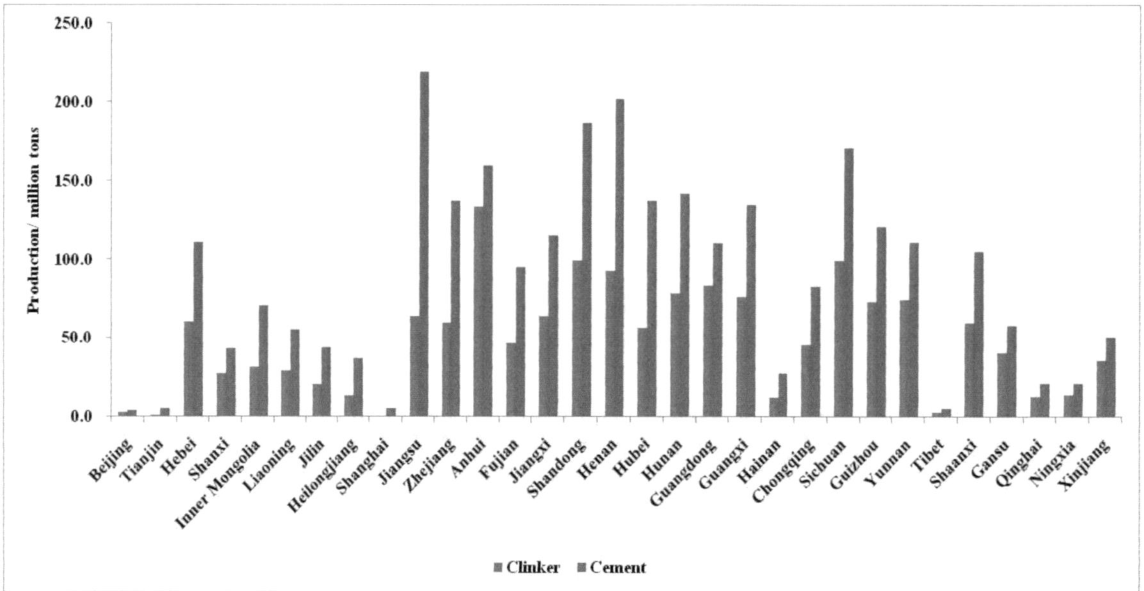


**Figure 4.** The production of clinker and cement in 2020.

PM pollution control can be considered from 3 perspectives: source control, process management and end-of-pipe treatment. Scenario 2 focused on end-of pipe treat (implementation of new emission standards) and process management (fugitive PM emission abatement), while Scenario 3 also considered source control. Under scenario 2, the PM emission would be reduced by 34% compared with BAU scenario, reaching 1.41 million tons. A further 0.14 million tons were expected under scenario 3. Implementation of higher emission standards could reduce organised PM emission effectively. The organised PM emission would decrease by about 40% since the new standard (GB4915-2013) took effect.

Fugitive PM emission is one of the important sources of particulate pollution in cement industry. Under the BAU scenario, fugitive PM emission contributed to over 40% of total PM emission. Actions must be taken to cut down fugitive PM emission. Firstly, advanced production technology should be widely adopted to prevent fugitive PM emission from the source. Secondly, fugitive PM can be collected and treated using dust removal facilities. Raw materials should be transported with closed belt conveyor and stored in closed warehouses. Thirdly, road dust resulted from raw materials and products transportation should be reduced. The fugitive PM emission would probably decrease by 25-30% by taking the above preventive measures.

## 4. Conclusions and Recommendations

In 2014, the PM emissions from cement industry in mainland China were 1.95 million tons. Organised emission was mainly from the clinker production process and fugitive emission was mainly from the cement grinding process. PM emissions from cement industry in the provincial administrative regions varied in a wide range due to the differences in the production scale and the particulate pollution prevention level.

According to the expectation of the economic growth during the 13th Five-year Plan period and the relevant industrial policies, the productions of cement and clinker were predicted to grow by 12% and 7%, respectively, compared with 2014. If the PM control remained at the same level as that in 2014, the quantities of PM emitted from cement industry in 2020 would increase by about 10%. Implementation of the new emission standards and comprehensive control of fugitive emission would probably reduce the PM emission by about 34%. Accelerating the elimination of outdated production capacities and strictly controlling the cement and clinker productions were highly likely to reduce the PM emission by a further 7%.

PM emission almost exists in every process of cement production. PM pollution prevention should be considered from three perspectives: source control, process management and end-of-pipe treatment. Meticulous management should be enhanced in cement industry.

## 5. References

[1] Daugher K 1973 Cement as an indicator of economicdevelopment *Cement and Concrete Research* **3** 163–83

[2] Stafford F, Raupp-Pereira F, Labrinca J2016 Life cycle assessment of the production of cement: A Brazilian case study *Journal of Cleaner Production* **137** 1293-9

[3] China NBS 2016 National Bureau of Statistics of China http://www.stats.gov.cn/

[4] Uwasu M, Hara K and Yabar H 2014 World cement production and environmental implications *Environmental Development* **10** 36-47

[5] Lei Y, Zhang Q, Nielsen C 2011 An inventory of primary air pollutants and CO2 emissions from cement production in China, 1990-2020 *Atmospheric Environment* **45** 147-154

[6] Lei Y, He K, Zhang Q 2008 Technology-based emission inventory of particulate matters (PM) from cement industry *Environmental Science* **29** 2366-71

[7] Hua S, Tian H, Wang K 2016 Atmospheric emission inventory of hazardous air pollutants from China's Cement plants: Temporal trends, spatial variation characteristics and scenario projections *Atmospheric Environment* **128** 1-9

[8] Sadavarte P and Venkataraman C 2014 Trends in multi-pollutant emissions from a technology-

linked inventory for India: I. Industry and transport sectors *Atmospheric Environment* **99** 353-64.

[9] Chen W, Hong J and Xu C 2015 Pollutants generated by cement production in China, their impacts, and the potential for environmental improvement *Journal of Cleaner Production* **103** 61-9

[10] Wei L and Liu Y 2013 Research progress of PM2.5 emissions in cement industry *Environmental Engineering* **31** 98-102

[11] Wang H, Liu Y, Wang F 2015 Analysis of waste gas volume and pollutants reduction potential for cement industry in China *Journal of Environmental Engineering Technology* **5** 241-6

[12] China MEP 2014 Technical Guideline for Compilation of Atmospheric Fine Particulate Matter Emission Inventory from Primary Sources (Trial).

[13] Zhang G, Liu J, Zhou J 2011 Measurement study of inhalable particulate from new type cement rotary kiln *Journal of China University of Metrology* **22** 1-5

[14] CCCPC 2016 Central Committee of The Communist Party of China, The 13th Five-Year Plan for Economic and Social Development of The People's Republic of China

IOP Conf. Series: Earth and Environmental Science **111** (2018) 012015 doi:10.1088/1755-1315/111/1/012015

# Microbial Fuel Cell Inoculated with *Ochrobactrum Tritici* KCC210 for Chromium (VI) Measurement in Electroplating Wastewater

**Jongtar Kuo, Juiling Kuo, Chiuyu Cheng, and Yingchien Chung***

Department of Biological Science and Technology, China University of Science and Technology, Taipei, Taiwan.
*Correspondence: ycchung@cc.cust.edu.tw; Tel.: +886-22782-1862; Fax: +886-22786-5456

**Abstract.** Many methods/techniques have been developed for Cr(VI) measurement, but they are often conducted offsite or/and cannot provide real-time for Cr(VI) monitoring. A microbial fuel cell (MFC) is a self-sustaining device and has great potential as a biosensor for in situ Cr(VI) measurement, especially for wastewater generated from different electroplating units. In this study, *Ochrobactrum tritici* KCC210, a facultatively anaerobic, Cr(VI)-reducing, and exoelectrogenic bacterium, was isolated and inoculated into the MFC to evaluate its feasibility as a Cr(VI) biosensor. The results indicated that *O. tritici* KCC210 exhibited high adaptability to pH, and temperature under anaerobic conditions. The maximum power density of the MFC biosensor was 17.5±0.9 mW/m$^2$ at 2,000 Ω. A good linear relationship was observed between the Cr(VI) concentration (10–80 mg/L) and voltage output. The stable performance of the MFC biosensor indicated its potential as a reliable biosensor system. Moreover, the developed MFC biosensor is a simple device that can accurately measure Cr(VI) concentrations in the actual electroplating wastewater generated from different electroplating units within 15 min with low deviations (–1.8% to 7.8%) in comparison with the values determined using standard method. Thus, the MFC biosensor can measure Cr(VI) concentrations in situ in the effluents and has potential as an early warning detection device.

## 1. Introduction

Chromium has been widely used in various industries, such as metal plating, wood preservation, tanning, and steel manufacturing, as well as in the military industry [1, 2]. Generally, chromium exists in water in two oxidation forms: hexavalent [Cr(VI)] and trivalent [Cr(III)] [3]. Cr(VI) is more toxic and mobile than Cr(III). Thus, Cr(VI) is considered toxic because it can cross cell membranes when the ambient pH exceeds 6 [4]. These toxic Cr(VI) compounds have been proved that cause health problems such as severe diarrhea, pulmonary congestions, and liver damage [5]. The U.S. Environmental Protection Agency (US EPA) identified Cr(VI) as one of the 17 chemicals posing the greatest threat to human health [6].

Removal of heavy metal treatment typically uses physical and chemical methods such as adsorption, precipitation, reverse osmosis, and ion exchange [7]. These methodologies exhibit high removal efficiency but may involve expensive equipment, and long treatment times [8]. The accidental release of industrial effluents without proper treatment would pose a serious threat to the environment during a long-term treatment process. Thus, the measurement of Cr(VI) concentrations in water or wastewater is crucial for environmental risk and health-related considerations.

Several chemical analysis methods, including ion chromatography (IC), atomic absorption spectrometry (AA), inductively coupled plasma mass spectroscopy (ICP-MS), and colorimetric methods, are currently being used for the measurement of chromium in water samples [9]. These techniques exhibit high accuracy and selectivity, however, they are usually complicated, time consuming, and expensive. Moreover, they often require sophisticated instrumentation and inadequate for use outside the laboratory. Hence, they have low applicability in routine measurements [10]. Recently, biosensors, such as amperometric enzyme-based sensors using cytochrome c3 or urease and cell-based sensors using bacteria or V79 cells, have been developed for Cr(VI) measurement [11]. They require less complex instrumentation and shorter measurement period times compared to classical methods. However, these methods are often limited to measuring very low Cr(VI) concentrations (e.g. μg/L). The dynamic ranges restrict their use to diluted samples only [9].

There are many scientific papers illustrate that pollutants can be removed from water bodies by aquatic plant, microalgae, and microorganisms [12-15]. A microbial fuel cell (MFC) is a device that uses plant, microalgae, or microorganisms as catalysts to generate electricity from organic fuels. Thus, it provides a potential approach for powering of electronic sensors or/and for the generation of renewable energy [16]. MFCs have been developed as powering electronic sensors to analyze pollutants (BOD, COD, toxins) and monitoring state variables for system control [17]. Thus, an MFC has great potential in measuring the Cr(VI) concentrations in water, if the appropriate bacteria are used. In this study, *Ochrobactrum tritici* KCC210, a facultatively anaerobic, Cr(VI)-reducing, and exoelectrogenic bacterium, was isolated from electroplating wastewater. It was inoculated into an MFC to evaluate its feasibility as a Cr(VI) biosensor. Crucial operating parameters were established to optimize the performance of the MFC. The relationship between the Cr(VI) concentration and voltage output was investigated. Cr(VI) concentrations in actual electroplating wastewater were in-situ measured using the MFC-based biosensor and ex-situ measured using standard colorimetric methods to illustrate the feasibility and accuracy of the self-sustaining device.

## 2. Materials and Methods

### 2.1. *Bacterial Strains, Cultivation, and Identification*

Sediments were collected from an electroplating wastewater treatment plant in Tainan County, Taiwan (22°47'N/120°15'E). The collected samples were centrifuged at 10,000*g* for 20 min and the obtained precipitates were inoculated in a 3 L working volume in a chemostat. Nutrient broth (Difco Co., USA) supplemented with $Na_2Cr_2O_7$ (99%, Sigma–Aldrich Co., USA), called NBCr medium, was continuously fed to the chemostat. The NBCr medium containing 30–120 mg/L of Cr(VI) was progressively fed into the system to acclimate the Cr(VI)-reducing bacteria under the anaerobic conditions. After a 30-day acclimation period, KCC210 bacterium predominated in the chemostat and was selected for subsequent experiments. To identify the isolated KCC210 bacterium, cell lysis, DNA extraction, 16S rRNA gene amplification, and sequencing were performed as described previously [18]. The KCC210 bacterium was identified as *Ochrobactrum tritici* (98.2% similarity).

### 2.2. *Factors Affecting Cr(VI) Removal by the Ochrobactrum Tritici KCC210*

To evaluate the effects of different operating conditions on Cr(VI) removal by the KCC210 bacterium, different pH values (4–10) and culture temperatures (15–45 °C) were used in anaerobic conditions. In this study, the initial cell number of KCC210 bacterium was $3.8\times10^7$ cfu/mL and initial Cr(VI) concentration was 50 mg/L. The Cr(VI) removal efficiency were analyzed after 24-h cultivation periods. All the experiments were conducted at least in triplicate to evaluate the accuracy and reproducibility of the obtained results.

### 2.3. *MFC Construction and Its Operation*

The schematic of the dual-chambered MFC is presented in Figure 1. The MFC was comprised of a 6 cm × 6 cm × 6 cm acrylic cube. The anode and cathode compartments had working volumes of 62 mL each. Anode and cathode compartments were physically separated using a PEM (Nafion 117, DuPont Co., USA) with a surface area of 25 $cm^2$. Graphite felt (25 $cm^2$ surface area) was used for the electrodes and an OK line connected the electrodes through a variable resistor. NBCr medium (50 mg/L Cr(VI)) containing $3.8 \times 10^7$ cfu/mL of the KCC210 bacterium was continuously recycled in the anode compartment of the MFC with a 1,000 Ω resistor for cell enrichment at 5-day retention time under anaerobic conditions. The anode compartment was kept anoxic by purging with nitrogen gas (99.5%, Sanfu Co. Taiwan). The catholyte consisted of 100 mM phosphate-buffered saline (Sigma–Aldrich Co., USA). When the potential of the inoculated MFC reached a steady state (after approximately 41 days of operation), the biofilm in the anode of the MFC was considered stable or mature.

To understand the operating characteristics of the MFC biosensor, 1/100 NB was used and the circuit was adjusted using variable resistance (700–8,000 Ω) to evaluate the relationship between current density and power density in a batch mode. In these batch experiments, the voltage and power density of the MFC were analyzed after 15 min operating periods. After obtaining the optimal external resistance, 1 mL Cr(VI) with a final concentration in the range of 1–100 mg/L was added to 1/100 NB to analyze the relationship between voltage output and Cr(VI) concentration.

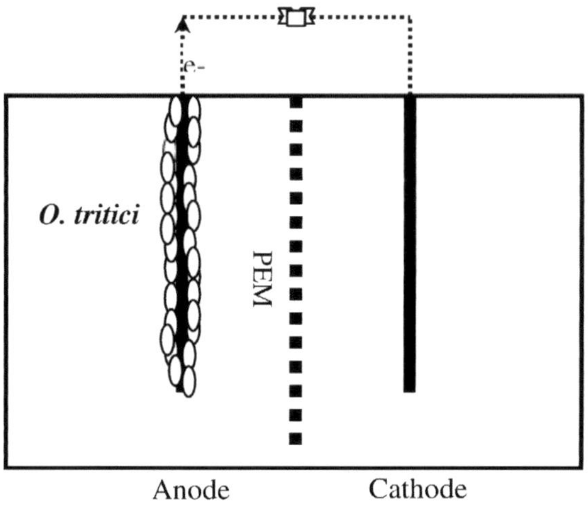


**Figure 1.** Schematic of the dual-chambered MFC.

### *2.4. Cr(VI) Measurement in Actual Electroplating Wastewater*

Cr(VI) concentrations in the actual electroplating wastewater were measured using the MFC biosensor and a standard colorimetric method [19]. Wastewater samples A–C were obtained from the effluents of different electroplating units. To measure Cr(VI) concentration in the actual wastewater, 61.5 mL of actual electroplating wastewater was mixed with 0.5 mL of 124/100 NB to maintain the NB concentration in the anolyte of the MFC. Based on the established s tandard curve between voltage output and Cr(VI) concentration, the Cr(VI) concentrations in the actual electroplating wastewater could be analyzed [16]. All the experiments were conducted using 4 separate MFCs and each analysis was conducted in triplicate.

### *2.5. Analysis*

The colorimetric method for Cr(VI) measurement was performed as described previously [19]. The MFC potential was measured using a multimeter (Model 2700, Keithley Instruments, Inc., USA). The measured voltage ($V$) was converted to current ($I$) according to the following equation: voltage ($V$) = current ($I$) × resistance ($R$). The power ($P$) was calculated as $P = I \times V$ and then normalized using the surface area of the anode.

To understand the changes in the bacterial community of the MFC, the biofilm at the anode was collected for bacterial community analysis using NGS. Cell lysis, DNA extraction, PCR amplification, and 454 pyrosequencing were conducted according to the methods described by Naz et al. [20].

## 3. Results and Discussions

### *3.1. Effects of PH and Temperature on Cr(VI) Removal of O. tritici KCC210*

As the pH in the wastewater is often variable, the effect of pH on the Cr(VI) removal efficiency of *O. tritici* KCC210 is evaluated. Figure 2A indicates that a pH of 5–10 led to >75% removal efficiency and an optimal pH range of pH 6–8, >92% removal efficiency could be achieved. This characteristic illustrates *O. tritici* KCC210 is alkalotolerant and favors the application of the MFC biosensor in measuring Cr(VI) concentration in different water bodies [21]. Figure 2B illustrates the effects of temperature on Cr(VI) removal efficiency of *O. tritici* KCC210. The results indicated that relatively high Cr(VI) removal efficiencies (82.5%–94.6%) were found in the range of 20–35 °C. However, at 15 °C and 45 °C, relatively low removal efficiencies of 46.1% and 32.5%, respectively, were observed.

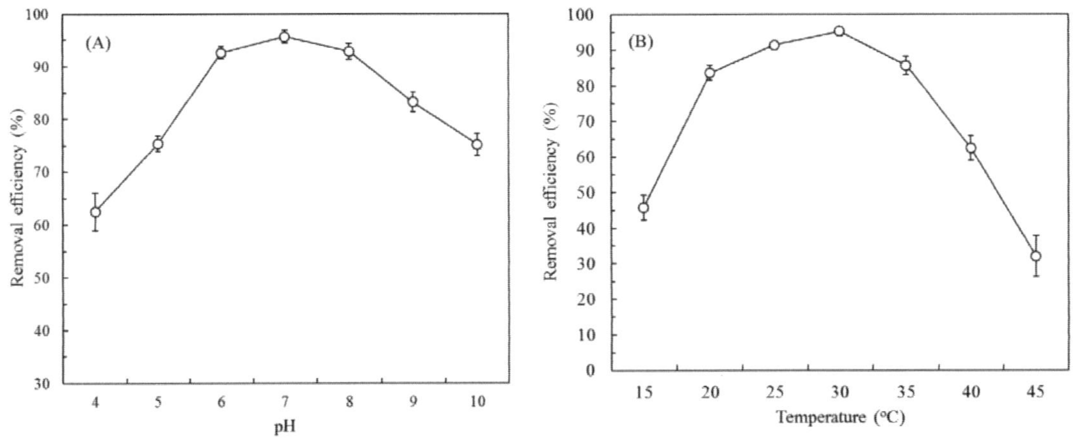


**Figure 2.** Effects of (A) pH (temperature: 30 °C) and (B) temperature (pH: 7) on the Cr(VI) removal efficiency of *O. tritici* KCC210 (Cr(VI) concentration was: 50 mg/L).

### *3.2. Operating Characteristics of MFC*

Figure 3 indicates the MFC potential/voltage decreased as the current density increased. Under such conditions, the maximum power density was 17.5 ± 0.9 $mW/m^2$, corresponding voltage was 325 ± 6.5 mV and the optimal external resistance was 2,000 Ω. Moreover, our results indicated that the decrease in MFC voltage had two significant slopes due to the activation loss and ohmic loss. However, the mass transport loss was not observed. These results illustrate that *O. tritici* KCC210 is an exoelectrogenic bacterium and the external resistance should be set at 2,000 Ω for subsequent experiments. The maximum power density of the MFC using *O. tritici* KCC210 was smaller than that of the Cr(VI)-MFC biosensor using *O. anthropi* YC152 [9]; thus, the MFC using *O. tritici* KCC210 may narrow the Cr(VI) concentration measurement range.

Figure 4 illustrates the relationship between the Cr(VI) concentration and the MFC voltage in the MFC in a batch mode. Generally, a 15 min reaction time was required to achieve the stable voltage output. A linear relationship could be observed when the Cr(VI) concentration was ranged from 10 mg/L to 80 mg/L. The regression equation was calculated as $y$ (voltage) = –1.3321$x$ (concentration) + 317.82. When the Cr(VI) concentration was <10 mg/L or >80 mg/L, the voltage variation was too slight or dramatic to establish their relationships. These results clearly illustrate operating characteristics of the developed MFC inoculated with *O. tritici* KCC210. Using the established calibration curve, the Cr(VI) concentration in the wastewater can be rapidly measured.

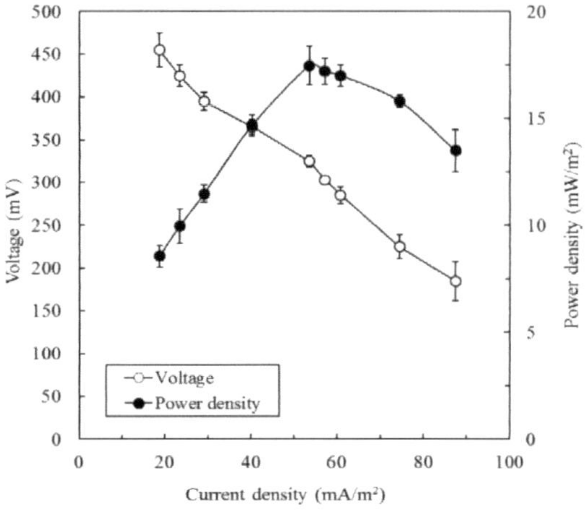


**Figure 3.** Polarization and power curves obtained from the MFC biosensor inoculated with *O. tritici* KCC210 during the stable phase of power generation.

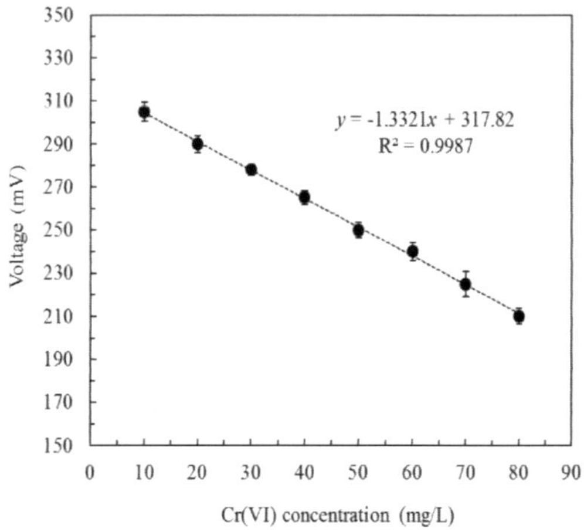


**Figure 4.** Relationship between Cr(VI) concentration and voltage output of the MFC biosensor inoculated with *O. tritici* KCC210.

### *3.3. Cr(VI) Measurement in Actual Electroplating Wastewater*

To evaluate the feasibility of using the developed MFC biosensor to measure Cr(VI) concentrations in actual wastewater, the Cr(VI) concentrations in the samples were measured using the MFC biosensor and colorimetric method. Figure 5 indicates the biosensor-based measurements on actual electroplating wastewater were accurate in comparison with standard colorimetric method. Moreover, both exhibited low deviations (−1.8% to 7.8%). Thus, the results clearly indicate that the biosensor can be used for accurate in situ measurement of Cr(VI) concentration in the electroplating wastewater.

To understand the changes in the bacterial community of the MFC after operation, the biofilm at the anode was analyzed using NGS. The results indicated the most abundant families found in the biofilm were *Brucellaceae* (98.1%) and then *Bacillaceae* (1.2%) after treating the actual electroplating wastewater with MFC for 152 days. Together, two families accounted for 99.3% of the total bacterial sequences obtained. The originally inoculated *O. tritici* KCC210 was most abundant species by further identification. The results clearly demonstrated that reliable MFC performance may be attributed to a stable bacterial community present during the treatment period.

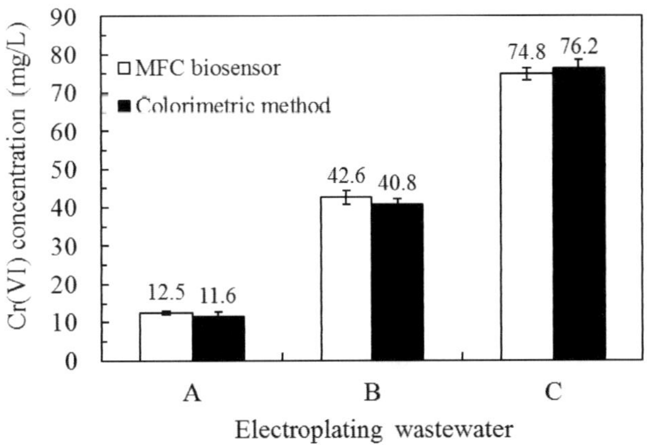


**Figure 5.** Cr(VI) measurement from actual electroplating wastewater by MFC biosensor and colorimetric method.

## 4. Conclusions

In this study, a MFC-based biosensor inoculated with *O. tritici* KCC210 for the in situ determination of Cr(VI) concentration in the electroplating wastewater was developed. *O. tritici* KCC210 is facultatively anaerobic, Cr(VI)-reducing, and exoelectrogenic; thus, it is suitable for application in an MFC to determine Cr(VI). The results illustrate that the developed MFC biosensor is a simple device for measuring a wide range of Cr(VI) concentrations with high accuracy. Although the detection limit of the biosensor was larger than the maximum limit of Cr(VI) in an effluent, the biosensor was not designed to assess whether the Cr(VI) concentration was in accordance with wastewater discharge standards or not, instead of being a warming function. Thus, the application of MFC biosensors as in situ devices for Cr(VI) determination in electroplating wastewater is promising.

## 5. Acknowledgments

The authors thank Min-Chi Hsieh for help with partially analytical measurements. This research was supported by MOST 104-2313-B-157-001-MY3.

## 6. References

[1] Thacker U and Madamwar D 2005 *World J. Microbiol. Biotechnol.* **21** 891–899
[2] Qamar M, Gondal M A and Yamani Z H 2011 *J. Mol. Catal. A-Chem.* **341**, 83–88
[3] Das K K, Dhundasi S A and Das S N 2011 *J. Basic Clin. Physiol. Pharmacol.* **22**, 3–10
[4] Quinteros F A, Poliandri A H B, Machiavelli L I, Cabilla J P and Duvilanski B H 2007 *Toxicol. Appl. Pharmacol.* **218**, 79–87
[5] Barrera-Diaza C E, Lugo-Lugo V and Bilyeu B 2012 *J. Hazard. Mater.* **223–224**, 1–12
[6] Cheung K H and Gu J D 2007 *Int. Biodeter. Biodegrad.* **59**, 8–15
[7] Abdel-Fattah T M, Mahmoud M E, Osmam M M and Ahmed S B 2014 *J. Environ. Sci. Health* A **49**, 1064–76
[8] Coelho C, Branco R, Natal-da-Luz T, Sousa J P and Morais P V 2015 *Chemosphere* **463**, 62–9
[9] Wang G H, Cheng C Y, Liu M H, Chen T Y, Hsieh M C and Chung Y C 2016 *Sensors* **16**, 1272

[10] Calvo-Pérez A, Domínguez-Renedo O, Alonso-Lomillo M and Arcos-Martínez M 2014 *Anal. Chim. Acta* **833**, 15–21

[11] Bohrn U, Mucha A, Werner C F, Trattner B, Backer M, Krumbe C, Schienle M, Stutz E, Schmitt-Landsiedel D, Fleischer M, Wagner P and Schoning M J 2013 *Sens. actuators. B Chem.* **182**, 58–65

[12] Rezania S, Taib S M, Din M F M, Dahalan F A and Kamyab H 2016 *J. Hazard. Mater.* **318**, 587–99

[13] Kamyab H, Din M F M, Hosseini S E,Ghoshal S K, Ashokkumar V, Keyvanfar A, Shafaghat A, Lee C T, Bavafa A A and Majid M Z A 2016 *Clean Technol. Environ. Policy* **35**, 2513–23

[14] Kamyab H, Din M F M, Ponraj M, Keyvanfar A, Rezania S, Taib S M and Majid M Z A 2016 Desalin. Water Treat. 57, 29118–25

[15] Gutiérrez-Corona J F, Romo-Rodríguez P, Santos-Escobar F, Espino-Saldaña A E and Hernández-Escoto H 2016 *World J. Microbiol. Biotechnol.* **32**, 191

[16] Hsieh M C and Chung Y C 2014 *Environ. Technol.* **35**, 2204–11

[17] Hsieh M C, Cheng C Y, Liu M H and Chung Y C 2016 *Sensors* **16**, 35

[18] Corby-Harris V, Snyder L A, Schwan M R, Maes P, McFrederick Q S and Anderson K E 2014 *Appl. Environ. Microbiol.* **80**, 7460–72

[19] Chen C Y, Cheng C Y, Chen C K, Hsieh M C, Lin S T, Ho K Y, Li J W, Lin C P and Chung Y C 2014 *J. Environ. Sci. Health Part* A **51**, 502–08

[20] Naz I, Hodgson D, Smith A, Marchesi J, Ahmed S, Avignone-Rossa C and Saroj D P 2016 *RSC Adv.* **6**, 104345–53

[21] Branco R, Alpoim M C and Morais P V 2004 *Can J. Microbiol.* **50**, 697–703

# Removal of Iron and Manganese from Natural Groundwater by Continuous Reactor Using Activated and Natural Mordenite Mineral Adsorption

**Y Zevi[1], S Dewita[1], A Aghasa[1,2] and D Dwinandha[1]**

[1]Department of Environmental Engineering, Faculty of Civil and Environmental Engineering, Bandung Institute of Technology (ITB), Jalan Ganesha 10, Bandung, West Java, Indonesia 40132
[2]Division of Environmental Science and Engineering, Pohang University of Science and Technology (POSTECH), 77 Cheongam-Ro. Nam-Gu. Pohang. Gyeongbuk. Korea 37673
Email: yz59@cornell.edu

**Abstract**. Mordenite minerals derived from Sukabumi natural green stone founded in Indonesia was tested in order to remove iron and manganese from natural groundwater. This research used two types of adsorbents which were consisted of physically activated and natural mordenite. Physical activation of the mordenite was carried out by heating at 400-600°C for two hours. Batch system experiments was also conducted as a preliminary experiment. Batch system proved that both activated and natural mordenite minerals were capable of reducing iron and manganese concentration from natural groundwater. Then, continuous experiment was conducted using down-flow system with 45 ml/minute of constant flow rate. The iron & manganese removal efficiency using continuous reactor for physically activated and natural mordenite were 1.38-1.99%/minute & 0.8-1.49%/minute and 2.26%/minute & 1.37-2.26%/minute respectively. In addition, the regeneration treatment using $NH_4Cl$ solution managed to improve the removal efficiency of iron & manganese to 1.98%/minute & 1.77-1.90%/minute and 2.25%/minute & 2.02-2.21%/minute on physically activated mordenite and natural mordenite respectively. Subsequently, the activation of the new mordenite was carried out by immersing mordenite in $NH_4Cl$ solution. This chemical activation showed 2.42-2.75%/minute & 0.96 – 2.67 %/minute and 2.66 – 2.78 %/minute & 1.34 – 2.32 %/minute of iron & manganese removal efficiency per detention time for chemically activated and natural mordenite respectively.

## 1. Introduction

The utilization of groundwater as a source of clean water by urban areas in Indonesia is still conducted by urban communities who still have not got better access to clean water supply through piping system. Iron and manganese are found simultaneously in groundwater with manganese concentrations which typically lower than the concentration of iron. According to [1], Iron and manganese pollution is a problem often found in groundwater access in Indonesia caused by a soil layer in Indonesia which is formed as the result of volcanic rock sedimentation containing high concentration of iron and manganese.

Iron and manganese are found simultaneously in groundwater with manganese concentrations which typically lower than the concentration of iron. Iron and manganese in groundwater can be removed effectively by using water treatment technique which could adapt to the form and concentration of iron and manganese contained in the water. In two previous studies conducted by [2] and [3] proved that the natural and activated mordenite were capable of removing iron and manganese ion content of artificial water sample. Meanwhile, this study was conducted to determine the ability of mordenite minerals to decreased iron and manganese in natural groundwater in accordance to the real condition on the field.

## 2. Materials and Methods

### *2.1. Preparation*
A tubular shaped reactor made from acrylic was used in this research. The adsorbent was mordenite minerals derived from natural green stone of Sukabumi, West Java. Adsorbent was cleaned using clean water then was activated physically by heating at a temperature of 400-600°C for two hours. Chemical activation was also conducted by immersing mordenite in $NH_4Cl$ solution for 24 hours. This research was conducted using groundwater from one of the residence wells in the settlement area around Bojongsoang, Bandung Regency, Indonesia.

### *2.2. Batch and Continuous Experiment*
Variation of conducted experiments were shown in **Table 1**.

**Table 1** Conducted Experiments

| Experiment | System | Adsorbent I | Adsorbent II |
|---|---|---|---|
| I | Batch | Physical Activation | Natural |
| II | Continuous | Physical Activation | Natural |
| III (Chemical Regeneration) | Continuous | Physical Activation | Natural |
| IV | Continuous | Chemical Activation | Natural |

### *2.3. Groundwater Sample Preparation*
This research was conducted using groundwater from one of the residence wells in the settlement area around Bojongsoang, Bandung Regency, Indonesia.

### *2.4. Measurement Parameters*

*2.4.1. Measurements of PH, Temperature and Turbidity.* The type of tools used in this research for pH and temperature measurement were pH-meter and Orion® digital thermometer for measurement in the laboratory and also portable pH-meter CT-6022 for field measurement which can also measure temperature. Meanwhile, turbidity measurement was done by using Lovibond® turbidimeter.

*2.4.2. Iron and Manganese Removal Efficiency.* To determine the removal efficiency of iron and manganese removal, concentrations of iron and manganese in groundwater and treated groundwater were measured using Flame Atomic Spectroscopy Adsorbent (Flame AAS) and then calculated using standad removal efficiency equation.

$$\text{Removal efficiency} = \frac{\text{Initial Concentration (ppm)} - \text{final concentration (ppm)}}{\text{initial concentration (ppm)}} \times 100\% \quad (1)$$

## 3. Result and Discussion

### *3.1. Initial Characteristics of Adsorbent*
Utilization of natural zeolite minerals for heavy metal adsorption can be done by using or without using treatment, such as physical and chemical activation which could enhance the adsorption capacity by producing a wider surface area through the formation of the porous structure and also eliminates polluting compounds. Physical activation was performed on mordenite by heating at a temperature of

400-600°C for two hours. Heating was used to reduce the water content and remove impurities inside zeolite pores. Mordenite has a stability to be heated up to 800°C in order to conduct physical activity in the form of heating without changing the structure of the zeolite and also avoiding the pore volume reduction. According to [3], heating of the zeolite at temperatures above 850°C could cause structural changes and also a decrease in surface area and zeolite pore volume. On the other hand, chemical activation was conducted by using $NH_4Cl$ solution. According to [2], There is an exchange of $NH_4^+$ with alkaline ion on mordenite. The exchange will reduce the mass percentage of some elements so that the selectivity of mordenite to iron and manganese were increases. **Figure 1** shows the result of SEM analysis on mordenite.

(a) Natural (b) Physical Activation (c) Chemical Activation

**Figure 1** Mordenite before Adsorption

*3.2. Groundwater Initial Characteristics*

Groundwater was collected from pipe directly connected to the wells. The results of groundwater characteristics analysis are shown in **Table 2.** It shows the high values of iron and manganese concentration which indicates that the groundwater must be treated before the usage while initial concentration of iron and manganese in groundwater used in the experiments are shown in **Table 3.**

**Table 2** Groundwater quality assessment results

| No | Parameter | Unit | Quality Standard | Methods | Result |
|---|---|---|---|---|---|
| 1 | Turbidity | NTU | 5 | SMEWW-2130-B | 55.9 |
| 2 | Color | Pt.Co | 15 | SMEWW-2120-B | 125 Colloid |
| 3 | Conductivity | uS/cm | - | SMEWW-2510 | 697 |
| 4 | Iron (Fe) | mg/L | 0.3 | SMEWW-3500-Fe-B | 36.4 |
| 5 | Fluoride (F) | mg/L | 1.5 | SMEWW-4500-F-D | 0.316 |
| 6 | Hardness ($CaCO_3$) | mg/L $CaCO_3$ | 500 | SMEWW-2340-C | 252 |
| 7 | Chloride (Cl) | mg/L | 250 | SMEWW-4500-Cl | 48.8 |
| 8 | Manganese (Mn) | mg/L | 0.4 | SMEWW-3500-Mn-B | 1.76 |
| 9 | Sulphate ($SO_4$) | mg/L | 250 | SMEWW-4500-$SO_4$-B | 1.13 |
| 10 | Total Bacteria Coliform | Units per 100 ml sample | 0 | J.P.T | 93 |

**Table 3** Initial characteristic of groundwater used in experiments

| Experiments | Source of Groundwater | Initial Concentration(mg/l) | |
|---|---|---|---|
| | | Iron | Manganase |
| I | Water Storage | 0.492 | 1.668 |
| II | Water Storage | 0.266 | 1.55 |
| III (regeneration) | Water Storage | 0.177 | 1.689 |
| IV | Water Storage | 2.626 | 1.439 |

### *3.3. Iron and Manganese Removal Efficiency*

Result of each experiments are shown in **Figure 2**. Iron and manganese removal efficiency from batch system in experiment I was fluctuated but tend to increase with the longer detention time. Iron removal was also higher than manganese same as in the previous study of [2] and [3]. According to [4], The presence of ions and other compounds in the solution may affect the adsorption of adsorbate while according to [5], heavy metal removal in batch adsorption process increases from the early stage until 60 minutes of detention time.

Continuous experiments were able to remove iron and manganese removal and were successful to meet the standard. Result of experiment II shows that there is no significant difference of iron and manganese removal efficiency between physically activated and natural mordenite. But chemical regeneration in experiment III was able to increase the efficiency after the adsorbent used for several days. Then it was continued to experiment IV that show significant difference of iron and manganese removal between chemically activated mordenite and natural mordenite. Different mechanism happened to affected the removal efficiency since activation could influenced the initial characteristic of the used mordenite. According to [2], chemical adsorption mechanism was slower than physical mechanism. Besides that, for the continuous process, the removal of iron and manganese were also influenced by the adsorbent characteristics which could be changed during the reactor operation. The iron and manganese removal in continuous reactor using down-flow system was not only involved in the mechanism of adsorption and ion exchange, but also involved in the filtration mechanism. According to [6], removal for heavy metals, such as $Fe^{3+}$, $Mn^{2+}$, $Zn^{2+}$, and $Cu^{2+}$ using zeolite adsorbents was not only involved in ion exchange mechanism, but also involved in the precipitation result of metal hydroxides from the solution. **Figure 3** shows that the increase of particle accumulation on mordenite surfaces proved that adsorption had taken place during the experiments.

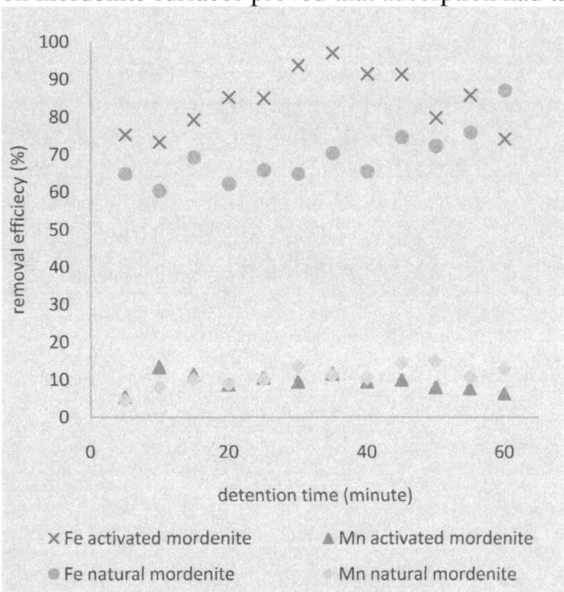


(a) Experiment I : Batch System

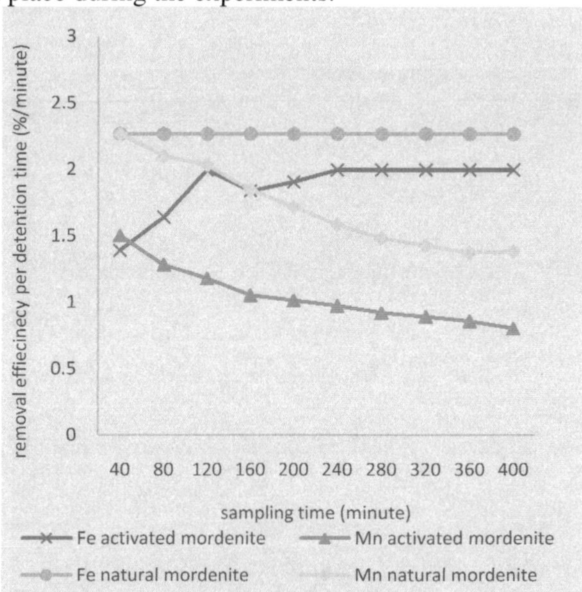


(b) Experiment II : Continuous System

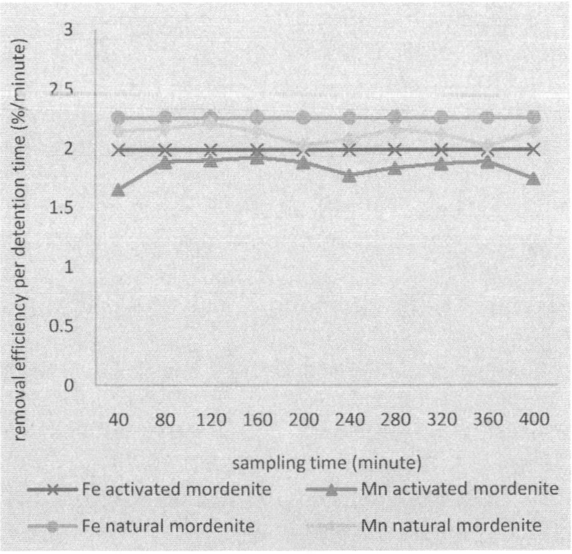


(c) Experiment III : Regeneration

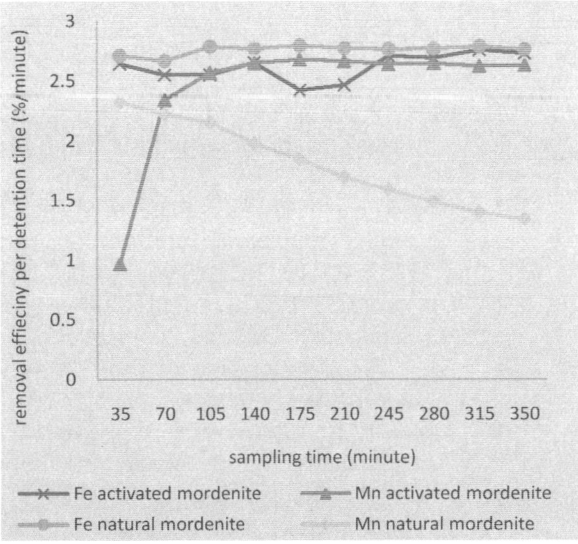


(d) Experiment IV : Continuous System

**Figure 2** Iron and Manganese Removal Efficiency

(a) Natural

(b) Physical activation

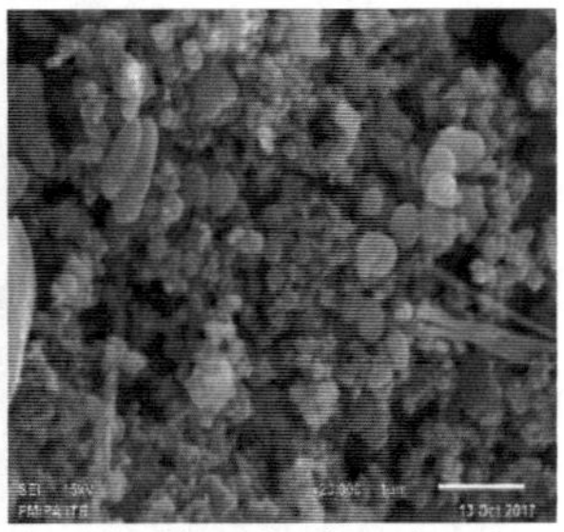

(c) Chemical activation

**Figure 3** Mordenite condition after adsorption process

## 4. Conclusions

The removal of iron and manganese from natural groundwater by using mordenite derived from Sukabumi green stone through the conducted experiments showed that iron removal was better than manganese. Even though the physical activation was already given in the process, natural mordenite could still remove iron and manganese better than mordenite with physical activation. On the other hand, chemical activation which was given to the mordenite using $NH_4Cl$ solution showed better iron and manganese removal than physical activation and natural mordenite. The maximum value of iron removal efficiency was obtained from natural mordenite which was 2.79 %/minute while manganese removal was 2.67 %/minute in chemically activated mordenite.

## 5. References

[1] Algamar K. 1989. *Activated Dry Sand Filter as an Alternatif to Reduce the Concentration of Fe and Mn in Water Treatment*. Australia: International Reference Centre for Community Water Supply and Sanitation (IRC) Library

[2] Novandy D. (2014). *Studi Kinetika dan Kapasitas Adsorpsi Ion Mangan dan Besi pada Adsorben Mineral Mordente dengan Sistem Batch*. Thesis. Bandung : Institut Teknologi Bandung

[3] Satria L. 2015. *Penyisihan Ion Besi dan Mangan dengan Mineral Mordenite yang Diaktivasi*

*dalam Fixed Bed Continuous Reactor*. Thesis. Bandung : Institut Teknologi Bandung

[4] Sharma S. 2001. *Adsorptive iron removal from groundwater*. UNESCO-IHE Institute for Water Education. Delft

[5] Lo S, Wang S, Tsai M, and Lin L. 2012. Adsorption capacity and removal efficiency of heavy metal ions by Moso and Ma bamboo activated carbons. Chemical Engineering Research and Design **90** (2012) 1397-1406

[6] Goel J, Kadirvelu K, Rajagopal C and Garg V. 2005. Removal pf lead (II) by adsorption using treated granular activated carbon: batch and column studies. Journal of Hazardous Materials **125** (2005) 211-220

[7] Cobzaru C. 2012. Modified Zeolite: Modification of Natural Zeolite for Catalytic Application. Handbook of Natural Zeolites 185-213 (**29**)

[8] Ali I. 2013. Water Treatment by Adsorption Columns: Evaluation at Ground Level. Separation & Purification Reviews. **43**: 175-205

[9] Motsi T, Rowson N and Simmons M. 2009. Adsorption of Heavy Metals from Acid Mine Drainage by Natural Zeolite. International Journal of Mineral Processing, **92**: 42-48

[10] Bordoloi S, Suresh K, Gogoi S and Dutta R. 2013. Arsenic and iron removal from groundwater by oxidation-coagulation at optimized pH : laboratory and field studies. Journal of Hazardous Materials **260** (2013) 618-626

[11] Amirnia S, Ray, Mad humita B and Margaritis A. 2016. Copper Ion Removal by Acer saccharum Leaves in a Regenerable Continuous-Flow Column. Chemical Engineering Journal **287** (2016) 755-764

# Numerical Simulation of the Effect about Groundwater Level Fluctuation on the Concentration of BTEX Dissolved into Source Zone

**Liqun Sun, Yudao Chen *, Lingzhi Jiang and Yaping Cheng**

College of Environmental Science and Engineering, Guilin University of Technology, Guilin 541004, China
Project (No. 41172229) supported by the National Natural Science Foundation
E-mail: cyd0056@vip.sina.com; 1106316393@qq.com

**Abstract:** The water level fluctuation of groundwater will affect the BTEX dissolution in the fuel leakage source zone. In order to study the effect, a leakage test of gasoline was performed in the sand-tank model in the laboratory, and the concentrations of BTEX along with water level were monitored over a long period. Combined with VISUAL MODFLOW software, RT3D module was used to simulate the concentrations of BTEX, and mass flux method was used to evaluate the effects of water level fluctuation on the BTEX dissolution. The results indicate that water level fluctuation can significantly increase the concentration of BTEX dissolved in the leakage source zone. The dissolved amount of BTEX can reach up to 2.4 times under the water level fluctuation condition. The method of numerical simulation combined with mass flux calculation can be used to evaluate the effect of water level fluctuation on BTEX dissolution.

## 1. Introduction

Monoaromatic hydrocarbons mainly including benzene, toluene, ethyl-benzene, and xylenes (referred to as BTEX) have several characteristics about solubility and mobility, which make it easy to penetrate through the soil vadose zone and stay in the aquifer. Additionally, it has some disadvantages about toxic, deformed, carcinogenic and hard to degrade. The water quality deterioration caused by oil leakage is a severe problem for humanity at present.

Monoaromatic hydrocarbons have a state of natural attenuation in the aquifer. The fluctuation of groundwater level will obviously improve the state, changing the migration of nitrate, dissolved oxygen and oxidation-reduction potential and so on [1]. The seasonal change can lead to the repeated eluviation at the smear zone. It also causes LNAPL (light non-aqueous phase liquid) to make vertical motion, and then influences the BTEX concentration [2]. At the same time, the fluctuation of water level around the smear zone may improve the biodegradation, but the increase amount of BTEX dissolved in groundwater is much more than consumed by degradation. Therefore, the amount of dissolved BTEX will increase significantly [3]. The trend of LNAPL pollution caused by water level fluctuation has been studied by some scholars, and they have made the simulation about natural attenuation by giving the adsorption conditions or electron acceptor for a certain organic pollutants or total petroleum hydrocarbon [4, 5]. There are few reports about the amount of BTEX dissolving in the source area using numerical simulation methods.

In order to determine the dissolved amount of BTEX caused by water level fluctuation, MODFLOW with its sub-module RT3D (multi-component interaction model of the migration) were used to simulate the natural attenuation process and the concentration changes of BTEX under the water level fluctuation condition. Combined with the effective dissolved amount of 63 mg/L of BTEX in gasoline with the steady water level, the BTEX dissolution was simulated under the condition of water level fluctuation and used to calculate BTEX mass flux. This work about BTEX dissolution under the condition of water level fluctuation can provide a reference for the actual contaminated site.

## 2. Materials and Methods

### *2.1. Experimental Situation*

The design of aquifer model used in this experiment is based on the generalized shallow sandy aquifers in the field. The dimension of the aquifer model is 5.8 m long, 2.9 wide and 1.30 high. Two narrow slots of 0.2 m wide is located on both east and west sides, respectively, and a 0.2 m impermeable wall divides the aquifer model into the north and south parts. The aquifer is 0.9 m thick, made of sand with 0.05-0.25 cm particle size. Groundwater pumped from nearby shallow aquifer was injected as supply water. The model has 40 sampling holes including A-E and 8 water level observation holes including W1-W8. The specific model structure is shown in Figure 1.

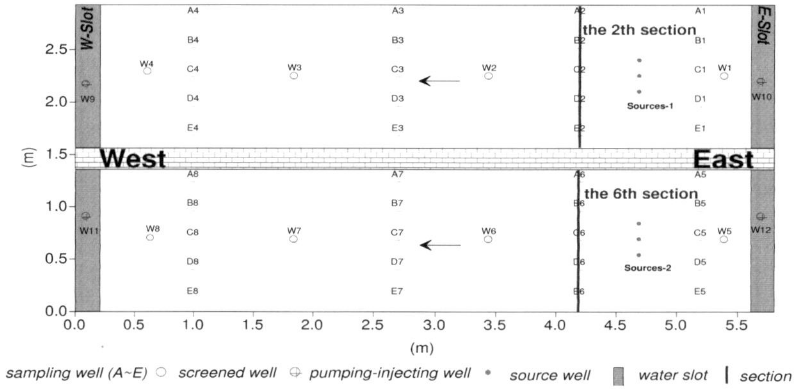


**Figure 1.** The plane figure of the aquifer model.

Before the experiment, the model was pumped and injected systematically to estimate hydrogeologic parameters by using fresh shallow groundwater. The water level was controlled at the height of 50 cm above the bottom of the model. After that, 3 L traditional gasoline was injected into the north part and equivalent of 10% ethanol gasoline was injected into the south part at the same time. The average flow rate of injection was about 500 mL/h, and the injection points was designed at the 45 cm height of the source wells of the model

After the experiment began, water level was relatively stable before the first 77 days. Water level occurred fluctuations in some extent, as shown in Table 1. During the period of experiment, BTEX concentrations were monitored A-E holes per month, detected by using an Agilent 6890 N gas chromatograph, and BTEX attenuation was paid attention. Significantly, The BTEX concentrations increased, correspondingly to each rise of the water level.

**Table 1.** The fluctuation of water level

| Period(day) | Water level changes | Water level range (cm) | Average water level (cm) |
|---|---|---|---|
| -76$^{th}$ | Stable | 48.1-51.4 | 49.8 |
| 77$^{th}$-78$^{th}$ | **Rise** | 50.4-64.3 | 57.4 |
| 78$^{th}$-128$^{th}$ | Fall | 64.3-50.8 | 56.1 |
| 129$^{th}$-147$^{th}$ | **Rise** | 50.8-63.4 | 57.0 |
| 148$^{th}$-167$^{th}$ | Stable | 63.4-64.1 | 61.3 |
| 168$^{th}$-173$^{th}$ | Fall | 64.1-53.1 | 54.1 |
| 174$^{th}$-237$^{th}$ | Stable | 53.1-49.3 | 50.9 |
| 238$^{th}$-285$^{th}$ | **Rise** | 49.3-61.5 | 57.4 |

### 2.2. *Mass Flux*

Mass flux refers to the total mass of a certain solute in a unit of time through a section perpendicular to the flow direction. In this paper, the in-site cross-section method and time-moment method were used to estimate the mass flux of BTEX concentration. The formulas are shown in the reference [6].

### 2.3. *Numerical Simulation*

The initial water level of the flow model was 0.5 m and the initial BTEX concentration was no detectable. Boundaries of the east and west sides belong to the first type of boundary condition, and other boundaries belong to zero flow condition. The top has evaporation boundaries. And set up the well boundary at east-west flume head observation well location and pollutant injection hole. The water flow model was corrected by parameter inversion, adjustment of hydraulic conductivity, water supply and water storage coefficient.

## 3. Results and Discussion

### 3.1. *The BTEX Concentration Changed with Water Level Fluctuation in the Source Area*

In the water quality model, a water injection well was set up at the pollution source, simulating the release of pollutants by setting a continuous boundary of point source injection. The pollutant flow rate was 500 ml/h. Since the model has electron acceptor, the initial concentration of each component is 4.5 mg/L dissolved oxygen, 16 mg/L sulfate, 3.8 mg/L nitrate. Shallow groundwater contains a large number of electron acceptors, so the influent sink boundary conditions was set up for the concentration of continuous injection on the boundary. In the shallow groundwater, the dissolved oxygen was 7.5 mg/L, nitrate was 3.5 mg/L and sulfate was 16 mg/L, and they were the average values of multiple measurements. The time step was calculated as 10 d [7]. By adjusting the effective solubility of BTEX within a reasonable range and fitting with the observation concentration data which changes caused by water level fluctuations to adjust parameters repeatedly. The model parameters are shown in Table 2.

Because the injection hole is located at a depth of 45 cm and the simulated water flow is similar to the one-dimensional flow. The central line at the water level of 45 cm is most impacted by the water level fluctuation. Therefore, the change of BTEX concentration is the most significant near the source region or within the source region affected by water level fluctuation, and it is more likely to be due to other factors at the downstream pore position. The model simulation results of pollution source area are shown in Figure 2.

**Table 2. Parameters that have been corrected**

| Model parameters | Value, unit | Source |
|---|---|---|
| Horizontal hydraulic conductivity | 41 m/d | measured |
| Vertical hydraulic conductivity | 4.1 m/d | [8] |
| Porosity | 0.3 | measured |
| Groundwater velocity | 0.207 m/d | measured |
| Bulk density | 1700 kg/m$^3$ | [9] |
| Specific yield | 0.2 | |
| Specific storage | 1.0×10$^{-5}$ (1/m) | [8] |
| Amount of evaporation | 0.1 mm/d | |
| Dispersion (N side model): | | |
| Longitudinal | 0.025 m | [10] |
| Transverse horizontal | 0.075m | parameter calibration |
| Transverse vertical | 0.003 m | parameter calibration |
| Dispersion (S side model): | | |
| Longitudinal | 0.025 m | [10] |
| Transverse horizontal | 0.012m | parameter calibration |
| Transverse vertical | 0.006 m | parameter calibration |
| BTEX transport velocity | 0.019 m/d | calculated |
| First-order reaction rate | 0.0002(1/day) | [11] |
| Partition coefficient ($K_d$) | 7.73×10$^{-4}$ m$^3$/kg | [12] |
| BTEX effective solubility (gasoline) | 44-112 mg/L | [13] |

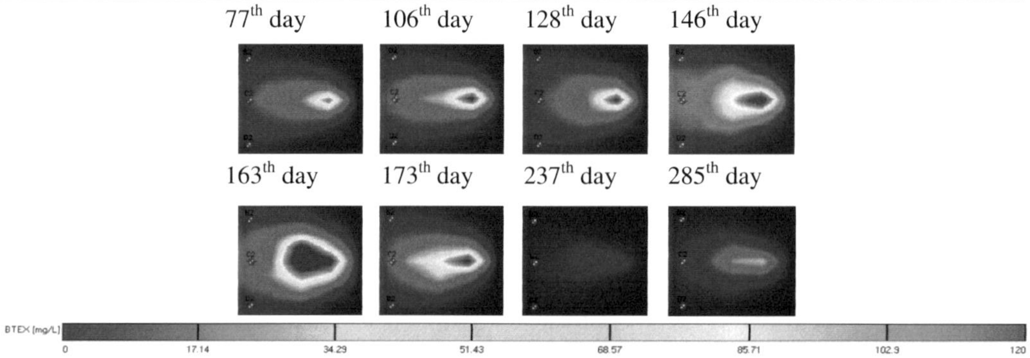


**Figure 2. The BTEX concentration changed along with the water level fluctuation at north source zone.**

The water level fluctuated 3 times in all. During the period with a stable water level before the 76th day, the concentration of BTEX in the source area first increased and then decreased. The concentration becomes higher when pollutants flow through the hole at this stage, and then decreased gradually with the natural attenuation effect. After a sudden increase in water level on day 77, the BTEX concentration increased significantly, and after the rise of the water level slowed down, the concentration decreased gradually. After a slow rise in water level at 129 days, the contaminant concentration increased. The water level gradually decreased from 148 days to 167 days after being stable for 20 days, and by day 182, it dropped to the initial level. At this point BTEX concentration also decreased gradually. At the end of the experiment, when the concentration of pollutants was very low and after the water level had risen, the BTEX concentration increased.

### *3.2. The Mass Flux of Source Area Fluctuates with Water Level Fluctuation in Downstream Section*

With the flux analysis can be seen that the difference of mass flux is obvious before and after the three water level fluctuations. Before the first water level fluctuation, the mean mass flux on both sides was

35.5 mg/d on day 71. After the fluctuation and on day 103, the mean mass flux on both sides was 78.1 mg/d. Before the second fluctuation, the mean mass flux on both sides was 75.1 mg/d on day 128. After the fluctuation and on day 163, the mean mass flux on both sides was 110.7 mg/d. Before the third water level fluctuation, the mean mass flux on both sides was 12.7 mg/d on day 237. After the water level fluctuation and on day 285, the mean mass flux on both sides was 50.4 mg/d. The observations, calculated results of mass flux are shown in Table 3. The mass flux results of observations are derived directly from the concentration observations into the relevant calculation formula, and the mass flux results of calculated is calculated by the value which is simulated by the MODFLOW software. Comparing the results showed that the simulation accuracy has improved.

**Table 3. The mass flux changed along with the times.**

| Time(d) | | | 71 | 103 | 106 | 118 | 128 | 142 | 163 | 173 | 237 | 285 |
|---|---|---|---|---|---|---|---|---|---|---|---|---|
| Mass flux (mg/d) | The 2$^{th}$ section | calculated | 26.3 | 54.9 | 73.4 | 71.0 | 61.1 | 55.0 | 98.9 | 79.7 | 5.5 | 53.8 |
| | | observations | 30.3 | - | 84.1 | 62.8 | - | 51.1 | - | 70.8 | 3.8 | - |
| | The 6$^{th}$ section | calculated | 39.7 | 101.3 | 146.3 | 112.4 | 92.0 | 93.4 | 122.4 | 111.9 | 19.9 | 47.0 |
| | | observations | 40.7 | - | 148.5 | 113.5 | - | 96.5 | - | 123.3 | 15.9 | - |

In summary, water level fluctuation will promote the dissolution of BTEX. After gasoline polluted the 45 cm layer of the aquifer, it will move with the flow of water. Most of the pollutants floating in the surface of groundwater are absorbed in the soil and rocks. Coupled with its own volatility, diffusion and dispersion, the concentration of BTEX in the smear zone increased greatly. Therefore, a sudden increase in water level will dissolve the pollutants in the water body, resulting in a substantial increase in the concentration of pollutants in the water body. When the water level drops, BTEX concentration will be reduced due to the water away from the source area. In real subsurface aquifers, away from the source area, the concentration of pollutants increases as the water level rises, due to the effect of dilution, which is also illustrated in the literature [14].

### *3.3. The BTEX Concentration Simulation Dissolved in Source Zone*

Water level fluctuations will have an effect of elution for LNAPL, and it will accelerate the dissolution of monocyclic aromatic hydrocarbons significantly. So the amount of dissolved pollutants increased sharply [6]. According to the literature, the effective solubility of BTEX is generally 63 mg/L.

**Table 4. The concentration gradient of contamination in source zone.**

| Period(d) | 0-77 | 77-78 | 78-128 | 128-148 | 148-167 | 167-173 | 173-237 | 237-285 |
|---|---|---|---|---|---|---|---|---|
| North side (mg/L) | 74.8 | 129.9 | 118.1 | 133.9 | 124.0 | 122.1 | 21.7 | 63 |
| South side (mg/L) | 63.0 | 126.0 | 144.2 | 145.7 | 149.6 | 129.9 | 25.6 | 68.9 |

After concentration correction, it was concluded that the continuous injection flow of the corresponding injection hole was 0.0036 $m^3$/d. The value we set can accurate simulate of BTEX dissolution relatively without water level fluctuation. Furthermore, through the calibration fitting with experimental results, the effective dissolved concentration of BTEX under every fluctuating condition was obtained with the stable injection flow. Finally, the model simulated the release concentration of BTEX pollutants caused by water fluctuation in the laboratory aquifer model by calibration parameters. And the concentration is the value of the point source boundary. The specific data are shown in Table 4, and it can provide data for the real reference to contaminated sites.

## 4. Conclusions

Fluctuation of water level will seriously affect the quality of groundwater in the source area. A rise in the water level will cause a lot more BTEX to dissolve into the groundwater sharply, resulting in a high BTEX concentration in the source area. A drop in the water level will alleviate the pollution in the source area. The method of numerical simulation combined with mass flux calculation can be used to evaluate the effect of water level fluctuation on BTEX dissolution. The BTEX dissolution under the condition of water level fluctuation can provide a reference for the actual contaminated site.

## 5. Acknowledgments

We gratefully acknowledge the National Natural Science Foundation of China (Project No. 41172229) and Guilin University of technology for their financial and technical support, and thank Li Liu-yue and Huang Jun-yu their assistance during fieldwork. We would like to thank the anonymous reviewers for their helpful comments and suggestions.

## 6. References

[1] Yuan Zhi-ye and Bai Shun-guo. Study on the transport of pollutants in soil under the condition of groundwater level rising and falling. Hebei: Hebei Agricultural University, 2013.
[2] Alan E, Kehew P M, Lynch, et al. Concentration trends and water-level fluctuations at underground storage tank sites. Environ Earth Sci, 2011, 62: 985–998.
[3] Richard D, Martin H, Schroth. Effect of water-table fluctuation on dissolution and biodegradation of a multi-component, light nonaqueous-phase liquid. Journal of Contaminant Hydrology, 2007, 94:235-248.
[4] Prommer H, Barry D A, Davis G.B. Modelling of physical and reactive processes during biodegradation of a hydrocarbon plume under transient groundwater flow conditions. Journal of Contaminant Hydrology, 2002, 59:113– 131.
[5] Jason N, Samuel W. A Case Study in the Use of 3-Dimensional Ground Water Modeling and Solute Transport Engines as a Tool in Site Assessment. Environment and Pollution, 2014, 3(2):55–64.
[6] Aris R. On the dispersion of linear kinematic waves. Mathematical and Physical Sciences, 1958. 245(1241):268-277.
[7] Chen X S, Brooks M C, Wood A L. The uncertainty of mass discharge measurements using pumping methods under simplified conditions. Journal of Contaminant Hydrology, 2014, 156:16-26.
[8] Wang Zhi-wei and Shan Xin-yu. Simulation of Aerobic Biodegration of BTEX Plume in Groundwater. Taiwan: National Chiao Tung University, 2012.
[9] Hudak P F. Chloride and Nitrate Distributions in the Hickory Aquifer, Central Texas, USA. Environment International. 2000, 25(4):393–401.
[10] John W H and Molson. Numerical Simulation of Hydrocarbon Fuel Dissolution and Biodegradation in Groundwater. Waterloo, Ontario, Canada, 2000.
[11] Suarez M P., and Rifai H S. Modeling Natural Attenuation of Total BTEX and Benzene Plumes with Different Kinetics. Ground Water Monitoring and Remediation, 2004, 24(3):53-68.
[12] Abdorreza V, Mohammad Z, Ezzat R, et al. Field-Scale Modeling of Benzene, Toluene, Ethyl-benzene, and Xylenes (BTEX) Released from Multiple Source Zones. Bioremediation Journal, 2012, 16(3):156–176.
[13] Johnson P C, Marian W K and James D C. Quantitative Analysis for the Cleanup of Hydrocarbon Contaminated Soils by In-Situ Soil Venting. Ground water, 1990, 28(3):413-429.
[14] Jamieson P D, Porter J R, Wilson D R. A test of computer simulation model ARC-WHEAT1 on wheat crops grown in New Zealand. Field Crops Research, 1991, 27(4):337-350.

IOP Conf. Series: Earth and Environmental Science **111** (2018) 012018 doi:10.1088/1755-1315/111/1/012018

# Characteristic Study of Shenmu Bituminous Coal Combustion with Online TG-MS-FTIR

**Guanfu Pan**[1, 2, 3]

[1] China Coal Research Institute Company of Energy Conservation Corporation Ltd..
[2] State Key Laboratory of Coal Mining and clean Utilization.
[3] National Energy Technology & Equipment Laboratory of Coal Utilization and Emission Control.
Address of all above: 5 Qingniangou east road, Hepingli Subdistrict, Beijing, China.
Email: panguanfubest@163.com

**Abstract.** The combustion characteristics of Shenmu bituminous pulverized coal (SBC) were comprehensively investigated with a combined TG-MS-FTIR system by considering the effect of particle size, heating rate and total flowrate. The combustion products were accurately quantified by normalization and numerical analysis of MS results. The results indicate that the decrease of the particle size, heating rate and total flowrate result in lower ignition and burnout temperatures. The activation energy tends to be lower with smaller particle size, lower heating rate and total flowrate. The MS and FTIR results demonstrate that lower concentrations of different products, such as NO, $NO_2$, HCN, $CH_4$ and $SO_2$ were produced with smaller particle size, slower heating rate and lower total flowrate. The decrease of particle size would lead to more contact area with oxygen and slower heating rate could provide more sufficient time for the diffusion. High total flowrate would reduce the oxygen adsorbability on the coal particle surface and shorten the residence time of oxygen, which makes the ignition difficult to occur. This work will guide to understand the combustion kinetics of pulverized coals and be beneficial to control the formation of pollutants.

## 1. Introduction

With the advantages of excellent thermal stability, high calorie and huge reserves, the Shenmu bituminous coal (SBC) has been adopted as one of the most widely used coal in China [1-3]. The investigation of the combustion characteristics is essential to understand the combustion process and control the pollutant products.

In recent years, SBC has been studied in terms of its burning and gasification properties. In 2003, Sun et al. [4] investigated the thermogravimetric (TG) characteristics of SBC and reported that vitrinite had higher volatile matter yield, maximum weight loss rate, lower initial decomposition temperature and peak temperature than that of inertinite. Compared to pressure and heating rate, the temperature has a more important impact on the devolatilization of SBC. With TG analysis, Zhao et al. [5] studied the ignition temperature ($T_{ig}$), maximum reaction temperature ($T_{max}$), burning rate, burnout temperature ($T_{burn}$) and combustion features of SBC coal and coke fine mixed sample. The results showed that with the increasing of the coke fine content, $T_{ig}$ of the coal and coke fine mixture, the caloric value of the sample and the $T_{max}$ would increase, while the burning rate would decrease. Yang et al. [6] studied the SBC ash's physicochemistry with temperature on the basis of TG-differential scanning colorimetry (TG-DSC) methods. The results indicated that the SBC ash could convert into a eutectic at low temperatures. Chang et al. [2] investigated the formation of nitrogeneous products from the gasification of SBC in a fluidized-bed/fixed-bed reactor. More recently, Yang et al. [7]

investigated the SBC's characteristics of combustion and nitrogen oxide (NOx) release in a fixed-bed reactor. The composition of the flue gas was analyzed to investigate the effect of sodium acetate on the combustion process and NOx emission. Sodium acetate was observed to reduce NOx emissions due to their special reactions with the nitrogen-containing species.

Although some works related to the combustion studies of SBC are available, a systematic study regarding the effect of particle size, heating rate and total flowrate on its combustion characteristics is scarce. Mass spectrometry (MS) and Fourier transform infrared spectroscopy (FTIR) are widely used in multicomponent analysis in energy chemical industry and are suitable to be combined with TG and gas chromatographic method [8]. It is efficient and reliable to use the TG-MS-FTIR system for numerical analysis of the combustion characteristics of coal. In the current work, the combustion characteristic of SBC was comprehensively studied with online TG-FTIR-MS system. The products formed in the combustion of SBC were accurately qualified and quantified by considering the electron impact ionization cross sections, ion flow intensities and the partial pressures of different species. The condition-characteristic relationship among the particle size, heating rate and total flowrate and ignition, burnout temperature as well as activation energy was addressed.

## 2. Materials and Methods

### *2.1. Coal Samples*

The SBC samples were bought from Shenmu Energy Developments Ltd Company and grinded into powder with different sizes. Before test, the sample was put into drying oven at the temperature of 105 °C for 90 min. Then the coals were sieved to a particle size of lesser than 40, 90-100, 128-180, 280-355 and 355-500 μm. The coal samples are mainly composed of $CaCO_3$, $SiO_2$ and Kaolinite, as shown in Figure. 1. The existence of carbonate could be the main reason to form the weak weight loss peak at the end of the combustion processes.

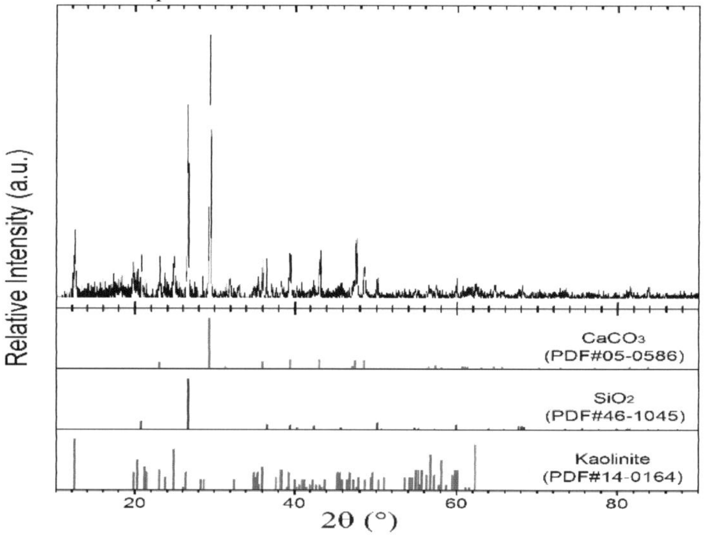


**Figure 1.** XRD patterns of SBC.

The elements of SBC samples were identified by the coal proximate analyzer as shown in Table I.

**Table 1.** The proximate analysis and ultimate analysis results of SBC.

| Proximate analysis | | | | | Ultimate analysis | | | | |
|---|---|---|---|---|---|---|---|---|---|
| $M_{ad}$ % | $A_{ad}$ % | $V_{ad}$ % | $FC_{ad}$ % | $Q_{ad.net}$ kJ/kg | $C_{ad}$ % | $H_{ad}$ % | $O_{ad}$ % | $N_{ad}$ % | $S_{ad}$ % |
| 6.53 | 10.41 | 32.40 | 50.66 | 26120 | 66.70 | 4.11 | 10.79 | 1.04 | 0.42 |

### *2.2. Comprehensive Test*

The tests of SBC combustion characteristics were performed with the TG-MS-FTIR system which comprises a TG analyzer (STA-449F3, Netzsch), a mass spectrometer (QMS403C, Aeolos) and an FTIR spectrometer (Tensor 27, Bruker). The combined system was controlled simultaneously and data were recorded by a computer, as shown in Figure. 2.

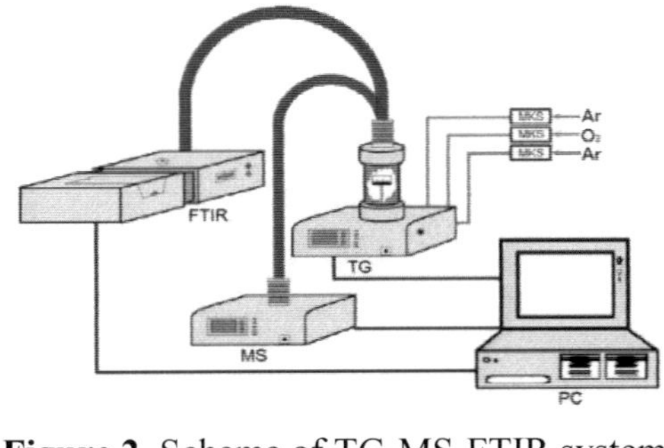


**Figure 2.** Scheme of TG-MS-FTIR system.

The tests were carried out in four steps. Firstly, about 10 mg sample was dispersed on a circular alumina pan and the air in TG chamber was replaced with a gas contained 20 % $O_2$ and 80 % Ar. So $O_2$ was more than the theoretically needed for the SBC combustion. Secondly, the samples were heated with a heating rate of 10 °C/min from 40 to 110 °C and kept for 30 min. Thirdly, the samples were heated to 1200 °C by four heating rate, namely 10, 20, 30 and 40 °C/min, respectively. Finally, the whole temperature-rising program was finished after an isothermal process of 10 min. Three total flowrates (50, 100 and 150 sccm) containing 20 % $O_2$ and 80 % Ar were sent into the TG chamber during the whole heating processes. The detailed experimental conditions are summarized in Table 2.

**Table 2.** Experimental conditions

| Condition | Mass (mg) | Flowrate (sccm) | | | Heating rate (°C/min) | Particle size (μm) |
|---|---|---|---|---|---|---|
| | | $O_2$ | Ar | Ar | | |
| 1 | 9.411 | 20 | 60 | 20 | 20 | 0-40 |
| 2 | 10.248 | 20 | 60 | 20 | 10 | 90-100 |
| 3 | 9.788 | 20 | 60 | 20 | 20 | 90-100 |
| 4 | 9.348 | 20 | 60 | 20 | 30 | 90-100 |
| 5 | 9.589 | 20 | 60 | 20 | 40 | 90-100 |
| 6 | 9.251 | 20 | 60 | 20 | 20 | 125-180 |
| 7 | 10.246 | 20 | 60 | 20 | 20 | 280-355 |
| 8 | 9.786 | 20 | 60 | 20 | 20 | 355-500 |
| 9 | 9.341 | 10 | 20 | 20 | 20 | 90-100 |
| 10 | 9.864 | 30 | 100 | 20 | 20 | 90-100 |

### *2.3. Data Treatment*

2.3.1. *Characteristic Temperature.* The ignition characteristic of SBC is analyzed based on $T_{ig}$. In the present work, $T_{ig}$ is determined by the commonly recognized TG-DTG methods [9]. The $T_{burn}$ is defined as the temperature with 98% total weight loss. The $T_{max}$, referring to the temperature at which combustion product own maximum volumetric flow, was also determined through numerical analysis.

2.3.2. *Kinetic Analysis.* The combustion kinetic parameters from TG data is calculated by the Coats-Redfern method [10]:

n=1
$$\ln\left[\frac{-\ln(1-\alpha)}{T^2}\right]=\ln\left[\frac{AR}{\beta E}\left(1-\frac{2RT}{E}\right)\right]-\frac{E}{RT} \quad (1)$$

n≠1
$$\ln\left[\frac{1-\ln(1-\alpha)^{1-n}}{T^2(1-n)}\right]=\ln\left[\frac{AR}{\beta E}\left(1-\frac{2RT}{E}\right)\right]-\frac{E}{RT} \quad (2)$$

where $\alpha=(m_0-m)/(m_0-m_\infty)$ is the weight loss ratio; m refers to the sample mass; $m_0$ and $m_\infty$ represent the initial mass and the final mass, respectively; $\beta$ is heating rate, K·min$^{-1}$; R is the gas constant; *E* stands for activation energy, *J·mol*$^{-1}$; A is the frequency factor, min$^{-1}$; n refers to the order of reaction. In general, the item $\frac{2RT}{E}<<1$ and $(1-\frac{2RT}{E})\approx 1$. As the first item of the right side of equations is nearly a constant, the two equations should result in a straight line of slope. *E* can be deduced through calculation with the slope.

*2.3.3. Products Analysis.* To obtain the volumetric flowrates of the products during the oxidation of SBC, both MS and FIIR were used. In order to give accurate quantitative analysis, a novel method of equivalent characteristic spectrum analysis was employed and the details of such method can be found elsewhere [11]. In the MS analysis, argon was used as a reference gas to calibrate the products by considering the electron impact ionization cross sections, ion flow intensities and the partial pressures of different species. The characteristic spectra and relative sensitivity were deduced. The effect of initial coal weight on the formation of products was excluded by normalization of the coal weight. The relative uncertainty of calculated MS results was estimated to be 0.052 [11].

## 3. Results and Discussion

### *3.1. TG/DTG Results*

Figure 3 presents the TG/DTG curves obtained in the combustion of SBC samples. As temperature increases, the coal sample could proceed with several steps, including devolatilization, coke formation and coke combustion. The peak of maximum weight loss moves towards high temperatures gradually and the peak value decreases slightly as the particle size increases. The combustion rate depends on the diffusion and reaction ability. During the rapid combustion processes, the major factor affecting the combustion is diffusion ability. As the particle size increases, the contact surface area decreases, which would inhibit the diffusion of oxygen and make the devolatilization and combustion difficult. So the devolatilization occurs at higher temperature and the value of weight loss peak decreases.

Compared to particle size, heating rate has more apparent effect on the characteristics of weight loss. As the heating rate increases, the peak of weight loss moves towards high temperatures. In addition, the peak values and residues decrease as well. Since diffusion ability is the dominant factor in affecting the combustion process, longer contact time of sample surface and oxygen could contribute to SBC conversion. Thus, the burnout time with low heating rate is longer. Sufficient contact of particles and oxygen is beneficial for the combustion process, and the maximum value of weight loss occurred at lower temperatures and the maximum weight loss shifted to larger values [12]. It is clearly that there is a small weight loss peak within 600~700 °C. This weak peak could come from the decomposition of carbonate in the coal samples. In addition, the TG/DTG results of different total flowrates are observed to be quite similar, as displayed in Figure. 4e and f. As the oxygen supplied is much more than the theoretical needed for the complete combustion, the weight loss characteristics are insensitive to the diverse total flowrate changed in experiments.

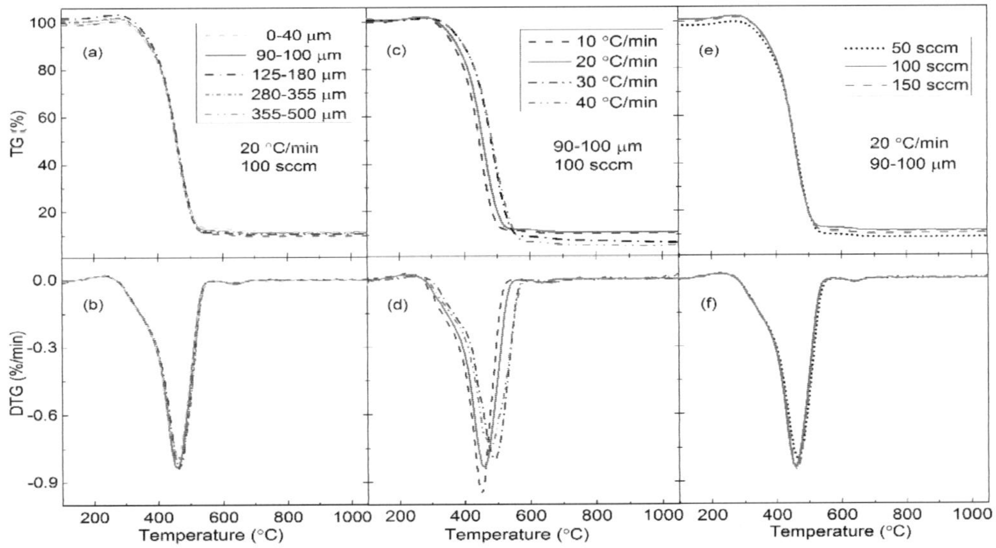


**Figure 3.** TG and DTG curve of SBC combustion with different particle size, heating rate and total flowrate.

## *3.2. Characteristic Temperatures*

The characteristic temperatures obtained in the combustion of SBC at different particle sizes, heating rates and total flowrates are shown in Figure. 4. As indicated in Figure. 4a, both $T_{ig}$ and $T_{burn}$ fall down as the particle size decreases, which is consistent with the results reported by Zhang et al. [13] for Yuanbaoshan and Datong coal and Zhou et al [14] for Lengshuijiang coal and lean coal. According to Zhang et al. and Zhou et al., the decrease of particle size could lead to larger surface area to expose to oxygen. The same behavior could occur to SBC and the larger surface area of SBC sample with smaller particle size would endow it with lower $T_{ig}$ and $T_{burn}$.

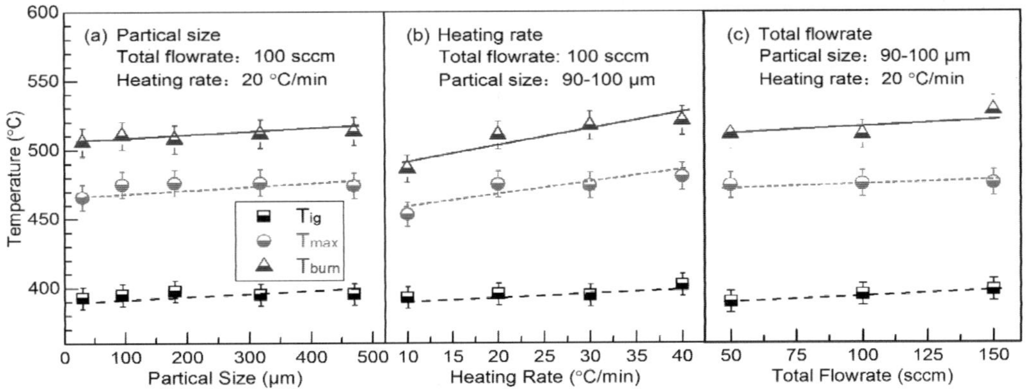


**Figure 4.** Characteristic temperatures of SBC combustion with different particle size, heating rate and total flowrate.

Figure 4b displays the characteristic temperature as a function of heating rate. The $T_{ig}$, $T_{max}$ and $T_{burn}$ tend to decrease slightly as the heating rate decreases, which agrees well with the conclusion drawn by Lu et al. [15] for Qilianta coal. This is due to the fact that the thermal conduction in coal particle at lower heating rate is better than that at higher ones. With the temperature increasing, the diffusion becomes better, which is helpful to the devolatilization and results in lower $T_{ig}$. Moreover, the reaction time is more sufficient at slower heating rate. As the plugging of the surface pores is avoided, it is more difficult to form ash cladding, which could give rise to a lower $T_{burn}$ [16].

For the total flowrate, a slight increase of $T_{ig}$ is observed with large total flowrate (see Figure. 4c). This slight increase mainly comes from shorter residence time of $O_2$ on the surface of coal particles at higher total flowrate. Even though more oxygen reached the coal surface, the higher speed of oxygen molecule makes the adsorption of oxygen more difficult. As the supplied $O_2$ was much more than the theoretically needed for the complete oxidation of the SBC sample, the influence of oxygen flow on the complete combustion and maximum combustion speed is not obvious. As a consequence, both $T_{max}$ and $T_{burn}$ exhibit similar values in the combustion of SBC at different total flowrates.

### *3.3. Activation Energy*

Figure 5 summarizes *E* of SBC with variation of particle size, heating rate and total flowrate. *E* tends to be low with the decrease of particle size, heating rate and total flowrate. As the particle size decreases, the surface area increases, which would result in more contact area with oxygen and better thermal reactivity. Thus, energy transfer from outside into inner becomes more rapidly, and *E* is getting low. Lower heating rate will make the energy transfer more evenly and slow down the rate of surface ash generating. When the total flowrate rises, the contact time of oxygen and sample surface will be shortened and result in increase of E. The lower energy barrier makes the oxidation occur more easily [17-19]. This finding also shows good agreement with the conclusion drawn by Lu et al. [15] for different kinds of coal. Similar to the tendency of $T_{ig}$, E decreased when the total flowrate turned lower. The adsorbability of oxygen on the coal particle surface is better with lower total flowrate, which could be responsible for the decrease of energy barrier.

### *3.4. MS and FTIR Results*

The volumetric flows of combustion products at different temperatures can be obtained by normalization and numerical analysis of MS results. The maximum volumetric flows of different combustion products as well as the corresponded temperatures ($T_{max}$) are used to explore the formation rule of emitted pollutants. Table 3 shows the maximum volumetric flows of $CH_4$, $NH_3$, NO, HCN, $NO_2$ and $SO_2$ as well as the corresponded temperatures ($T_m$). By comparing the results obtained with different particle size, the listed species tend to be less produced with smaller particle size. In view of the heating rate, lower amounts of these products were measured at lower heating rate. As far as the total flowrate is concerned, the quantity of these species becomes lower when the total flowrate is smaller. In general, the $T_m$ is insensitive to the particle size. However, the increase of the total flowrate normally results in smaller Tm. With the increase of the heating rate, $T_m$ of NO, $NH_3$ and HCN tends to be higher, while Tm of SOx is getting lower. To summarize, lower concentrations of $CH_4$, $NH_3$, NO, HCN, NOx and $SO_2$ were observed at smaller particle size, slower heating rate and bigger total flowrate.

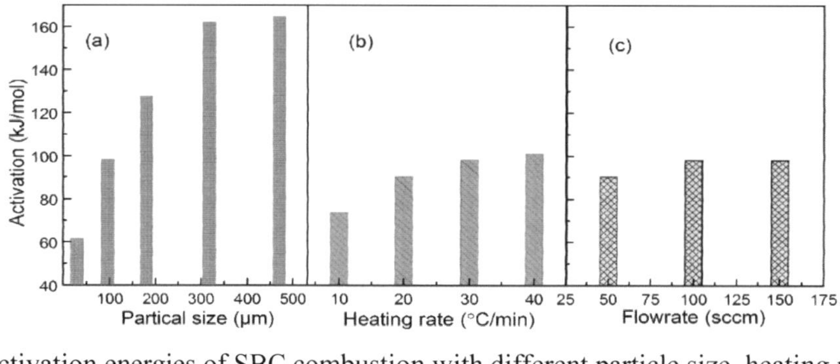


**Figure 5.** Activation energies of SBC combustion with different particle size, heating rate and total flowrate.

Figure 6 shows the volumetric flowrates of species formed under condition 4 (see Table 2). In the combustion of SBC, both $NH_3$ and HCN exhibited two-peak shapes. The peak before $T_{ig}$ comes from the devolatilization, whereas the peak after $T_{ig}$ results from the decomposition of nitrogen components in volatile and char. Further increase of the temperature led the conversion of NO and $NO_2$.

$NO_2$ exhibited a shoulder peak and an asymmetric single peak. The shoulder peak is due to the weak release of $NO_2$ during devolatilization. The asymmetric single peak could come from the conversion of nitrogen-containing heterocyclic species at the end of devolatilization and the beginning of char combustion. However, the other nitrogenous components enriched in char will release $NO_2$ during the rapid combustion of char. Less $NO_2$ emissions were observed at smaller particle size, slower heating rate and higher total flowrate. Similarly, the emissions of $SO_2$ are relatively less than those formed with slower heating rate and higher total flowrate. However, the amount of $SO_2$ is quite similar for the investigated particle size range, indicating that it is insensitive to the particle size.

**Table 3.** Maximum volumetric flow of representative combustion products

| No. | $CH_4$ | | $NH_3$ | | NO | | HCN | | $NO_2$ | | $SO_2$ | |
|---|---|---|---|---|---|---|---|---|---|---|---|---|
| | Value (slm) | $T_m$ (°C) | Value (slm) | $T_m$ (°C) | Value (slm) | $T_m$ (°C) | Value (slm) | $T_m$ (°C) | Value (slm) | $T_m$ (°C) | Value (slm) | $T_m$ (°C) |
| 1 | 4.08E-05 | 442 | 3.61E-04 | 318 | 4.72E-05 | 493 | 5.05E-04 | 452 | 6.14E-05 | 431 | 6.38E-06 | 400 |
| 2 | 3.53E-05 | 427 | 2.41E-04 | 309 | 3.56E-05 | 458 | 3.68E-04 | 443 | 4.17E-05 | 438 | 4.52E-06 | 402 |
| 3 | 5.91E-05 | 440 | 2.81E-04 | 305 | 7.69E-05 | 492 | 3.09E-04 | 451 | 6.41E-05 | 440 | 7.73E-06 | 388 |
| 4 | 6.51E-05 | 437 | 2.17E-04 | 307 | 6.46E-05 | 494 | 4.94E-04 | 453 | 9.84E-05 | 429 | 8.37E-06 | 379 |
| 5 | 5.06E-05 | 430 | 3.41E-04 | 337 | 1.40E-04 | 507 | 3.97E-04 | 512 | 1.29E-04 | 439 | 1.24E-05 | 350 |
| 6 | 7.27E-05 | 453 | 4.23E-04 | 338 | 4.60E-05 | 494 | 6.74E-04 | 463 | 7.10E-05 | 442 | 4.55E-06 | 400 |
| 7 | 7.94E-05 | 453 | 4.89E-04 | 306 | 6.97E-05 | 504 | 5.19E-04 | 453 | 7.43E-05 | 442 | 6.33E-06 | 410 |
| 8 | 7.96E-05 | 432 | 3.73E-04 | 317 | 4.95E-05 | 503 | 4.52E-04 | 463 | 7.20E-05 | 442 | 6.20E-06 | 390 |
| 9 | 9.75E-05 | 453 | 7.72E-04 | 327 | 9.44E-05 | 504 | 6.53E-04 | 463 | 1.47E-04 | 453 | 1.01E-05 | 400 |
| 10 | 3.76E-05 | 430 | 1.58E-04 | 316 | 4.32E-05 | 482 | 3.03E-04 | 472 | 4.99E-05 | 440 | 5.75E-06 | 398 |

Note: $T_m$ refers to the temperature at which the peak flow is obtained.

By comparing the formation of $NO_2$ and $SO_2$ with different particle sizes, heating rates and total flowrate shown in Figure. 7, it is obvious that the amount of $SO_2$ production is much lower than those of NO and $NO_2$. To summarize, the SBC combustion with smaller particle size, slower heating rateand bigger total flowrate is beneficial to controlling the formation of pollutants. This finding is consistent with the results reported for Hegang, Tiefa and Zhungeer coal [20] and Heshan sulfer coal. [18, 21].

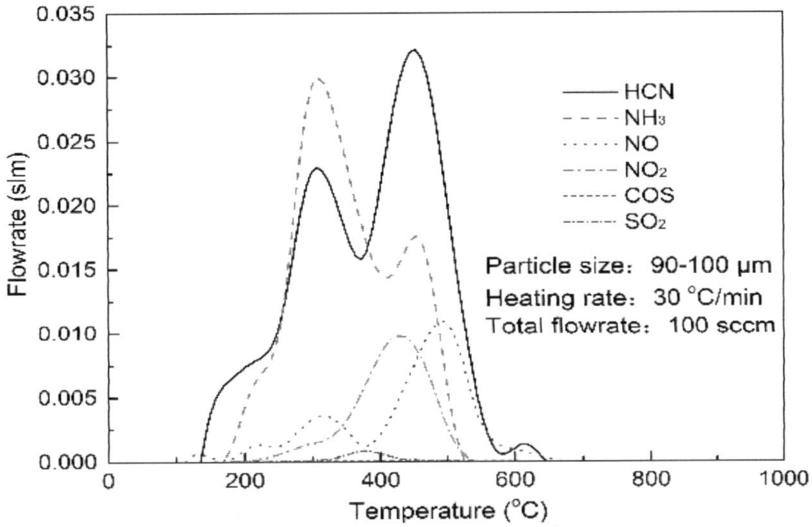


**Figure 6.** Flowrates of major products.

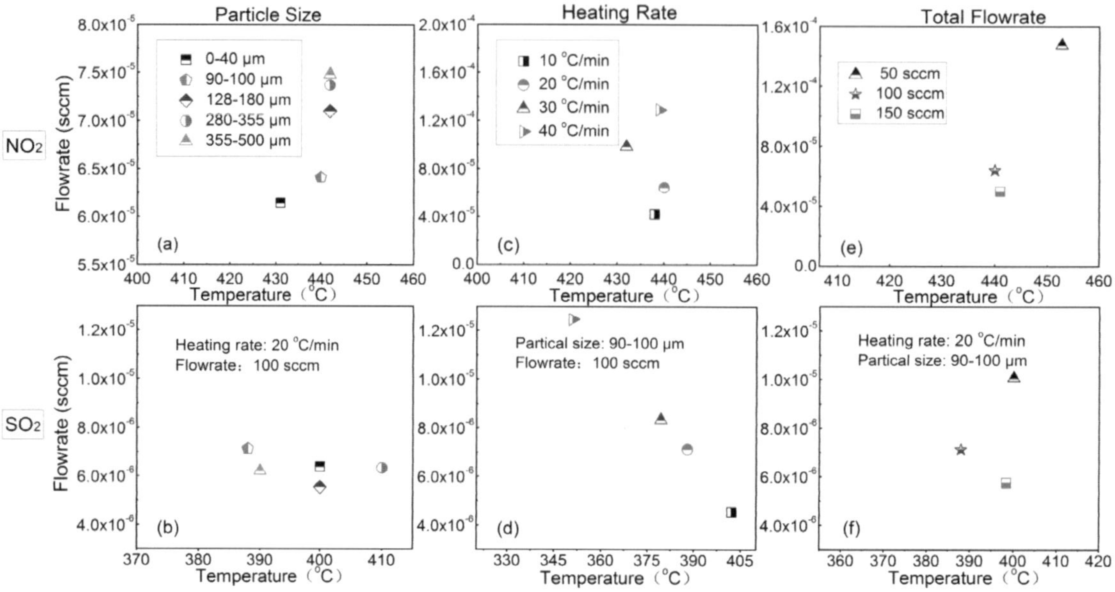


**Figure 7.** Production of $NO_2$ (a, c, e) and $SO_2$ (b, d, f) with different particle size, heating rate and total flowrate.

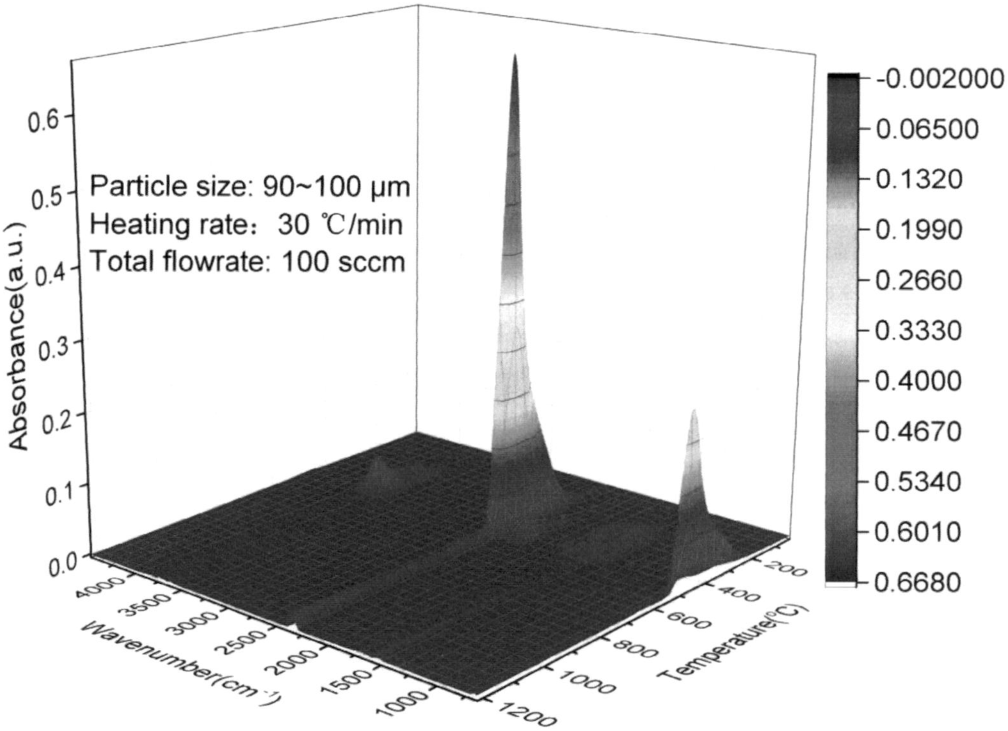


**Figure 8.** FTIR spectra with different wavenumber and temperature.

To visualize the formation of the major species and avoid the effect of ion fragmentation in MS, FTIR with spectral range of 500-5000 $cm^{-1}$ was involved. Figure 8 presents a representative three-dimensional spectrum measured with particle size of 90-100 μm and heating rate of 30 °C/min. The maximum absorbance peak is located at the wavenumber of 2380 $cm^{-1}$ corresponding to the anomalistic vibration of $CO_2$. Another obvious peak at 700 $cm^{-1}$ belongs to the bending vibration of $CO_2$. The peaks at 1375 and 3600 $cm^{-1}$ corresponds to $SO_2$ and $H_2O$, respectively. By comparing the standard spectral peaks of different species, overlapped peaks were observed. For instance, the peak between 1500-1800 $cm^{-1}$ was the overlap of $H_2O$ and NO [22, 23]. In Figure. 8, $CO_2$ becomes detectable at around 140 °C. Up to 200 °C, a slow increase was observed. Another turning point was measured at about 375 °C, which also agrees well with the wave valley in the profiles of HCN and NH3. Besides $CO_2$, the fast production of the major species was achieved within the temperature range of 400-500 °C, which is consistent with the rapid combustion temperature revealed in the MS analysis.

## 4. Conclusion

The combustion of SBC was comprehensively studied with online TG-FTIR-MS system in terms of characteristic temperatures as well as qualitative and quantitative analysis of products. Five particle sizes ranging from 0-500 μm, four heating rates from 10 to 40 °C/min and three total flowrate from 50-150 sccm were used. To avoid the influence of overlap peaks, signal drift and dynamic response delay in ion current spectra during MS analysis, argon was used as a reference gas to calibrate the products by considering the electron impact ionization cross sections, ion flow intensities and the partial pressures of different species. The results indicate that the decrease of the particle size, heating rate and flowrate lead to lower ignition and burnout temperatures, and the activation energy tends to be lower with smaller particle size, slower heating rate and lower flowrate. The decrease of particle size could lead to more contact area with oxygen and better thermal reactivity. Slower heating rate could provide more sufficent time for the reaction. Moreover, higher total flowrate would reduce the oxygen adsorbability on the coal particle surface at higher flow speed. From MS and FTIR analysis, lower concentrations of different products were observed to be formed at smaller particle size, slower heating rate and higher total flowrate. These findings will guide to understand the combustion kinetics of SBC and be beneficial to control the formation of pollutants.

## 5. Acknowledgments

The author thanks the financial support from youth fund of China Coal Technology & Engineering Group Corp (No. 2018QN004).

## 6. References

[1] J Cheng, JH Zhou, JZ Liu, XY Cao, KF Cen. 2009 *Energy Sources Part a-Recovery Utilizatio and Environmental Effects*. 31p 956-66.
[2] LP Chang, KC Xie, CZ Li. 2004 *Fuel Processing Technology*. 85 p 1053-63.
[3] XM Wang, YQ Jiao, LQ Wu, H Rong, *XM Wang, J Song. Ordos Basin.* 2014 Fuel. 136p 233-9.
[4] Ql Sun, W Li, HK Chen, BQ Li. 2003 *Journal of China University of Mining & Technology*. 32p 664-9.
[5] SY Zhao. 2007 *Coal Science and Technology*. 35 p 80-2.
[6] JG Yang, FR Deng, H Zhao, KF Cen. 2007 *Asia-Pacific Journal of Chemical Engineering*. 2 p 165-70.
[7] WJ Yang, JH Zhou, MS Liu, ZJ Zhou, JZ Liu, KF Cen. 2007 *Energy & Fuels*. 21p 2548-54.
[8] T Dijkmans, MR Djokic, KMV Geem, GB Marin. 2015 *Fuel*. 140p 398-406.
[9] QM Yu, YJ Pang, HG Chen. 2001 *North China Electric Power*. p 9-10+50.
[10] JP Redfern, AW Coats. 1964 *Nature Biotechnology*. 201p 68-9.
[11] HD Xia, K Wei. 2014 Thermchimica Acta, 602 15-21.
[12] SQ Wang, YG Tang, HH Schobert, GD Mitchell, FR Liao, ZZ Liu. 2010 *International Journal of Coal Geology*. 81p 37-44.
[13] CQ Zhang, LJ Yu, ZG Cui, XM Jiang. 2005 *Journal of Chemical Industry and Engineering*. 56 p 2189-94.
[14] Z hou, L Liu, YL Wang, H Liu. 2007 *Thermal Power Generation*. 3 p 35-8+47.
[15] HB Lu, HJ Xu, CX Jia, DL Zhang. 2006 *Power System Engineering*. 22 p 11-2+5.
[16] MS Nyathi, M Mastalerz, R Kruse. 2013 *International Journal of Coal Geology*. 118 p 8-14.

[17] DM Fan, ZP Zhu, YJ Na, QG Lu. 2012 *Journal of Thermal Analysis and Calorimetry*. 113p 599-607.
[18] XM Jiang, JB Li, JR Qiu. 1999 *Journal of China Coal Society*. 24 p 643-7.
[19] S Lu, GJ Lu, XG Jiang, Y Chi, JH Yan, KF Cen. 2014 *Journal of Coal Science & Engineering*. 3 p 554-61.
[20] LH Wei, XM Jiang, TH Yang, YJ Li, L Wang. 2006 *Acta Scientiae Circumstantiae*. 11 p 1780-4.
[21] XM Jiang, H Liu, JB Li, CG Zheng, DC Liu. 2002 *Environmental Science & Technology*. 23 p 126-8.
[22] T Ahamad, SM Alshehri. 2012 *Journal of Hazardous Materials*. 199 p 200-8.
[23] S Silvera. 2013 *Journal of Analytical and Applied Pyrolysis*. 104 p 95-102.

# Characteristic Study of Some Different Kinds of Coal Particles Combustion with Online TG-MS-FTIR

**Guanfu Pan** [1, 2, 3]

[1] China Coal Research Institute Company of Energy Conservation Corporation Ltd..
[2] State Key Laboratory of Coal Mining and clean Utilization.
[3] National Energy Technology & Equipment Laboratory of Coal Utilization and Emission Control.
Address of all above: 5 Qingniangou east road, Hepingli Subdistrict, Beijing, China.
Email: panguanfubest@163.com

**Abstract.** Four kinds of pulverized coal samples from China and Indonesia were studied by thermogravimetry coupled with mass spectrometry and fourier transform infrared spectroscopy (TG-MS-FTIR). The thermal behaviors and gaseous emissions of these coals were analyzed in this work. The results indicate that the relative lower values of H/C ratios, which normally represent the degree of aromatization and ring condensation in coal samples, could lead to the relative more intense thermal reaction. The time-evolved profiles of some typical gas products (i.e., CO, $SO_2$, $CH_4$, NO, $NO_2$, $NH_3$ and etc.) were provided by TG-MS-FTIR, and their variations are different. For all the samples, the releases of $SO_2$ and COS can be found at lower temperature than those of NO and CO. As the temperature increases, the possible conversion of $NO_2$ and $NH_3$ to NO is deduced in this work.

## 1. Introduction

Coal is an important and promising energy resource for electricity production due to its huge reserve in the world. At present, most power plants utilize the combustion of pulverized coal with fine particles below 100 μm to produce electricity [1]. In China, coal is the main source of primary energy and it accounts for about 70% of the total energy consumed [2]. Due to its huge reserves and lack of sufficient natural gas or oil reserves, coal will still be the main energy resource in China in the next decades. However, coal combustion leads to the increase of greenhouse gas emissions as well as air pollutants, such as nitrogen oxides ($NO_x$), $SO_2$ and etc. Thus, understanding the kinetics and mechanisms during the combustion of pulverized coals is crucial to control the product composition.

Coal combustion involves lots of chemical reactions, including the cleavage of bridge bonds in the aromatic ring systems, the loss of heteroatom functional groups, and fragmentation of the macromolecular network [3]. The combustion of coal is affected by a series of factors, such as heating rate, temperature and coal natures (i.e., coal type, particle size, chemical composition and etc.) [4-6]. Among these factors, coal properties are often considered as the key one before real application. Some traditional techniques and characteristic parameters, such as proximate and ultimate analysis, ash fusion temperatures and ash chemical composition, are commonly used to assess the behavior of pulverized coal during combustion. However, the test conditions of the above techniques are generally different from the real condition within a boiler or reactor, which sometimes would cause a large error [7]. To make good use of the pulverized coal, understanding of their thermal behavior is essential. The thermal analysis coupled to the measurement of evolved gas is very important to study the mechanism underlying coal combustion [8]. Recently, thermogravimetry coupled with mass spectrometry and Fourier transform infrared spectroscopy (TG-MS-FTIR) has been successfully used to study the thermal behavior and gaseous emission characteristic during the pyrolysis and combustion of specific coals [9-11]. The results indicate that TG-MS-FTIR is a very useful method, which integrates the advantages of TG, MS and FTIR. However, the systematic study on the effect of coals characteristic on their combustion behavior and evolved gases is rare.

The present work is oriented to study the thermal behaviors and gaseous emissions of four selected coal samples from different regions (SQ, YQ and DT are from China, YN is from Indonesia) by TG-MS-FTIR. The ignition and burnout temperature as well as the activation energy of the samples were deduced. Moreover, the information of the evolution of typical gaseous compounds during the combustion of the samples was revealed.

## 2. Experimental Methods

### *2.1. Samples*

The coal samples were bought from Shenmu Energy Developments Ltd. Company and grinded into powder with particle size of 90~100 μm. The results of proximate and ultimate analysis of the samples used in this work are listed in Table 1.

The $M_{ad}$, $A_{ad}$, $V_{ad}$, and $FC_{ad}$ represent the moisture, ash, volatile and fixed carbon in air-dried basis, respectively. The $C_{ad}$, $H_{ad}$, $O_{ad}$, $N_{ad}$ and $S_{ad}$ represent carbon, hydrogen, oxygen, nitrogen and sulfur elements in air-dried basis, respectively. H/C and O/C refer to the H/C and O/C atomic ratio, respectively.

**Table 1** The proximate and ultimate analysis results of four coal samples

| Coal | Proximate analyses (wt.%) | | | | Ultimate analysis (wt.%, daf) | | | | | H/C | O/C |
|---|---|---|---|---|---|---|---|---|---|---|---|
| | $M_{ad}$ | $A_{ad}$ | $V_{ad}$ | $FC_{ad}$ | $C_{ad}$ | $H_{ad}$ | $O_{ad}$ | $N_{ad}$ | $S_{ad}$ | | |
| SQ | 6.50 | 18.98 | 27.13 | 47.39 | 59.26 | 3.29 | 10.34 | 0.86 | 0.77 | 0.67 | 0.13 |
| YN | 1.13 | 52.51 | 16.37 | 29.99 | 35.26 | 2.23 | 7.52 | 0.67 | 0.68 | 0.76 | 0.16 |
| DT | 0.88 | 34.90 | 6.39 | 57.83 | 59.80 | 1.20 | 1.67 | 0.92 | 0.63 | 0.24 | 0.02 |
| YQ | 0.60 | 14.42 | 2.70 | 82.28 | 80.71 | 1.16 | 0.92 | 1.13 | 1.06 | 0.17 | 0.01 |

### *2.2. TG-MS-FTIR Test*

The experiments were performed with a TG analyzer (STA-449F3, Netzsch) coupled to a quadrupole mass spectrometer (QMS403C, Aeolos) and a FTIR spectrometer (Tensor 27, Bruker). About 10 mg of each sample was heated with a heating rate of 10 °C/min from 40 to 110 °C and kept for 30 min. Then, the samples were heated to 1070 °C at the heating rate of 20 °C/min in an atmosphere of 20 sccm $O_2$ and 80 sccm Ar.

## 3. Results and Discussion

### *3.1. TG Analysis*

The TG and DTG curves of four coal samples are displayed in Figure. 1. As shown in Figure. 1a, the initial slopes of TG curves of SQ and DT are steeper than those of YQ and YN, showing that the thermal decomposition processes of SQ and DT occurred at lower temperatures than those of YQ and YN. From Figure. 1b, the rates of weight loss of SQ, DT, and YN are less than that of YQ. Therefore, the SQ and DT show more intense thermal reaction than YQ and YN. As can be seen from Table 1, the H/C atomic ratios of YN, SQ and DT are higher than that of YQ. It has been reported that the high H/C atomic ratio normally generate a low degree of aromatization and ring condensation in coal samples [3]. Thus, the SQ and DT are liable to suffer from thermal decomposition at lower temperatures than YQ who has relatively higher degree of aromatization and ring condensation. However, for the YN coal with the highest value of H/C atomic ratio, less intense thermal reaction was observed than SQ and DT, which is most likely caused by its highest ash content and O/C ratio among the four coal samples.

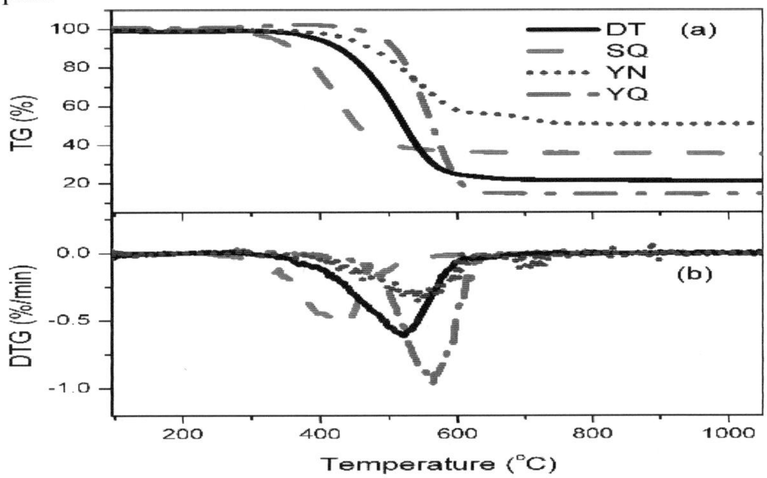


**Figure. 1** TG and DTG curves of four coal samples.

The characteristic temperatures and activation energies of the four coal samples are shown in Figure. 2. In this work, the ignition temperature ($T_{ig}$) is determined by the commonly-used method [12]. The maximum reaction temperature ($T_{max}$) refers to the temperature at which combustion products reach maximum volumetric flow. The burnout temperature ($T_{burn}$) is defined as the temperature at 98% total weight loss. The activation energy is calculated by the Coats-Redfern method [13,14]. For the four coal samples, the orders of $T_{ig}$ and $T_{max}$ are consistent with that of activation energies. The $T_{ig}$ and $T_{max}$ as well as activation energies of SQ and DT are less than those of YN and YQ. The $T_{burn}$ of YN is higher than that of other samples, which could be due to its highest ash contents. In fact, high ash contents will lead to the plugging of the surface pores and ash cladding, which would result in a higher burnout temperature.

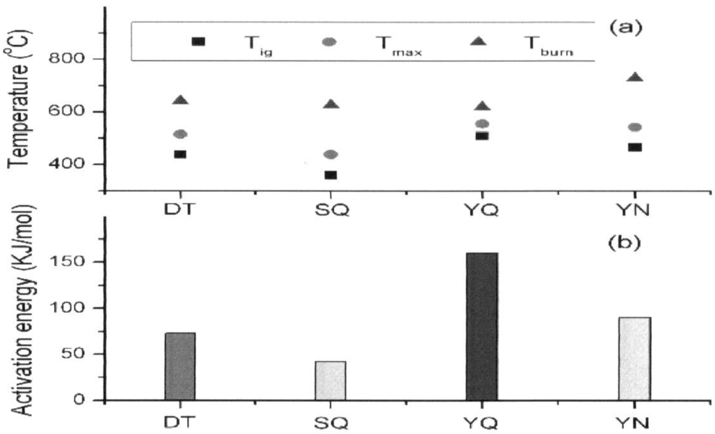


**Figure. 2** Characteristic temperatures and activation energies of four coal samples.

### *3.2. MS Analysis*

Figure. 3 summarizes the time-evolved flowrates of some selected products during the combustion process of the four coal samples. The data was analyzed through calculation by taking into consideration the electron impact ionization cross sections, ion flow intensities and the partial pressures of different species [15].

For $CO_2$, it appears in the temperature range of 250 to 650 °C. At low temperature, $CO_2$ comes mainly from the emission of adsorbed gas in the coal surface. With temperature increasing, $CO_2$ is possibly from aromatic carboxyl or aliphatic groups [3]. At around 700 °C, a small amount of $CO_2$ can be detected during YN coal combustion, which is possibly from oxygen-containing groups in YN coal [16]. The temperature range of CO is similar to that of $CO_2$, while the CO production is likely from the decomposition of carbonyl structures and phenolic groups in coal samples as temperature increases [17]. Among the four coal samples, the amounts of $CO_2$ and CO emission from YQ are the highest, while those of YN are the lowest, which is consistent with the amount of carbon contents in the four coal samples (see Table 1).

The maximum temperatures of $SO_2$ and COS emission are lower than each $T_{max}$, and the releases of $SO_2$ and COS can be found at lower temperature than those of NO and CO. The curve of $SO_2$ exhibits double peaks during YN and YQ coals combustion, which may result from the oxidation of aliphatic sulfur and aromatic sulfur components, respectively [18]. According to literatures [18-20], the formation of COS is mainly from the reaction of sulfur and CO generated in the combustion process.

For all samples, the formation of $CH_4$ is detectable from 300 °C to above 650 °C. In fact, the temperature corresponding to the max flowrate is irregular for different samples. As shown in Figure. 3e, the temperature of max flowrate of YQ is higher than that of the other three samples, especially for SQ. The observation agrees with the previous discussion on the relationship of H/C atomic ratio and degree of aromatization. The $CH_4$ produced at lower temperature results from the C-C bond cleavage between methyl and aliphatic chains or aromatic side chains [21]. At higher temperature, most of $CH_4$ comes from the breakage of aryl-methyl group or aryl-alkyl ether bonds [3,22].

The $NH_3$, NO and $NO_2$ are most likely from the decomposition of nitrogen components, such as nitrogen-containing heterocyclic species. The amounts of $NH_3$, NO and $NO_2$ emissions from YQ are the highest among the four samples, which is due to its highest nitrogen contents. For all samples, it should be noted that the maximum temperature of $NO_2$ and $NH_3$ emission are lower than $T_{max}$ of each sample, while the maximum temperatures of NO emission are higher than each $T_{max}$. Thus, the possible conversion of $NO_2$ and $NH_3$ to NO can be deduced as temperature increases during the combustion processes of the coal samples.

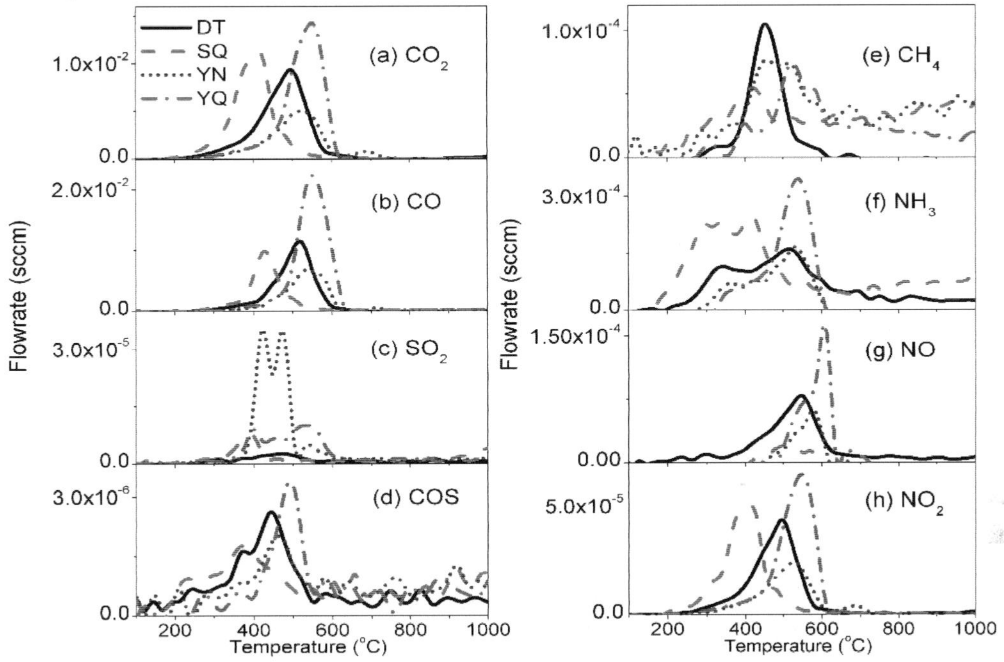


**Figure. 3** The time-evolved flowrates of some selected products during the combustion process of the four coal samples.

### *3.3. FTIR Analysis*

To visualize the variations of the combustion products directly, FTIR analysis was performed. As the four samples exhibit similar behavior, only the three dimensional FTIR spectra during DT coal combustion (Figure. 4) is given here as a representative. The $T_{max}$ is consistent with the DTG curve of DT coal shown in Figure. 1b, which can also be observed during the combustion of the other coal samples (not shown).

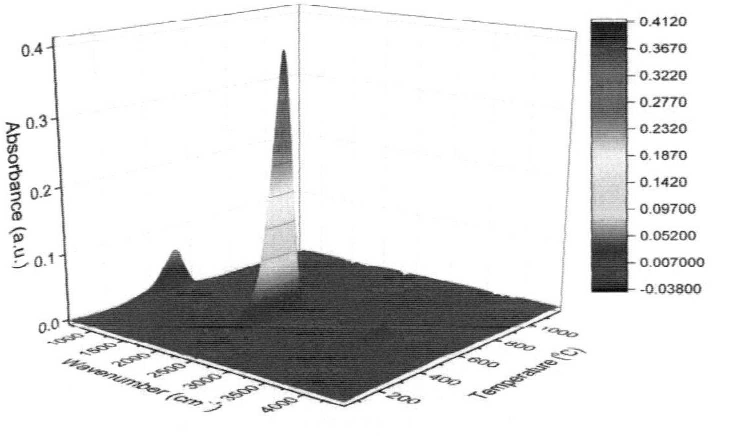


**Figure. 4** Three dimensional FTIR spectra during DT coal combustion.

The FTIR spectra during the combustion processes of four coal samples at different temperatures are shown in Figure. 5. The max absorbance bands at 2315-2380 $cm^{-1}$ come from $CO_2$ [23]. CO was confirmed by the appearance of bands at 2180 $cm^{-1}$ and 2120 $cm^{-1}$ [24]. The bands at 1320-1400 $cm^{-1}$ and 1350-1500 $cm^{-1}$ are from $SO_2$ and HCN, respectively [25]. The presence of $H_2O$ is verified by the bands at 3550-3850 $cm^{-1}$ and 1500-1800 $cm^{-1}$, which can hide the bands of some gases, such as NO (bands at 1500-1900 $cm^{-1}$), $NO_2$ (bands at 1500-1700 $cm^{-1}$), and $NH_3$ (bands at 1450-1800 $cm^{-1}$) [25]. At the temperature of 400 °C, the emissions of $CO_2$, CO, and $H_2O$ from SQ are higher than those from other samples. As the temperature increases to 600 °C, the releases of $CO_2$, CO, and $H_2O$ from YQ are the highest, while those from SQ are the lowest among the four coal samples, which can confirm the results discussed in the section of MS analysis.

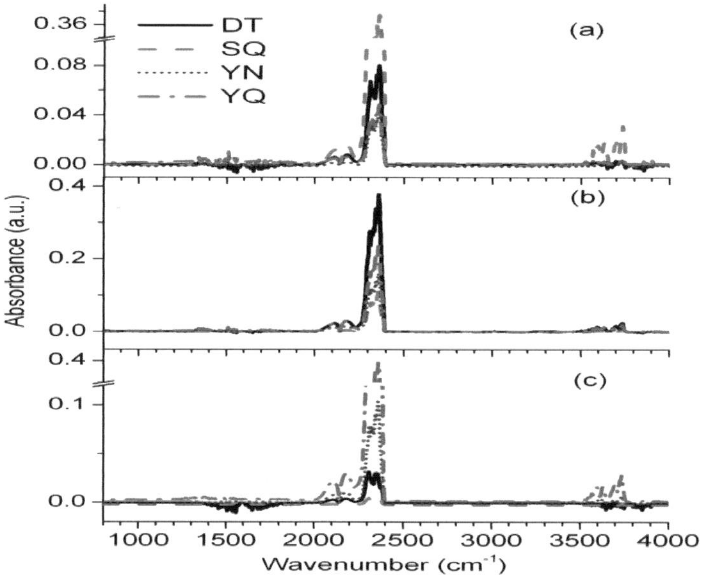


**Figure. 5** FTIR spectra during the combustion processes of four coal samples at different temperatures: (a) 400 °C, (b) 500 °C, and (c) 600 °C.

## 4. Conclusions

The thermal behaviors and gaseous emissions of four selected coal samples from different regions were studied by TG-MS-FTIR. The results show that the SQ and DT exhibit more intense thermal reaction than YQ due to the relative lower degree of aromatization and ring condensation in SQ and DT than YQ, which normally can be deduced by the relative higher values of H/C ratios in SQ and DT than that in YQ. However, for the YN coal with the highest value of H/C ratio, less intense thermal reaction was observed than SQ and DT, which is most likely caused by its highest ash content and O/C ratio among the four coal samples. The information on the evolution of gaseous compounds, such as $CO_2$, CO, $SO_2$, COS, $CH_4$, $NH_3$, NO, and $NO_2$, during the combustion of the coal samples were simultaneously obtained in this work. For all the samples, the releases of $SO_2$ and COS can be found at lower temperature than those of NO and CO. As the temperature increases, the possible conversion of $NO_2$ and $NH_3$ to NO is deduced in this work. For coal samples with higher values of H/C atomic ratio, the gaseous products normally reach the max flowrates at lower temperatures. Thus, the release of gaseous compounds produced in the combustion processes is related to the different chemical structures and compositions in the coal samples.

## 5. Acknowledgments
The authors thank the financial support from the Natural Science Foundation of Beijing, China(Grant No. Z161100004816012).

## 6. References
[1] S.M. Hwang, K. Oomagari, F.T. Akamatsu, M. Katsudi, R. Kurose, H. Tsuji, H. Makino, Combustion Science and Technology in Asia-Pacific Area: Today and Tomorrow, 2003, pp. 414-417.
[2] R. Li, G.C.K. Leung, Energ. Policy 40 (2012) 438-443.
[3] S.Q. Wang, Y.G. Tang, H.H. Schobert, Y.N. Guo, W.C. Gao, X.K. Lu, J. Anal. Appl. Pyrol. 100 (2013) 75-80.
[4] P. Ghetti, Fuel 65 (1986) 636-639.
[5] S. Kizgut, S. Yimaz, Fuel Process. Technol. 85 (2004) 103-111.
[6] W. Zhu, W. Song, W. Lin, Energy Fuels 22 (2008) 2482-2487.
[7] D.I. Barnes, Appl. Therm. Eng. 74 (2015) 89-95.
[8] W. Xie, W.P. Pan, J. Therm. Anal. Calorim. 65 (2001) 669-685.
[9] A. Arenillas, F. Rubiera, J.J. Pis, Environ. Sci. Technol. 36 (2002) 5498-5503.
[10] A. Arenillas, C. Pevida, F. Rubiera, R. Garcia, J.J. Pis, J. Anal. Appl. Pyrol. 71 (2004) 747-763.
[11] T. Kaljuvee, M. Keelman, A. Trikkel, V. Petkova, J. Therm. Anal. Calorim. 113 (2013) 1063-1071.
[12] L.M. Lu, Thermal Analysis in Practice, Donghua University Press, 2010.
[13] A.W. Coats, J.P. Redfern, Nature 201 (1964) 68-69.
[14] A.W. Coats, J.P. Redfern, J. Polym. Sci. Pol. Lett. 3 (1965) 917-920.
[15] H.D. Xia, K. Wei, Thermochim. Acta, 602 (2014) 15-21.
[16] A. Arenillas, F. Rubiera, J.J. Pis, J. Anal. Appl. Pyrol. 50 (1999) 31-46.
[17] K.H. Vanheek, W. Hodek, Fuel 73 (1994) 886-896.
[18] W.H. Calkins, Energy Fuels 1 (1987) 59-64.
[19] M. Fatemibedi, A.W. Scaroni, Abstr. Pap. Am. Chem. Soc. 195 (1988) 27.
[20] P.A. Montano, P.P. Vaishnava, J.A. King, E.N. Eisentrout, Fuel 60 (1981) 712-716.
[21] A. Arenillas, F. Rubiera, J.J. Pis, M.J. Cuesta, M.J. Iglesias, A. Jimenez, I. Suarez-Ruis, J. Anal. Appl. Pyrol. 68-69 (2003) 371-385.
[22] W. Hodek, J. Kirschstein, K.H. Vanheek, Fuel 70 (1991) 424-428.
[23] P.L. Gendreau, F. Vitaro, Can. J. Publ. Health-Rev. Can. San. 96 (2005) 167-172.
[24] T. Ahamad, S.M. Alshehri, J. Hazard. Mater. 199 (2012) 200-208.
[25] P.J. Linstrom, W.G. Mallard, NIST Chemistry Webbook. National Institute of Standard and Technology, Number 69, Gaithersburg, MD, http://webbook.nist.gov/, 2005.

IOP Conf. Series: Earth and Environmental Science **111** (2018) 012020 doi:10.1088/1755-1315/111/1/012020

# Pollution Assessment and Sources Identification of Heavy Metals in Surface Sediments from the Nantaizi Lake, Middle China

**Jinlong Ma, Fei Li[1], Xiaolin Jia and Jingdong Zhang**

Research Center for Environment and Health, Zhongnan University of Economics and Law, Wuhan 430073, China
E-mail: [1]lifei@zuel.edu.cn

**Abstract.** The total contents of heavy metal elements including Cr, Cd, Cu, Zn, Pb and As were investigated in sediments from the Nantaizi Lake in Hanyang district of Wuhan. The heavy metal pollution level of Nantaizi Lake was calculated by potential ecological risk index and the main sources of pollutants were researched by correlation analysis and principal component analysis. The results show that heavy metal concentration of Nantaizi Lake sediments is within the Chinese Environmental Quality Standard for Soils (GB 15618-1995) level-II standard limitation. According to the result of potential ecological risk index, ecological hazard rank of heavy metal element of Nantaizi Lake sediments is: Cd>Cu>As>Pb>Zn>Cr, and whole water environment of lake is slightly polluted. Through correlation analysis and principal component analysis, it is found that industrial sewage and domestic wastewater in human activities are the main contributors to heavy metal sources of Nantaizi Lake, and chemical processes, such as endogenous microbial activities of lake etc., also affect heavy metal sources in sediments simultaneously.

## 1. Introduction

Heavy metals can enter into river system through hydrological cycle. The vast majority of heavy metal pollutants entering into waters will transform from water phase to solid phase through physical, chemical and biological effect etc. of a certain period, and combine with suspended solids and sediments in waters to become water pollution endogenesis, which causes lasting damage to water environment and may enter into human dining-table, directly affecting human health [1-2]. Main sources of heavy metal of lake environment are divided into human activities and natural sources [3-4]. In recent years, with rapid development of urbanization and industrialization, artificial sources, such as wastewater discharge, mining and fossil fuel combustion etc., and natural sources, such as geological structure activity, erosion and weathering etc., have caused entry of a great deal of heavy metal pollutants into waters, and made them gathered at lake sediments, which causes huge damage to urban lakes and threatens urban resident health [5-8]. As important constituent part of urban ecological system, urban lakes play an irreplaceable role in many respects [9]. Therefore, analyzing heavy metal pollution source by researching heavy metal pollution distribution of urban lake environment sediments is of great importance to protect urban lake environment and maintain urban ecological safety.

Wuhan is called as "city of hundreds of lakes", and its lake area occupies 25% of total area. Because of ecological pressure caused by population explosion and pollution by-product generated by rapid development of industry, many lakes of Wuhan are polluted in different degree, which arouses the attention of many scholars. Research of Qiao Shengying et al. [10] shows that the bottom mud heavy metal content of lakes within Wuhan is greater than that of lakes at suburb. Yu Jinsong et al. [11] researched the migration of heavy metal of main lake environment of Wuhan. Zhi Fengyang et al. [12] made analysis aimed at sediment heavy metal pollution degree of Wuhan Yangtze river basin and lake environment. Zhang Hailin et al. [13] analyzed 8 main lakes of Wuhan by utilizing lake eutrophication model. Through monitoring analysis to 6 representative sample sites of river system of Nantaizi Lake, and based on potential ecological risk index model, correlation analysis principle and principal component analysis, the author evaluates heavy metal pollution in Nantaizi Lake environment sediments and judges its main pollution source, to provide related data and evidences proving pollution condition of Nantaizi Lake to Wuhan Municipal Government.

## 2. Methods and Materials

### *2.1. Study Area*

Nantaizi Lake (114°10′E to 114°13′E, 30°29′N to 30°30N) is located within Hanyang District of Wuhan. Hanyang is the earliest industrial area in Wuhan, and has been developed into the largest industrial park of Wuhan currently [14]. In July of 2015, Nantaizi Lake of Wuhan Economic and Technological Development Zone was suddenly polluted, with lake water presenting black brown and a great deal of dead fish floating on lake. Within one month, gross of dead fish appearing turned to be 75000kg roughly. Before this, large area of dead fish has appeared in water area of Nantaizi Lake because of serious pollution for 4 continuous years, which causes serious economic losses to local fishermen and threatens health of adjacent residents simultaneously.

### *2.2. Sampling and Analysis*

According to distribution characteristics of sewage pipe of Nantaizi Lake, 6 representative sample sites are selected for stationing sampling in Figure 1. ZYA-0204 sludge sampler is used to collect sediments sample, 0~10 cm sediments at surface layer of lake is collected as sample, and 3 samples are collected at each sample site, and placed into clean PVC bag after blending

Sediment samples are naturally dried at laboratory with residual body of flora, fauna and stone removed. Then, sediments were grinded and sieved through 200-mesh nylon mesh. Next, all sediments were kept in plastic bottles and were numbered before analysis.

Weigh and take 0.2500 g of screened sediments sample correctly with TE124S electronic scales and place it into polytetrafluoroethylene digestion tank, and then prepare it into solution to be measured after digestion with $HNO_3$-HF-$HClO_4$. Samples were handled under the reference preprocessed steps according to NY/T 1121.10-2006. The total contents of heavy metal Cr, Cd, Cu, Zn and Pb were all detected by atomic absorption spectrometer (ZEEnit-700P). In addition, the concentration of As were measured by full-automatic dual-channel atomic fluorescence photometers

(AFS-9730). Guarantee accuracy and correctness of experimental result through repeated analysis and standard sample analysis to sample. Take 10% of sample as duplicate samples, and when analytical error of duplicate sample is lower than 5%, analysis result would be reliable. Experimental process conforms to the Chinese Technical Specifications for Soil Environmental Monitoring HJ/T 166-2004.

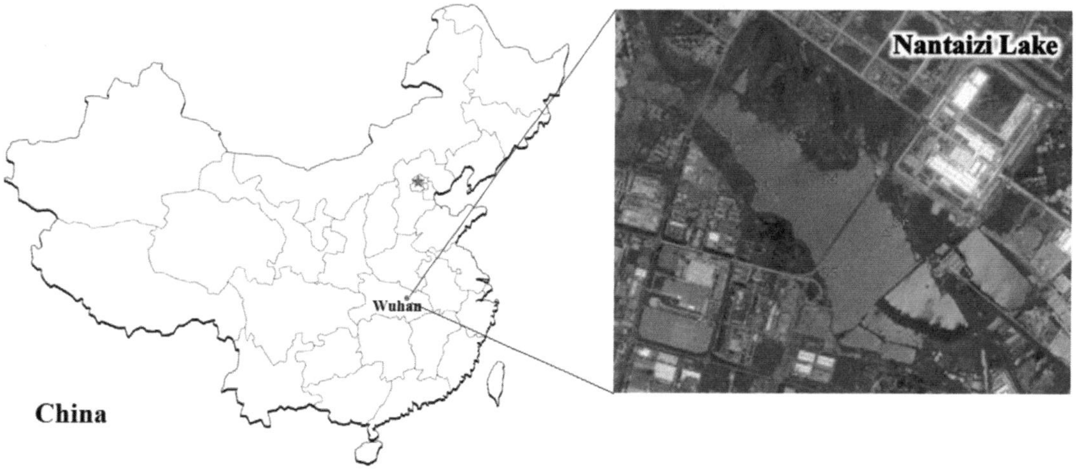


**Fig. 1** The location of Nantaizi Lake and sampling sites

## *2.3. Potential Ecological Risk Index*

Swedish scientist Hakanson [15] proposed potential ecological risk index method in 1980. This method is widely applied to evaluation of sediment heavy metal pollution, which mainly makes evaluation from perspective of toxicological effect of heavy metal, and its computational formula is as follows:

$$E_r^i = T_r^i \times C_f^i = T_r^i \times C^i / C_i \quad (1)$$

In the formula, $C_f^i$ is enrichment factor of overlying sediment heavy metal $i$; $C^i$ is measured content of sediment heavy metal element $i$, mg•kg$^{-1}$; $C_i$ is reference value of element $i$, mg/kg, and the highest background value of heavy metal in sediments before industrialization is taken as reference value, and see Table 1; $T_r^i$ is toxicity response coefficient of heavy metal $i$ in the Table 1; $E_r^i$ is potential ecological risk coefficient of heavy metal $i$.

**Table 1.** Reference value ($C_n^i$) and toxicity coefficient ($T_r^i$) of heavy metal in sediments

| Element | Cr | Cu | Zn | Pb | Cd | As |
|---|---|---|---|---|---|---|
| $C_n^i$(mg/kg) | 86 | 30.7 | 83.6 | 26.7 | 0.17 | 12.3 |
| $T_r^i$ | 2 | 5 | 1 | 5 | 30 | 10 |

Integrated potential ecological risk degree of numerous heavy metals in sediments, Hakanson [15] method, is represented by ecological risk index *RI*:

$$RI = \sum_{i=1}^{n} E_r^i \qquad (2)$$

Corresponding pollution level can be judged according to potential ecological risk index value obtained, which is as shown in Table 2.

**Table 2.** Division on level of ecological pollution degree of sediment heavy metals

| Scope of potential ecological risk index of single heavy metal | Potential ecological risk degree of single heavy metal | Scope of potential ecological risk index (*RI*) of Hakanson | Potential ecological risk degree of heavy metal |
|---|---|---|---|
| <40 | Level 1, low risk | <150 | Level 1, low risk |
| 40~80 | Level 2, intermediate risk | 150~300 | Level 2, intermediate risk |
| 80~160 | Level 3, considerable risk | 300~600 | Level 3, considerable risk |
| 160~320 | Level 4, high risk | >600 | Level 4, high risk |
| >320 | Level 5, very high risk | | |

## 3. Results and Discussion

### *3.1. Heavy Metal Distribution of Sediments in Nantaizi Lake*

Table 3 shows heavy metal content of sediments on the surface of Nantaizi Lake. According to the results of Table 3, average content values of Cr, Cu, Zn, Pb, Cd and As in Nantaizi Lake surface sediments are respectively 40.11, 27.86, 86.21, 20.03, 0.23 and 5.19 mg/kg; referring to Environmental Quality Standard for Soils (GB15618-1995), 6 kinds of heavy metal contents in sediments shall be within class II standard limit. According to arrangement condition of heavy metal concentrations in sample sites, it can be found out clearly that heavy metal contents in sediments shall increase from half northern part and half southern part of lake to central area of lake.

**Table 3.** Detection result for total heavy metals in Nantaizi Lake sediments

| Items | | Concentration/(mg/kg) | | | | | |
|---|---|---|---|---|---|---|---|
| | | Cr | Cu | Zn | Pb | Cd | As |
| Sample sites | S1 | 7.94 | 20.45 | 60.53 | 17.61 | 0.13 | 2.50 |
| | S2 | 55.43 | 24.18 | 78.38 | 20.87 | 0.30 | 6.07 |
| | S3 | 58.66 | 35.61 | 118.78 | 22.64 | 0.38 | 7.35 |
| | S4 | 56.64 | 29.43 | 91.49 | 21.02 | 0.20 | 7.19 |
| | S5 | 34.89 | 30.89 | 100.43 | 21.70 | 0.19 | 4.65 |
| | S6 | 27.09 | 26.60 | 67.67 | 16.34 | 0.19 | 3.39 |
| | Average value | 40.11 | 27.86 | 86.21 | 20.03 | 0.23 | 5.19 |
| Class II standard limit (GB 15618-1995) | | 350 | 100 | 300 | 350 | 0.6 | 20 |

### *3.2. Potential Ecological Hazards Assessment*

In order to characterize the ecological pollution degree of heavy metals in Nantaizi Lake sediments, the results calculated by potential ecological risk index were in Table 4, it can be seen clearly that ecological pollution degree in all sample sites of Nantaizi Lake sediments is slight. Analyzed from all sample sites that ecological risk in sample site S3 index RI is higher than other sample sites; analyzed from all aspects of heavy metals, heavy metal element Cd is main contributor of ecological harm index and its ecological harm index is far higher than that of other elements, where Cd ecological harm coefficient of sample site S2 and S3 exceeds 40, which belongs to middle ecological pollution and shows that ecological harm degree of Cd in Nantaizi Lake is relatively high. 6 kinds of heavy metal elements in Nantaizi Lake were sorted according to ecological harm coefficient of average value of heavy metal elements: Cd>Cu>As>Pb>Zn>Cr.

**Table 4.** Potential risk Index evaluation results for heavy metals in Nantaizi Lake sediments

| $E_r^i$ | Cr | Cu | Zn | Pb | Cd | As | *RI* | Ecological pollution degree |
|---|---|---|---|---|---|---|---|---|
| S1 | 0.18 | 3.33 | 0.72 | 3.30 | 23.68 | 2.03 | 33.25 | Slight |
| S2 | 1.29 | 3.94 | 0.94 | 3.91 | 52.56 | 4.94 | 67.57 | Slight |
| S3 | 1.36 | 5.80 | 1.42 | 4.24 | 66.92 | 5.98 | 85.72 | Slight |
| S4 | 1.32 | 4.79 | 1.09 | 3.94 | 35.03 | 5.84 | 52.01 | Slight |
| S5 | 0.81 | 5.03 | 1.20 | 4.06 | 32.82 | 3.78 | 47.72 | Slight |
| S6 | 0.63 | 4.33 | 0.81 | 3.06 | 33.45 | 2.75 | 45.04 | Slight |
| Average value | 0.93 | 4.54 | 1.03 | 3.75 | 40.75 | 4.22 | 55.22 | Slight |

## *3.3. Pollution Sources Analysis*

According to the correlation analysis between all heavy metals in Nantaizi Lake sediments listed in Table 5. Table 5 shows that there is relatively remarkable positive correlation between heavy metal elements in sediments: Cu-Zn, Pb-Zn and Pb-As, which shows that there is similar geochemical property in all heavy metals in Nantaizi Lake environment sediments Cu, Zn, Pb and As, which may have certain homology. Besides, there is significant correlation in As and Cr, which shows that they are closely related.

**Table 5.** Pearson's correlation matrix of heavy metals in Nantaizi Lake sediments

| Items | Cr | Cu | Zn | Pb | Cd | As |
|---|---|---|---|---|---|---|
| Cr | 1 | | | | | |
| Cu | 0.648 | 1 | | | | |
| Zn | 0.706 | 0.945** | 1 | | | |
| Pb | 0.769 | 0.701 | 0.884* | 1 | | |
| Cd | 0.775 | 0.619 | 0.690 | 0.646 | 1 | |
| As | 0.972** | 0.695 | 0.774 | 0.827* | 0.743 | 1 |

** Significant correlation (both sides) under $P < 0.01$.

* Significant correlation (both sides) under $P < 0.05$.

Principal Component Analysis (PCA) is to reflect weight of all factors by dividing variables factors by utilizing dimension reduction technology and by representing correlation degree between different factors with different main components based on displaying original data to the maximum [16].Urban inland lake waters has reflected that lake waters are polluted by heavy metals, of which contents and sources have recorded influence of human activity and natural change on environment [17]. Pollution source of heavy pollution can be divided into main source and secondary source according to sources. Based on PCA, by simplifying and extracting original variables factors, independent heavy metal elements can be combined to form a new variable, so as to judge and analyze heavy metal sources in water environment sediments [16]. Therefore, PCA has been implemented for characteristics values of heavy metals of 6 sample sites of Nantaizi Lake water system. It can be shown from results of PCA (Table 6) that characteristics value of the first principal component is 4.824 and its variance contribution rate reaches 80.392%; characteristics value of the second principal component is 0.591 and its variance contribution rate is 9.848% and accumulated contribution rate of the first principal component and the second principal component occupies for 90.239% of total variance, which shows that extracted principal component can better reflect large part of information of all data.

**Table 6.** Principal component analysis of heavy metal data in Nantaizi Lake sediments

| Principal | Initial factor solution | | | Extracted original solution | | |
|---|---|---|---|---|---|---|
| component | Total | % of variance | Cumulative % | Total | % of variance | Cumulative % |
| 1 | 4.824 | 80.392 | 80.392 | 4.824 | 80.392 | 80.392 |
| 2 | .591 | 9.848 | 90.239 | .591 | 9.848 | 90.239 |
| 3 | .341 | 5.691 | 95.931 | | | |
| 4 | .223 | 3.715 | 99.646 | | | |
| 5 | .021 | .354 | 100.000 | | | |
| 6 | -7.02E-17 | -1.17E-15 | 100.000 | | | |

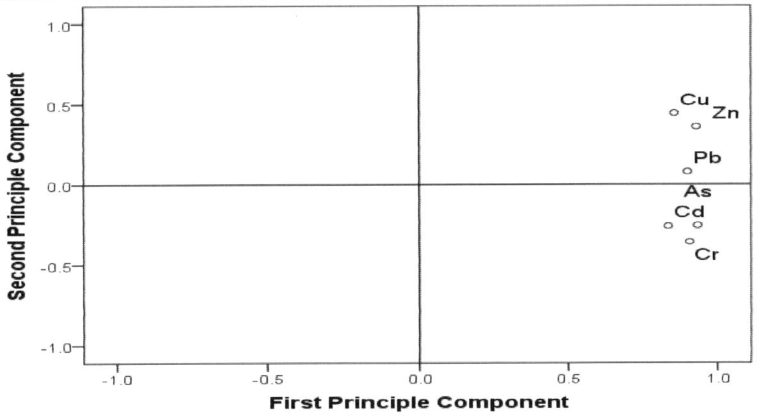


**Fig. 2** Figure of main component load

Based on Table 6 and Figure 2, the contribution rate of the first principal component in the total variance is far higher than other principal components, where Cu, Zn, Pb, As, Cd and Cr are relatively higher than positive load. According to research of Tang Zhenwu [18], heavy metal elements Cd, Cu, Pb and Zn, etc. in the first principal component of heavy metals in lake sediments in Hanyang industrial area have relatively large load and Cd, Cu, Pb and Zn contents are higher than that of other lakes, which is basically consistent with result of the experiment. Nantaizi Lake is located in Hanyang High-new Technical Park, where there are many industrial enterprises, such as chemical industry, metallurgy and plating, etc.; at the same time, because of rapid development of Wuhan economy, transfer of industry and enterprises drive removal of population, which leads to closed communities and buildings near Nantaizi Lake [19]. By referring to relevant materials on line, the Author has known that sewage from Longyang Lake and Moshui Lake, etc. have been discharged to Nantaizi Lake for several times because disposal capacity of sewage disposal factory is limited within Hanyang Industrial Park, which has caused severe damage to Nantaizi Lake environment. It is judged comprehensively that Cu, Zn, Pb, As, Cd and Cr may be related to human activities. Therefore, according to analysis of 6 kinds of heavy metals Cd, Cu, Pb, Zn, Cr and As and combined with relevant materials, it can be known that the first principal component represents influence of industrial sewage and domestic sewage on lake environment. It can be known from Table 5 that there is relatively strong correlation between contents of heavy metals Cd, As and Cr; therefore, it can be known that the first principal component mainly reflects sources of heavy metals Cd, As and Cr in lake sediments and at the same time, it also reflects sources of Cu, Pb and Zn.

Heavy metals, of which the second principal component is positive load are Cu, Zn and Pb. Content of the 3 kinds of heavy metals in crust is not high, mainly in mineral [20]. There is a lot of rainfall in plum rain season in Wuhan annually, which brings quantities of street rubbish and residual tailings to lake and will accumulate in lake sediments. Under certain condition, metabolic activity of microorganism in lake environment will lead to that heavy metal in sediments will be released and generate "secondary pollution" [21]. It can be inferred that Cu, Zn and Pb may come from the release of inner source pollution of lake sediments. Therefore, the second principal component has mainly reflected influence of inner source pollution of lake on heavy metal contents.

## 4. Conclusion

Average values of Cr, Cd, Cu, Zn, Pb and As contents in Nantaizi Lake sediments are within GB 15618—1995 class II standard limit. According to potential ecological risk index, except for that heavy metal Cd content in lake sediments belong to middle ecological pollution, rest heavy metal contents belong to slight ecological pollution. 6 kinds of heavy metal elements of Nantaizi Lake shall be ranked according to ecological harm coefficient of mean value of heavy metals: Cd>Cu>As>Pb>Zn>Cr. Therefore, general ecological pollution degree of Nantaizi Lake belongs to slight pollution. Based on correlation analysis and principal component analysis, the sources of heavy metals on Nantaizi Lake surface mainly comes from two main components: the first principal component is industrial pollution discharge in human activities and point-source pollution generated by domestic sewage, which has decided the sources of Cd, As and Cr and has partly dominated the sources of Cu, Pb and Zn; the second principal component is inner source pollution generated by microbial degradation of minerals in natural environment and the sources of Cu, Pb and Zn mainly depend on second principal component.

## 5. Acknowledgments

This study was financially supported by the Science and Technology Research Project of Hubei Provincial Education Department (B2017601).

## 6. References

[1] Li F, Huang J H, Zeng G M, Yuan X Z, Li X D, Liang J, Wang X Y, Tang X J and Bai B 2013 *J.*

*Geochem. Explor* **132** 75-83
[2] Ajah K C, Ademiluyi J, Nnaji C C 2015 *J. Environ. Health. Sci. Eng* **13** 15
[3] Zhang Y, Liu Y, Niu Z, Jin S 2017 *Environ. Sci. Pollut. R* **241** 2301-12311
[4] Islam M S, Ahmed M K, Raknuzzaman M, Habibullah-Al-Mamun M, Islam M K 2015 *Ecol. Indic* **48** 282-291
[5] Dundar M S, Altundag H, Eyupoglu V, Keskin S C, Tutunoglu C 2012 *Environ. Monit. Assess* **184** 33-41
[6] Zhang W, Jin X, Di Z, Zhu X, Shan B 2016 *Environ. Sci. Pollut. R* **23** 25364-25373
[7] Zhong Z, Tan S K, Tang W W 2015 *Chinese. Geogr. Sci* **25** 775-790
[8] Wokhe T 2015 *J. Sci. Res. Rep* **6** 157-164
[9] Ma Z, Chen K, Yuan Z, Bi J, Huang L 2013 *J. Environ. Qual* **42** 341
[10] Qiao S Y, Jiang J Y, Xiang W, Tang J H 2005 *Res. Environ. Yangtza. Basin* **14** 353-357
[11] Deng W B, Xu-Xiang L I 2015 *J. Earth. Environ*
[12] Zhi F Y, Ying W, Zhen Y S, Jun F N, Zhen W T 2008 *J. Hazard. Mate* **166** 1186-119
[13] Zhang H L, He B Y 2006 *Chin. J. Oceanol. Limnol* **24** 285-290
[14] Zhang C, Yang Y, Li W, Zhang C, Zhang R, Mei Y, Liao X, Liu Y 2015 *Environ. Monit. Assess* **187** 556
[15] Hakanson L 1980 *Water. Res* **14** 975-1001
[16] Liu H, Li L, Yin C, Shan B 2008 *Chinese J. Environ. Sci* **20** 390-397
[17] Martins M V A, Zaaboub N, Aleya L, Frontalini F, Pereira E, Miranda P, Mane M, Rocha F, Laut L, Bour M E 2015 *Plos. One* **10**
[18] Jiang J, Xu Y, Peng H 2014 *Fresen. Environ. Bull* **23** 502-507
[19] Guo S 2014 *J. Suzhou. Univ. Sci. Technol*
[20] Qu M, Li H, Li N, Liu G, Zhao J, Hua Y, Zhu D 2017 *Chemosphere* **168** 1515-1522
[21] Kastratović V, Željko Jaćimović, Bigović M, Đurović D, Krivokapić S 2016 *Environ. Monit. Assess* **188** 1-15

IOP Conf. Series: Earth and Environmental Science **111** (2018) 012021 doi:10.1088/1755-1315/111/1/012021

# The Effect of the Wastewater of Penicillin and Amoxicillin on the Antioxidant Indexes of Limnodrilus Hoffmeisteri

**Yuxin Dong[1] and Hongyan Shen[1,a*]**

[1]College of Environmental Science and Engineering, Hebei University of Science and Technology, Pharmaceutical Molecular Chemistry Laboratory of Hebei Province, Hebei University of Science and Technology, Shijiazhuang, China.
E-mail: [a]1178611805@qq.com

**Abstract**. In order to obtain some basic data for ecological risk assessment, an exposure experiment was carried out to study the effect of different volume of mixed penicillin and amoxicillin wastewater on Superoxide Dismutase (SOD) and Malondialdehyde (MDA) in Limnodrilus hoffmeisteri. Limnodrilus hoffmeisteri was exposed to different volume (10%, 20%, 30%, 40%, and 50%) of mixed wastewater for 15 days. According to the experimental data, the MDA contents and SOD activities in Limnodrilus hoffmeisteri muscle tissue had seen significant change significantly during the period of exposure. The results showed that the trend of SOD activities in low concentration groups (10% and 20%) increased at first, then decreased and increased at last. As exposure concentration increased (40% and 50%), SOD activities were inhibited in the early days (3 d), and was gradually induced in the later phase. The change of MDA content in muscle tissue of Limnodrilus hoffmeisteri was further investigated. It was found to be negatively correlated with the activity of SOD, and the high concentration group (50%) was in a remarkable induction state and reached the maximum at 6 d. According to the experimental data, the MDA contents and SOD activities in Limnodrilus hoffmeisteri muscle tissue had changed significantly and caused the oxidative damage.

## 1. Introduction

At present, China is one of the world's largest antibiotics producer. In the process of producing antibiotics, it will discharge a lot of wastewater. It is found that an increasing number of drinking water, surface water, groundwater, soil, sludge, and river sediment be detected of different levels of antibiotics. Aquatic sediments are important part of aquatic ecosystems and provide habitat for many benthic organisms [1]. Some studies have shown that concentrations of pollutants in sediments are usually hundreds or even thousands of times the amount of water in solution [2]. The research was conducted using Limnodrilus hoffmeisteri, which is a typical bottom-dwelling species and widely used in biotoxicity experiments [3].

In this study, the mixed wastewater of penicillin and amoxicillin was used to determine their effect on the muscle tissue of *Limnodrilus hoffmeisteri* in cell level, and the ecological effects of the mixed wastewater on aquatic organisms were determined, which provided scientific basis for its ecological risk assessment.

## 2. Materials and methods

### 2.1. Experimental materials

Instruments and reagents: UV-visible light spectrophotometer(UV-2550); hypothermia

refrigerator(BBC-226STV); high speed centrifuge(TG16-WS); aeration pump; dissolved oxygen meter(JPBJ-608); whirlpool mixer (XW-80A); enzyme standard instrument; glass homogenizer.
SOD and MDA kit were purchased from a biological engineering research institute.
The water samples were taken from the mixed wastewater treatment plant of penicillin and doxycycline in a pharmaceutical enterprise. The water sample was tested by the analytical test center, which contained penicillin (0.39 g/L) and amoxicillin (198 μg/L).
Limnodrilus hoffmeisteri were obtained from aquafarm in Shijiazhuang.

### *2.2. Experimental methods*

SOD and MDA samples: 18 Limnodrilus hoffmeisteri were sampled from the corresponding concentration groups before each measurement, anatomized, and fetch about 0.2 g of muscle tissue block, rinsed in cold saline, accurately weighing the determinant, put in a glass homogenate, use a liquid move adding the physiological saline nine times with the ratio of weight (g) : volume (ml) = 1:9, homogenizing with ice water bath, rotating stamp stem up and down for 10 times, crushing and making the tissue homogenate fully. The 10% tissue homogenate was centrifuged at 3000 r / min for 10 min with an ordinary centrifuge. Activities of superoxide Dismutase (SOD) and Malondialdehyde (MDA) content in samples were estimated by the method of SOD and MDA kit.

### *2.3. Data processing*

The mean value of the three parallel groups is obtained. The statistical results are expressed by the mean ± standard deviation (Mean ± SD). One-way ANOVA was used to analyze the experimental data by SPSS17.0 statistical software. The difference between the two groups was analyzed by the least significant difference (LSD). The difference was significant when $P < 0.05$; $P < 0.01$ indicates that the difference is extremely significant.

## 3. Results and discussion

### *3.1. Effect of the mixed wastewater on the SOD activities in Limnodrilus hoffmeisteri muscle tissue*

Effect of the mixed wastewater on the SOD activities in Limnodrilus hoffmeister muscle tissue was shown in table.3-1.

**Table.3-1.** Effect of the mixed wastewater on the SOD activities in Limnodrilus hoffmeisteri muscle tissue

**Unit :**(u/mgprot)

| **Experimental (%)** | **Expour time(d)** | | | |
|---|---|---|---|---|
| | ***3*** | ***6*** | ***9*** | ***12*** |
| control | 46.73±4.131 | 43.52±5.371 | 49.03±2.732 | 45.27±4.111 |
| 10 | 33.06±5.931** | 45.60±1.534 | 48.64±4.918 | 50.30±2.469* |
| 20 | 43.23±5.982 | 47.39±2.976 | 46.55±6.187 | 54.59±1.772** |
| 30 | 39.87±8.235* | 28.26±1.271** | 49.30±3.304 | 41.29±6.996 |
| 40 | 41.66±3.496 | 31.22±3.416** | 23.17±4.028** | 40.59±2.932 |
| 50 | 40.29±3.972 | 32.89±1.972* | 21.03±1.469** | 48.44±4.073 |

All the data in the same day are compared with control group, *p <0.05, ** p<0.01.
As shown in table.3-1, the SOD activity in the groups exposed to 20% was slightly different from that of the control group, when Limnodrilus hoffmeisteri were exposed to the mixed wastewater for 3 days and other exposure groups were subjected to different degrees of inhibition. Three groups (20%, 40%, and 50%) of SOD activities were negatively correlated with the mixed wastewater concentration, it shows that the higher concentration of the mixed wastewater, the greater the stress of the body to produce oxygen free radical, the more consumption of SOD. High concentration group (50%) do not fit this pattern, this may be due to the antioxidant defense system in body besides SOD, which can withstand too much oxygen free radicals, there are many other antioxidant enzymes involved in the oxidation process. At 6 d, in the low concentration groups (10% and 20%) SOD activity was significantly higher than those of the control group (P < 0.05). The significant inhibition of high concentration groups (30%, 40%, and 50%) were more severe than those in 3 d (P<0.01), 30% concentration group reached the minimum SOD activity in the whole experiment (28.26 U/mgprot). At 9 d, there was no significant difference in SOD activity between the low concentration groups (10%, 20%, and 30%) and the control group (P>0.05). on 12 d, the mixed wastewater with medium-low concentrations were not much different from those in the control group, and the oxygen free radical elimination was still in dynamic balance. It is shown that the body's antioxidant capacity is still in the limit of free adjustment; and the external source pollutants do not cause severe oxidative damage to the body. The trend was the same as the study of the exposure to 20 days of Brocarded Carp in low-concentration streptomycin wastewater [4].

### *3.2. Effect of the mixed wastewater on the MDA contents in tissue of Limnodrilus hoffmeisteri*

**Table.3-2.** Effect of the mixed wastewater on the MDA contents in tissue of Limnodrilus hoffmeisteri

**Unit (u/mgpro)**

| **Experimental (%)** | **Expour time(d)** | | | |
|---|---|---|---|---|
| | ***3*** | ***6*** | ***9*** | ***12*** |
| control | 10.02±1.072 | 9.94±0.937 | 7.951±0.233 | 7.68±1.224 |
| 10 | 10.68±0.241 | 8.21±2.323 | 6.25±0.166** | 5.72±0.107** |
| 20 | 7.81±0.070** | 9.85±0.273 | 7.23±2.323 | 5.86±0.432** |
| 30 | 9.37±0.518 | 7.49±0.572* | 9.32±0.913** | 7.19±1.937 |
| 40 | 12.32±0.981* | 9.31±1.241 | 8.91±0.274* | 8.48±1.703 |
| 50 | 13.19±1.131 | 15.11±2.638** | 7.57±1.935 | 14.32±1.661** |

All the data in the same day are compared with control group, *p <0.05, ** p<0.01.
As shown in table.3-2, in the early exposure (3 d), the MDA content of the low concentration groups (10%, 30%) compared with that of control group were both decreased and affected by different degree of inhibition. This may be owing to the participation of antioxidant defense system components (SOD) in the Limnodrilus hoffmeisteri, reducing the large body active oxygen concentration. The MDA content of high concentron groups (40%, 50%) significantly increased (p<0.01). It indicates that the body produces excess of oxygen free radicals due to high concentration of exogenous contaminants. Moreover, the ability of eliminate free radical in the body's antioxidant system is limited [5, 6]
In the 6d; in addition to the 50% concentration group, there was no significant difference between the exposure groups and the control groups. However, 50% concentration group was significantly increased (P<0.01) compared with the control groups and reached the maximum (15.11nmol/mL). It shows that

high conceration of the oxogenous pollutants cause futher oxidative damage. During 9~12d period, only the MDA content of 50% concentranion were significantly induced, and other concentration groups were same as that of control group($p>0.05$). So it is clear that body is constantly adjusting its antioxitant defense system when resisting the exogenous pollutants of oxidative stress to make a dynamic equilibrium between consumption of oxygcn free radicals and protection against oxidative damage in the body. This result is consistent with shen's discovery, the cefotaxime sodium of 125mg/L caused severe lipid peroxidation in the Zebrafish at the early stage of exposure, and the damage eventually recovered completely with the extension of time[7]. 50% concentration group has been in a significantly induced condition. This suggests that the antioxidant system of Limnodrilus hoffmeisteri has been subjected to a large degree of damage. They combined with unsaturated fatty acids, producing the lipid peroxidation. But the increasing in MDA contents cannot be digested, eventually leading to a serve damage of membrance system.

## 4. Conclusion

The high concentration of mixed wastewater of exposure can cause significant decrease in SOD activities in Limnodrilus hoffmeisteri. But within 15 days of the low-concentration exposure cell can recover itself after damaged, at the end of exposure, SOD has basically returned to the control level.The effect of the mixed wastewater on MDA content in Limnodrilus hoffmeisteri muscle tissue was significantly related to the "concentration-time-effect" relationship.The MDA content in different concentration groups had the same trend of "rising-decreasing-increasing". The activity of SOD in each concentration group was negatively correlated with MDA content.The change of SOD activities and MDA contents of Limnodrilus hoffmeisteriwere significant in the exposure to mixed wastewater of penicillin and amoxicillin, indicating that the mixed wastewater had oxidative damage to the Limnodrilus hoffmeisteri.

## 5. Acknowledgment

This study was supported by the National Natural Science Foundation of China (41373096), National Environmental Protection Project (201509041-05), Hebei Province Natural Science Fundation (B2014208068), Hebei Province Pharmaceutical Molecular Chemistry Laboratory Open Foundation, Hebei Province Environmental Protection Public interest Project, and Hebei Province key disciplines construction fund project

## 6. References

[1] Huang yao,Wu qunhe Research and progress in the quality benchmarking of water sediments. Environmental technology, 2003, (S1) : 24-27.

[2] Salomons W, Stigliani W. Biogeodynamics of pollutants in soils and sediments.Heidelberg: Springer-Verlag,1995:352.

[3] Fu rongshu Acute toxicity effects of lead and cadmium pollution on the water silk worms. Journal of shandong normal university (natural science edition). 200812 (23), 93-95

[4] Hongyan Shen Chenhong Wu, Streptomycin wastewater on the antioxidant enzymes and MDA content of koi fish [J]. The influence of safety and environmental engineering, 2013, (6) : 65-68

[5] guo-xia zhang. Of nitro aromatic hydrocarbons in water induced oxidative stress of fish biomarker research. [master of hebei university of science and technology thesis]. Shijiazhuang: hebei university of science and technology, 2012

[6] Shen Hongyan, wang bing, zhao month, etc. Ofloxacin on the brocade carp antioxidant system and the effect of DNA damage. Environmental science and technology, 2015, 38 (12) : 59-66

[7] Hongyan ShenXiaohui Jiao,Cefotaxime sodium of zebrafish SOD activity and MDA content and the effect of DNA damage [J]. Journal of environmental science, 2015, (8) : 2626-2632.

# Determination of Trace lead (Ⅱ) by Resonance Light Scattering Based on Pb (Ⅱ)-KI-MG System

**Ninghua Chen[1], Yingchun Yang[1, a]*, Shuai Hao[1], Yangmin Li[1]**

[1]Chengdu University of Information Technology, Department of Resources and Environment, Chengdu, Peoples Republic of China
E-mail: [a]yangyingchun@cuit.edu.cn

**Abstract**. In pH=3.0 weak acidic solution, it is found that $Pb^{2+}$ can react with $I^-$ to form $[PbI_4]^{2-}$, and it further reacted with MG to form ion-association complex. As a result, the new spectra of RLS appeared and their intensities enhanced greatly. Accordingly, a new method developed for the determination of Pb (II).The appropriate reaction conditions were optimized through experiments. The results show that a strong and stable resonance scattering spectra emerge at the wavelength of 338 nm. The resonance light scattering strength ($\Delta I_{RLS}$) has good linear relationship with the concentration of Pb (II) in the range of 0.2 μg/mL ~ 1.0 μg/mL. The detection limits (LOD) is 0.0155 μg/mL. The relative standard deviation (RSD) is 3.61% (n=11) for the determination of 0.6 μg/mL Pb (II) standard solution. And this method was successfully applied to the determination of three environmental water samples (nongfu spring, tap water, laboratory wastewater). Results illustrate that the addition standard recovery are 80%~107% with relative standard deviation (RSD) between 1.8% to 4.6%.

## 1. Introduction

As a strong neurotoxin heavy metal element, lead has been considered as one of the most serious environmental pollutants in human civilization. Lead is widely used in many fields, such as industry and transportation, and lead pollution in living environment seriously endangers human health. [1] According to relevant medical report, children's blood lead levels above or equal to 100 g/L will have an effect on children's intellectual development, leading to a decline in children's intelligence. The blood lead content of children was significantly negatively correlated with IQ (IQ). When blood Pb (II) levels increased by 100 g/L, the average IQ decreased by 1~3 points. Thus, the detection of Pb (II) content, especially in environmental water samples, remains a key challenge in the implementation of environment measuring and controlling. [2]

At present, the methods of determination of Pb (II) include atomic absorption spectrometry (AAS), [3-6] spectrophotometry [7-10] and paleography [11-13]. Among them, the AAS is fast and accurate, but the Pb(II) content in the water is usually low, and the detection limit of this method can hardly meet the requirements of trace analysis. On the other hand, such samples must be preseparated and enriched before used. Spectrophotometry is one of effective detection methods that can realize simultaneous detection of large samples, while the low degree of accuracy limits its practical application. In addition, paleography is another convenient method, but suffers from its large errors. Therefore, the development of simple and rapid determination methods of Pb (II) with high detection limit and accuracy are crucial. The determination of Pb (II) by RRS is a new method, which have the advantages of higher sensitivity, simplicity and easy operation. It has been widely used to the research of biological macromolecule, pharmaceuticals, nanoparticals, surfactants, metal and non-metal ions in the environmental samples.[14,15] Compared with the traditional method, such RRS, has the characteristics of using nontoxic substances, saving reagents and samples, and its operation is simple and easy.

In this work, We investigate pH = 3, Pb(II) with too much $I^-$ ions form $[PbI_4]^{2-}$ complex anion, and further form ion association with MG, so that the system of the resonance light scattering intensity is enhanced, thus establishing a kind of spirit high lightness, good selectivity, fast, accurate determination of Pb(II).

## 2. Experiment section

### 2.1. Apparatus and reagents

A RF-5301 spectrofluorophotometer (Shimadzu Company, Japan) was used for recording and measuring the RLS and fluorescence spectra. A UV-2550 spectrophotometer (Shimadzu Company, Japan) was used for recording the absorption spectra. A PHS-3C acidity meter (Shanghai Precision & Scientific Instrument Co. Ltd., Shanghai) was used for adjusting pH values.

The concentration of Pb (II) stock solution is 1.0 mg $mL^{-1}$ and the working solution has been prepared by diluting the stock solution to 1.0 μg $mL^{-1}$.$1.0\times10^{-3}$ mol $L^{-1}$ of malachite solution is used as stock solution, and it is diluted to $1.0\times10^{-4}$mol $L^{-1}$ working solution. The stocking solution concentration of potassium iodide and hydrochloric acid are 10% (mass concentration) and $1.0\times10^{-3}$ mol $L^{-1}$, respectively.

All reagents are analytical reagent grade and come from Chengdu Kelong Chemical Reagent Factory. The milli-Q ultrapure water is used throughout.

### 2.2. General procedure

According to the experimental order, suitable amount of Pb(II) wording solution,0.8 mL HC1 solution, 2.6mL malachite working solution, 0.6mL of potassium iodide solution(10%)were injected into a 10.0 mL calibrated glass tube. The resulting solution was diluted to the mark with ultrapure water and mixed thoroughly, and then it was set aside for 15 min before determination. The RLS spectra of the system were recorded with synchronous scanning at λ ex = λ em and RLS intensity ($I_{RLS}$) for sample solution and $I^0_{RRS}$ for reagent blank at 338 nm were measured accordingly, $\Delta I_{RLS} = I_{RLS} - I^0_{RLS}$. The selected excitation and emission-slit widths are 10nm.

## 3. Results and discussion

### 3.1. RLS spectra

The RLS spectra of Pb (II)-$I^-$-MG system are shown in Fig.3-1 and Fig.3-2. From Fig.3-1, it can be seen that the RLS intensities of single Pb (II), MG solution, and I- or the mixture solution of any two of them are very weak. However, when Pb (II), $I^-$ and MG mixed each other, the intensities of RLS spectra are enhanced greatly due to the formation of a ternary ion association complex. And four obvious scattering peaks located at the wavelength of 338 nm, 468 nm, 589 nm and 660 nm, respectively. From Fig.3-2, the shape and stability for the peak of 338 nm are relatively good, and the added value of scattering intensities (ΔI) increase with the accretion of Pb (II) concentration. There is a good linear relationship between ΔI and Pb (II) concentration, which can be applied to the determination of Pb (II). Therefore, 338 nm was chosen as the quantitative measuring wavelength in further experiments.

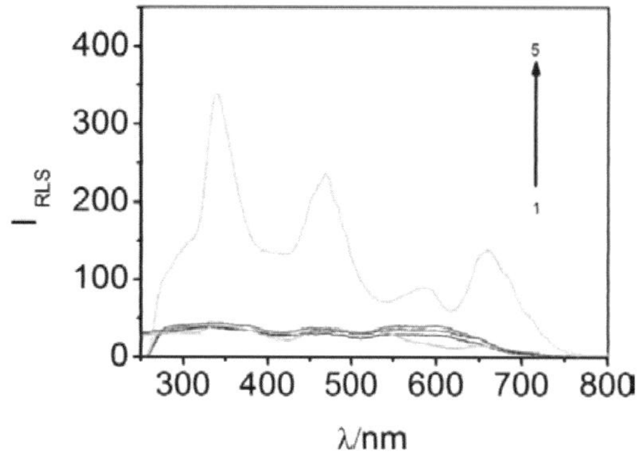


**Figure.3-1.** The resonance light scattering spectra of various system.1-5:Pb

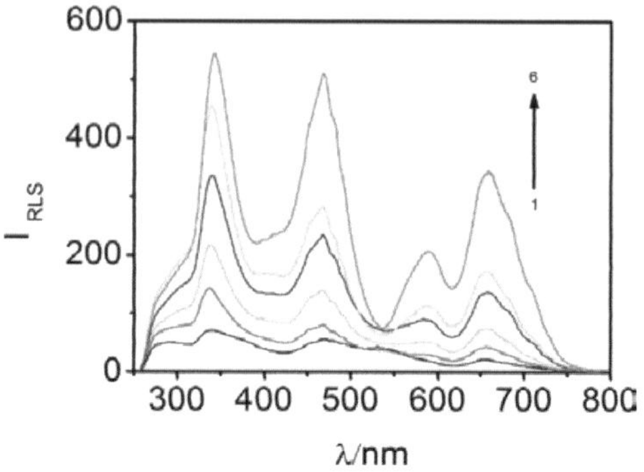


**Figure.3-2.** The resonance light scattering spectra of Pb(II)-I--MG reaction system.1-6(cPb2+):0.2, 0.4, 0.6, 0.8, 1.0μgmL-1; cKI:10%(mass concentration); cMG: 1.0×1035molL-1; pH=3.0

*3.2. Optimum Reaction Conditions*

*3.2.1. Effect of acidity.* Various kinds of acid and buffer solutions, such as sulfuric acid, hydrochloric acid, nitric acid, phosphoric acid, HC1-NaAc buffer solution and BR buffer solution, were used to investigate the effects of acidity on the intensity of RLS. The results indicate that $1.0\times10^{-3}$ mol $L^{-1}$ hydrochloric acid is the best. When hydrochloric acid was used as the reaction medium, different volume of $1.0\times10^{-3}$ mol $L^{-1}$ hydrochloric acid was tested. The results suggest that 0.8 mL dosage of hydrochloric acid is the best. When the addition value of hydrochloric acid was lower or higher than 0.8mL, $\Delta I_{RLS}$ decreased obviously. Thus, $1.0\times10^{-3}$ mol $L^{-1}$ hydrochloric acid was chosen as the reaction medium in the following experiments, and the appropriate addition amount of 0.8 mL was used in view of the strength and stability of RLS signal.

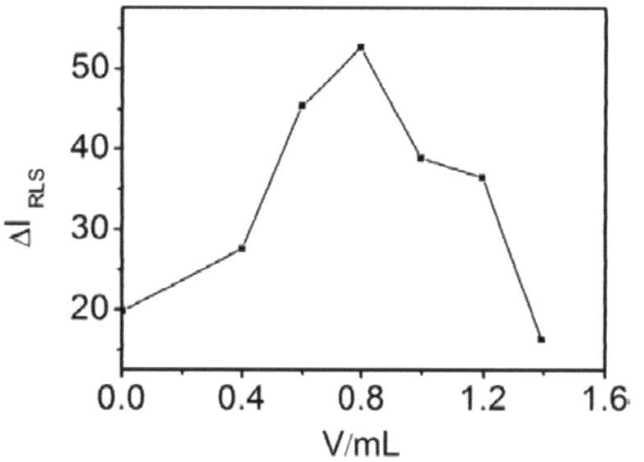


**Figure.3-3.** Effect of HCl dosage on $\Delta I_{RLS}$

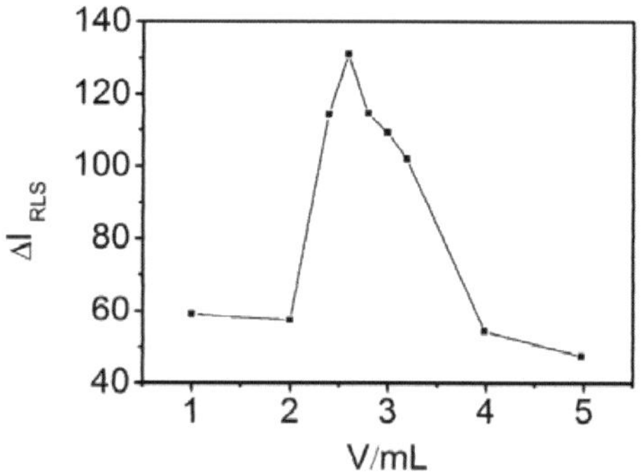


**Figure.3-4.**Effect of MG dosage on $\Delta I_{RLS}$ $cPb^{2+}$: 100.0μg $L^{-1}$; cKI: $2.8\times10^{-3}$mol $L^{-1}$; pH=3.0

*3.2.2. Effect of MG concentration.* Effects of MG concentration on RRS were examined. 1 to 5 mL of $1.0\times10^{-4}$ mol/L MG solution were added into the system respectively. The influences on RLS intensity of various dosage of MG are shown in Fig.3-4. When the dosage of MG is in the range of 1 to 2.6 mL, the relatively $\Delta I_{RLS}$ intensity increased accordingly. However, when the dosage is higher than 2.6 mL, the $\Delta I_{RLS}$ intensity continually decreased. The maximum $\Delta I_{RLS}$ intensity occurred at the dosage of 2.6mL. Consequently, the addition of $1.0\times10^{-4}$mol/L MG was selected as 2.6 mL in the next experiments.

*3.2.3. Effect of potassium iodide concentration.* Generally, excessive potassium iodide is required for fully formation of $[PbI_4]^{2-}$ from trace amount of Pb (II) with iodide. Therefore, effects of the concentration of potassium iodide on RRS intensities ($\Delta I_{RLS}$) were tested using higher concentration (10% mass concentration) of potassium iodide. The results are shown in Fig.3-5. It is can be seen from Fig.3-5, when the addition of potassium iodide was 0.6 mL, $\Delta I_{RLS}$ reached maximum and stable. Therefore, 0.6mL of potassium iodidc was chosen in the subsequent experiments.

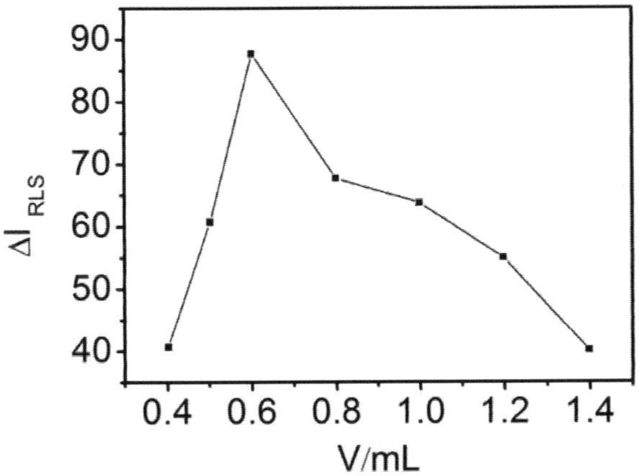


**Figure.3-5.**Effect of KI dosage on $\Delta I_{RLS}$

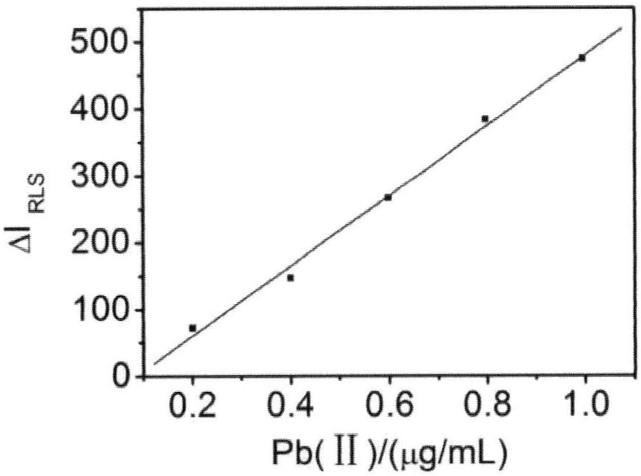


**Figure.3-6.**Calibrate curve

*3.2.4. Other influence factor.* Different kinds of surfactants were inspected in this scattering system, which included arabic gum, cetylpyridinium bromide, Triton X-100, hexadecyltrimethyl ammonium bromide and laurel sodium sulfate. The results indicated that there are not influences on the system. And the influence of temperature through 10°C to 60°C was examined. It is no distinctive influence. Accordingly, room temperature and no surfactants were selected in the following experiments.

*3.2.5. Reaction speed and the stability of scattering system.* At room temperature (15-25°C), the RLS intensity ($\Delta I_{RLS}$) reached maximum within 15 min, and it could remain constant within 70 min free from the straight irradiation of sunlight. After that, the RLS intensity ($\Delta I_{RLS}$) gradually reduced. So the experiment should be accomplished within 70 min. normally, we performed the whole determination within 15-20min. Moreover, there is not obviously influence on the magnitude of RLS intensity ($\Delta I_{RLS}$) of different mixing sequence of reaction reagents.

*3.2.6. Effect of ionic strength.* The influences of ionic strength on RLS intensity ($\Delta I_{RLS}$) were investigated by introducing NaCl into the mixed solution. Results show that the RLS strength will decrease when the concentration of NaCl is higher than 0.05 mol $L^{-1}$, but there is no obvious influence on scattering strength of the system as the concentration of NaCl is lower than 0.05 mol $L^{-1}$. The reason is maybe the electrostatic screening effect by excessive NaCl in solution, which interfered with the combination of MG with $[PbI_4]^{2-}$. Thus, other inorganic salt should not be introduced under normal circumstances.

### *3.3. Sensitivity of methods*

Under the selected optimum reaction conditions, the linear regression equations were performed. The specific information was displayed in Fig.3-6. It can be seen from Fig.3-6 that the intensity of $\Delta I_{RLS}$ has good linear relationships to the concentration of Pb (II) in the range of 0.2-1.0μg $mL^{-1}$. The correlation coefficient is 0.9986, and the detection limits is 0.016μg $mL^{-1}$.

### *3.4. Effect of coexistent substances*

Effects of more than 20 kinds of coexisting substances were tested by using RLS method. While the content of Pb (II) is 200 μg $L^{-1}$, relative error within ±5%, the allowing multiples of coexisting substances were investigated. The results show that most common ions do not interfere with the determination of Pb (II). However, there are bigger interferences of $Ag^{+}$, $Cd^{2+}$ and $Bi^{3+}$. In the determination of real environmental water samples, the disturbances of them could be removed by adding complexion II and dimercaptopropanol according toliterature.

### *3.5. Analytical application*

According to the abovementioned experimental methods, concentration of Pb (II) in three water samples were determined after necessary pretreatment steps. The results show that the recoveries of the method are between 80% and 107%; the relative standard deviation is between 1.8% and 4.6%. This method can fast and accurate determination of trace levels of Pb (II) in the water samples.

**Table.3-1.** Determination of Lead (II) in Real Samples (n=3)

| Sample | *Found* ($\mu g{\cdot}mL^{-1}$) | *Added* ($\mu g{\cdot}mL^{-1}$) | *Total Found* ($\mu g{\cdot}mL^{-1}$) | *Recovery* (*%,n=3*) | *RSD* (*%,n = 3*) |
|---|---|---|---|---|---|
| No.1 | ND | 0.40 | 0.35 | 87.50 | 1.8 |
| | ND | 0.60 | 0.64 | 106.68 | 2.9 |
| | ND | 0.80 | 0.69 | 86.25 | 3.1 |
| No.2 | ND | 0.40 | 0.32 | 80.00 | 4.6 |
| | ND | 0.60 | 0.55 | 91.66 | 3.4 |
| | ND | 0.80 | 0.65 | 81.25 | 3.1 |
| No.3 | 0.21 | 0.40 | 0.58 | 92.50 | 2.2 |
| | 0.33 | 0.60 | 0.89 | 93.33 | 1.8 |
| | 0.25 | 0.80 | 0.95 | 87.50 | 4.3 |

ND: Not detected

## 4. Conclusion

In acidic medium, neutral ion-association complex $[PbI_4]^{2-}$ reacted with malachite green resulted in the obvious enhancement of RLS. The concentration of Pb (II) is proportion to the increment of intensity of RLS in a certain range. Based on this, a new method for the determination of Pb (II) was established. The developed method has the advantages of simplicity, rapidity, good selectivity and high sensitivity.

## 5. Acknowledgements

The Scientific Research Fund of Sichuan Provincial Education Department (No. 16ZA0217) supports this work.

## 6. References

[1] X. Liu, Z. Zhu, H. Li, D. He, Y. Li, H. Zheng, Y. Gan, Y. Li, N. Belshaw and S. Hu, Anal. Chem., 2017, 89, 6827–6833.
[2] W. Kang, X. Pei, C. Rusinek, A. Bange, E. Haynes, W. Heineman and I. Papautsky, Anal. Chem., 2017, 89, 3345–3352.
[3] D. Citak and M. Tuzen, Food Chem. Toxicol, 2010, 48, 1399–1404.
[4] Y. Xiao, Y. Feng, H. Lu and G. Zhang, J. Anhui Normal Univ., 2010, 33, 48.
[5] H. Wang, L. Ou, Y. Suo and H. Yu, Anal. Chem., 2011, 83, 1557–1563.
[6] G. Huang, Y. Fang, Y. Hang and X. Tang, Acta Tabacaria Sinica, 2013, 19, 5–9.
[7] G. Alpdogan and S. Zor, J. Aoac. Int., 2017, 100, 1524–1530.
[8] Q. Wu, H. Zhang, S. Liu, X. Wang and J. Gao, Shandong Chem. Industry, 2017, 46, 120–123.
[9] T. Zhang, H. Sun and B. Zhao, Guangzhou Chem. Industry, 2017, 45, 96–97.
[10] R. Zhou, D. Fu, D. Yuan, X. Li and X. Guo, App. Chem. Industry, 2017, 46, 601–603.
[11] L. Zhang, Mining and Metallurgy, 2017, 26, 91–93.
[12] Q. Liang, Technology and Economic Guide, 2017, 2, 78.
[13] Y. Fu, Science and Technology, 2016, 26, 154.
[14] W. Ren, Y. Zhang, Y. Fan, J. Dong, N. Li and H. Luo, J. Hazard. Mater., 2017, 336, 195–201.
[15] X. Hu, D. Xu, S. Liu, Z. Liu, C. Li and P. Chen. Anal. Lett., 2010, 43, 2125–2133.

# Testing Method of Degrading Heavy Oil Pollution by Microorganisms

**Qi Wu[1,a], Lin Zhao, Aijin Ma[2,b*]**

[1]Institute of Food and Agriculture Standardization, China National Institute of Standardization, Beijing 100191, China
E-mail: [a]wuqi@cnis.gov.cn, [b*]maaj@cnis.gov.cn

**Abstract.** With the development of human society, we are more and more relying on the petrochemical energy. The use of petrochemical energy not only brings us great convenience, but is also accompanied by a series of environmental pollution problems, especially oil pollution. Since it is impractical to restore all pollution problems, the proper use of some remedial measures, under the guidance of functional orientation, may be sufficient to minimize the risk of persistent and diffusing pollutants. In recent years, bioremediation technology has been gradually developed into a promising stage and has played a crucial role in the degradation of heavy oil pollution. Specially, microbes in the degradation of heavy oil have made a great contribution. This paper mainly summarizes the different kinds of microorganisms for degrading heavy oil and the detection method for degradation efficiency of heavy oil pollution.

## 1. Introduction

We begin to pay more and more attention to a variety of toxic organic compounds that have been brought into our life intentionally or unintentionally, among which petroleum hydrocarbons, as the most important energy resources and industrial raw materials, bring great convenience to our life, but produce a lot of environmental pollution problems. Currently, there are no better substitutes for petroleum hydrocarbons which make the solution of pollution problems very important. The bioremediation technology refers to the method [1] for eliminating or reducing the concentration of harmful substances in polluted areas by take advantage of the potency of biology, especially microorganisms. As a pollution-free and low-cost method, it is superior to other physical and chemical methods in degradation of heavy oil pollution.

The heavy oil pollution has seriously damaged the balance of the soil ecosystem- reducing the soil fertility, affecting the normal growth of plants, and even in filtrating underground and causing serious pollution to the groundwater. All of these have seriously affected our life. Currently, there are two kinds of main methods for controlling the heavy oil pollution in soil: physical-chemical methods and biological methods; physical-chemical methods involves in excavation and landfill, isolation control, oxidation, extraction, adsorption, and so on. However, all of these methods have drawbacks, like they are temporary relieving methods, and prone to bring secondary pollution. In the contrast, microbiological methods, such as plant cultivation, microbiological method, have become a hot topic in dealing with the problems on oil pollution in soil for their advantages, for example, low cost, better effective and free of secondary pollution .[2-3]

Microorganisms have the following advantage on the degradation of heavy oil pollution: There are widely distributed numerous microorganisms with a strong reproductive capacity, adaptability and faster metabolic rate. The studies on degrading heavy oil pollution by microorganisms have shown that the degradation efficiency of heavy oil pollution has been able to reach about 70% under certain conditions[4-5].In addition, it is possible to adjust the degradation ability of microorganisms to heavy oil pollution by adjusting the living environment of microorganisms, such as adjusting the pH, adding mineral elements, adjusting the temperature and so on.[6-7]Currently, the microorganisms used for degrading heavy oil pollution are mainly from contaminated soils near the oil fields, oil-contaminated sea, forest soil, etc.[8-10]. , There area wide variety of microorganisms in different environment and different regions. .

There are various testing methods for the evaluation of heavy oil pollution degradation by microorganisms. Most of these methods are mainly based on gas chromatography (GC) [11-13]. GC is generally simple and accurate and can be combined with the use of mass spectrometry, flame ionization detector, and so on for more accurate results [10, 14]. Beside GC, there are also reports based on infrared absorption spectrum [15], UV spectro-photometer [16], acid-base titration [17], gravimetric analysis [18], and so on.

## 2. Microorganisms

There are many kinds of microorganisms that degrade oil pollution, and they are mainly screened from different oil-contaminated environment. Generally, several or dozen kinds of different bacterial species can be obtained in the same environment. Therefore, in the existing environment, multiple bacterial species are producing combined effects of degradation on oil pollution. Reports showed that Gram-negative bacteria perform better and dominate in the degradation process, [19] when compared with Gram-positive bacteria. The most common bacteria are bacilli, of which bacillus and pseudomonas are found more in the process of study. The following table.2-1. shows several common microorganisms used for degrading heavy oil pollution.

**Table.2-1.** Microorganisms used for degrading heavy oil pollution

| Bacterial strain | References |
|---|---|
| *Pseudomonas* | 4, 6, 15, 18,20,21,22 |
| *Bacillus* | 4, 15,21,23,24 |
| *Ochrobactrum* | 5,16,24 |
| *Acinetobacter* | 4,25 |
| *Proteobacteria* | 19,26 |
| *Firmicutes* | 19,26 |
| *Alcaligenes* | 5,19 |
| *Microbacterium* | 5 |
| *Chryseobacterium* | 15 |
| *Micrococcus* | 27 |
| *Nitratireductor* | 22 |
| *Actinobacteria* | 26 |
| *Penicillium* | 28,29 |
| *Aspergillus* | 28 |
| *Ganodermalucidum* | 7 |
| *Candida bombicola* | 23 |

Bacteria have been found to be the microorganisms actively involved in the degradation of organic pollutants, and many bacteria mentioned above are capable of effectively reducing the polycyclic aromatic hydrocarbons. Benzopyrene in petroleum hydrocarbons is considered to be the most carcinogenic and toxic. Studies have shown that a variety of strains have the ability to degrade benzopyrene. [30] For example, 10 strains of bacteria were isolated from various contaminated sites (oil, engine oil, wood), by Aitkin et al. [30], and there are at least three kinds of pseudomonas, as well as agrobacterium, bacillus and so on. It is generally idea that the common coordination effect of different bacterial species is better than that of single strain because different bacterial species have different sensitivity for different kinds of hydrocarbons. In the context that the heavy oil pollution is often a mixture of multiple hydrocarbons, the coordination between bacterial species can be more effective for

its degradation. Daane et al. [31-32] isolated and characterized the petroleum hydrocarbon degradation bacteria associated with the salt marsh plant roots. They divided the isolated bacteria into three major bacterial groups: Gram-negative pseudomonas, Gram-positive bacteria and bacillus. They found that the fraction with rich phenanthrene was able to take advantage of more polycyclic aromatic hydrocarbons than fraction with rich naphthaline. They confirmed that bacteria in the genus bacillus isolated from petroleum-contaminated sediments and salt-marsh roots could use naphthaline or phenanthrene as the sole carbon source and therefore could degrade polycyclic aromatic hydrocarbons.
At present, the degradation efficiency of bacteria on oil pollution can be significantly improved by adjusting the living environment of microorganisms. Jingxiu Wang et al. [15]found that the optimum degradation conditions for crude oil with the concentration of 1000 mg/L by the three kinds of bacteria including chryseobacterium, bacillus and pseudomonas were 10 °C temperature, pH = 7 and salinity of 10g*L-1. The degradation efficiencies of these three kinds of bacteria were 62.3%, 61.6% and 60.9% after 8 days under these conditions.
In addition, some fungi microorganisms can also effectively degrade oil pollution. For example, Fuad Ameen et al. [28] extracted 45 kinds of fungi, mostly aspergillus, and other 9 species, from the beach near the Red Sea. The whole degradation efficiency on crude oil can reach 55.5%.There are many algae which also made a great contribution to the degradation ofocean oil pollution.
In general, studies on degradation of heavy oil pollution by bacteria are most popular, and the kind of involved bacteria are diverse. In the current stage, the most challenge for bioremediation technology is that, microbial populations have their own living environment and limitation of survival time, thus a certain field of microorganisms cannot be applied in other fields. In other words, there are space limitation and application restrictions needed to be overcome.

## 3. Test method

There are various methods for testing the oil pollution degradation efficiency in bioremediation of heavy oil pollution. Currently, we mainly rely on spectrum, chromatography and other methods to examine the content of residual oil pollution and then to calculate the degradation efficiency. The most common test methods are as below.

### *3.1. Gas chromatography-Mass spectrometry (GC-MS)*

Gas chromatography - Mass spectrometry (GC-MS) is a method which combines the features of gas chromatography and mass spectrum for identifying different substances in the samples, and it is the most widely used method for testing oil pollution degradation. The main components of petroleum hydrocarbon are alkenes and polyaromatic hydrocarbon. Firstly, alkenes fraction and polycyclic hydrocarbons fraction are separated roughly through column chromatography. Then, both fractions can be detected effectively and respectively by GC-MS. By this way, the content and degradation efficiency of each component can be determined; the total oil pollution degradation efficiency can also be deduced. For example, Jingxiu Wang et al. [15] conducted research on the degradation of oil pollution by three bacterial strains in cold climate with GC-MS. According to their research, the small and medium molecular weight in the oil pollution after microbial action increases significantly and large molecular weight decreases obviously, which means that macromolecules are gradually degraded into small molecules after microbial degradation.
Besides, by detecting degradation process with GC-MS, it can be found out that which hydrocarbon compounds microorganism is more sensitive to and has more degradation ability and so on. These series of detailed information could greatly help to further study of microorganism's degradation ability. For example, SERGI DI'EZ et al. [33] conducted a research on the bioremediation of two microbiomes with GC-MS. The results show that, n-alkane, alkyl cyclohexane, alkylbenzene, diaromatic hydrocarbon and triaromatic hydrocarbon are degraded obviously, and the biodegradation efficiency of higher polycyclic hydrocarbon is limited under the initial action. In addition, by analyzing and comparing the molecular species at different time quantum, the degradation tendency of oil pollution is given: non-ring isoprene > disilane > C27- gonane > bb- gonane > pentacyclic triterpene > monoaromatics> gonane > triarylstane.

### *3.2. Gas chromatograph - Flame ionization detector (GC-FID)*
Similar to GC-MS, the concentrations of different components, such as alkanes, aromatic compound, can be detected by using gas chromatograph (GC) and flame ionization detector (FID) jointly, as well as the overall petroleum hydrocarbon and the degradation efficiency can be determined finally. Since FID has a high sensitivity, good stability and rapid response to organic compounds and tolerance to a wider temperature range, volatile hydrocarbon can be detected more conveniently. FID is widely used in organic compounds detection. Eman Koshlafet al. [14] detected the impact of peas on the degradation of petroleum hydrocarbon in polluted soil through GC-FID. Their results showed that the optimum degradation efficiency could reach 96.1%.

### *3.3. Microbiological respiration (output of carbon dioxide)*
The degradation of oil pollution by microorganism mainly relies on metabolism process using hydrocarbon compound as the carbon source. Large quantity of carbon dioxide will be produced during the metabolism. The determination of $CO_2$ content can be used for indication of the metabolism activity of microorganism indicated, and further used for the determination of degradation efficiency of oil pollution. Acid-base titration, gas chromatographic method, barium peroxide absorption method and other methods can be adopted for the determination of $CO_2$ content. For example, OBIRE, O et al. [34] determined the degradation efficiency of microorganism by detecting the output of $CO_2$ with Cornfield (1961) method. Cornfield method was firstly reported in 1961.In this method, barium peroxide solutionis added in airtight reactor producing barium carbonate after absorbing $CO_2$.Then the content of absorbed $CO_2$ is determine by acid base titration. Cornfield method is relatively complicated. With the development of detection method the method of detecting $CO_2$ output becomes convenient and rapid now. For example, Samiha Al Kharusiet al. [35] made the analysis of $CO_2$ contents in the 250 milliliter gas absorbed from airtight reaction bulb in different time stages with gas chromatographic method, the degradation ability of microorganism can be determined by comparing the $CO_2$ output produced at different time stages or under different conditions.
Though it is easy to determine the degradation ability of microorganism by detecting $CO_2$ output with microbiological respiration, the microbe activity and degradation can only be concluded by comparison, the degradation efficiency cannot be obtained clearly, which is the most disadvantage of this method.

### *3.4. Quality method*
Quality method is a simple and convenient method for determining the degradation efficiency of microorganisms, that is, to determine the quality of reaction samples before the degradation, and then to determine the quality of remaining reaction samples after a period of degradation. The quality of the degraded oil pollution can be obtained by comparing the difference between the two qualities, and the degradation efficiency can be determined by simple calculations. However, this method can be only applied to the preliminary determination, with a relatively large error. And the quality method can only be used for figuring out the changes of total oil pollution, but not the determination of degradation efficiency of each component. Besides, the quality method cannot be used for detection the degradation process of long-chain alkenes to short-chain alkenes. By using this method, Qu Lina et al. [36] determined the degradation efficiency of microorganisms for oil pollution in soil.

### *3.5. Ultraviolet spectrophotometer method*
Similar to the quality method, the determine the degradation efficiency by ultraviolet spectrophotometer analysis is also based on determination of the total oil pollution content. This method determines the degradation efficiency of the microorganism for oil pollution by detecting the change of absorbance of the mixed oil pollution solution. For example, Chen Lihua et al. [37] applied this method to study the degradation efficiency of microorganisms in heavy oil pollution soil. Firstly, they used the absorbance of oil pollution solution with known concentration at 225nm to draw the standard curve, and then they determined the absorbance of oil pollution solution before and after degradation and calculated the

concentration of oil pollution solution before and after degradation. Finally, the degradation efficiency was obtained. The issue with this method is that the standard curve of oil pollution containing different components is different. Therefore, it is necessary to draw a different standard curve for different oil pollution degradation, which is slightly cumbersome.

### *3.6. Infrared spectroscopy*
The main component of oil pollution is hydrocarbon. The C-H bond of the alkanet compound has tensile vibration at 2930 $cm^{-1}$. The concentration of the alkanet compound can be obtained by determining the absorption spectrum of the infrared spectroscopy, and then the degradation efficiency can be determined by comparing the oil pollution concentration. Compared with the UV spectrophotometer, this method is easier to operate. A. M. Zyakun et al. [38] determined that the degradation efficiency of microorganisms is in the range of 38±5% by this method. It was found that the results were consistent with that measured by GC-MS.
There are a wide variety of methods for determination of the oil pollution degradation efficiency. For the moment, gas chromatography is the most widely used method. Gas chromatography is mainly used with the combination of a series of other detection methods. Among these methods, gas chromatography-mass spectrometry is most widely used. This method featured with the advantages of high precision, high sensitivity and simple operation steps. Most importantly, oil pollution, as a series of mixtures of hydrocarbon compounds, can be effectively separated by this method, and then the degradation efficiency of different components can be determined. In addition, UV-visible absorption, infrared-visible absorption, quality method, microorganisms' respiration method and other methods are generally used as an auxiliary method together with GC-MS method and can also be used as the verification of result. For example, Martínez ÁlvarezLM et al [39] determined the concentration of hydrocarbons by GC-FID and IR, and have determined the oil pollution degradation efficiency of microorganisms by these two methods.

## 4. Conclusion
Large number of organic chemicals, such as petroleum hydrocarbons, halogenated and nitro aromatic compounds, phthalates, etc., have seriously polluted the soil and aquatic environment. The petroleum hydrocarbon contaminants pose a serious threat to human health and the environment because of its low reactivity and resistance to degradation. The conventional restoration method for these contaminated sites based on either physical or chemical methods is technically and economically challenging. In contrast, bioremediation is relatively effective, cost-effective and the most promising technology. Therefore, bioremediation technology is deemed as an effective and economical alternative over physical and chemical methods. We use microorganisms for the degradation of heavy oil pollution, which can avoid the secondary pollution and be at lower costs for its treatment. However, the existing bioremediation method is subject to many limitations, including poor ability of microbial communities, low bioavailability in space and time scales, and no reference values for functional test of bioremediation. At present, the lower microorganism degradation efficiency of heavy oil pollution under natural conditions is still the most serious problem to be overcome. It is of particular importance to use microorganisms to achieve efficient degradation of oil pollution and to determine its degradation efficiency in a simple and convenient manner.

## 5. Acknowledgment
This work was supported by National Key Research and Development Program (No. 2016YYF0202300).

## 6. References

[1] YU Yang, ZOU Li, SUN Tingting, et al. The Mechanism of Oil Degradation by Microbe and Influencing Factors [J]. Journal of Anhui Agriculture Science, 2014(16):5198-5200.
[2] Wang Meng, Zhang Peng, a Li-guo, et al. Biodegradation of petroleum pollution by microorganisms [J]. Science & Technology Information, 2009(2):14-15.
[3] Varjani S J. Microbial degradation of petroleum hydrocarbons [J]. Bioresource Technology, 2016, 223:277-286.
[4] Bento F M, Camargo F A O, Okeke B C, et al. Comparative bioremediation of soils contaminated with diesel oil by natural attenuation, biostimulation and bioaugmentation[J]. Bioresource Technology, 2005, 96(9):1049-1055.
[5] Tahseen R, Afzal M, Iqbal S, et al. Rhamnolipids and nutrients boost remediation of crude oil-contaminated soil by enhancing bacterial colonization and metabolic activities [J]. International Biodeterioration & Biodegradation, 2016, 115:192-198.
[6] Hesnawi R M, Mogadami F S. Bioremediation of Libyan Crude Oil-Contaminated Soil under Mesophilic and Thermophilic Conditions [J]. Apcbee Procedia, 2013, 5(5):82-87.
[7] Ting W T E, Yuan S Y, Wu S D, et al. Biodegradation of phenanthrene and pyrene by Ganoderma lucidum [J]. International Biodeterioration & Biodegradation, 2011, 65(1):238–242.
[8] Sal nitro J P, Dorn P B, † M H H, et al. Crude Oil Hydrocarbon Bioremediation and Soil Ecotoxicity Assessment [J]. Environmental Science & Technology, 1997, 31(6):1769-1776.
[9] Hazen T C, Prince R C. Marine Oil Biodegradation [J]. Environmental Science & Technology, 2015, 50(5):2121-2129.
[10] Liu J, Bacons H P, Liu Z. Potential Environmental Factors Affecting Oil-Degrading Bacterial Populations in Deep and Surface Waters of the Northern Gulf of Mexico:[J]. Frontiers in Microbiology, 2016, 7(1325), 02131.
[11] Xu J, Kong F, Song S, et al. Effect of Fenton pre-oxidation on mobilization of nutrients and efficient subsequent bioremediation of crude oil-contaminated soil.[J]. Chemosphere, 2017, 180:1-10.
[12] Chaîneau C H, Rougeux G, Yéprémian C, et al. Effects of nutrient concentration on the biodegradation of crude oil and associated microbial populations in the soil [J]. Soil Biology & Biochemistry, 2005, 37(8):1490-1497.
[13] Haghollahi A, Fazaelipoor M H, Schaffie M. The effect of soil type on the bioremediation of petroleum contaminated soils [J]. Journal of Environmental Management, 2016, 180:197-201.
[14] Koshlaf E, Shahsavari E, Aburto-Medina A, et al. Bioremediation potential of diesel-contaminated Libyan soil [J]. Ecotoxicology & Environmental Safety, 2016, 133:297-305.
[15] Wang J, Wang J, Zhang Z, et al. Cold-adapted bacteria for bioremediation of crude oil-contaminated soil [J]. Journal of Chemical Technology & Biotechnology, 2016, 91(8):2286-2297.
[16] Ibrahim H M. Biodegradation of used engine oil by novel strains of Ochrobactrum anthropi HM-1 and Citrobacter freundii HM-2 isolated from oil-contaminated soil [J]. Biotech, 2016, 6(2):226, 1-13.
[17] 17. WU Xiao-lin, HOU Zhao-wei, CHEN Jian, et al. The microbial degradation of Daqing crude oil by selected bacteria [J]. Journal of Daqing Petroleum Institute, 2004, 28(6):14-16.
[18] Varjani S J, Upasani V N. Biodegradation of petroleum hydrocarbons by oleophilic strain of Pseudomonas aeruginosa NCIM 5514[J]. Bioresour Technol, 2016, 222:195-201.
[19] Shahi A, Aydin S, Ince B, et al. Reconstruction of Bacterial Community Structure and Variation for Enhanced Petroleum Hydrocarbons Degradation Through Biostimulation of Oil Contaminated Soil[J]. Chemical Engineering Journal, 2016, 306:60-66.
[20] Ruchi G, Anshu G, Khare S K. Lipase from solvent tolerant Pseudomonas aeruginosa strain: production optimization by response surface methodology and application. [J]. Bioresource Technology, 2008, 99(11):4796-4802.
[21] Wu M, Li W, Dick W A, et al. Bioremediation of hydrocarbon degradation in a petroleum-contaminated soil and microbial population and activity determination[J]. Chemosphere, 2017, 169:124-130.

[22] Hanafy A E M E, Anwar Y, Mohamed S A, et al. Isolation and identification of bacterial consortia responsible for degrading oil spills from the coastal area of Yanbu, Saudi Arabia [J]. Biotechnology & Biotechnological Equipment, 2015, 86(1):1763-1775.
[23] Minucelli T, Ribeiro-Viana R M, Borsato D, et al. Sophorolipids Production by Candida bombicola, ATCC 22214 and Its Potential Application in Soil Bioremediation[J]. Waste & Biomass Valorization, 2016:1-11.
[24] Li X, Zhao L, Adam M. Biodegradation of marine crude oil pollution using a salt-tolerant bacterial consortium isolated from Bahia Bay, China. [J]. Marine Pollution Bulletin, 2016, 105(1):43-50.
[25] Roberto L, Vazquez S C, Cormack W P M. Effectiveness of the natural bacterial flora, biostimulation and bioaugmentation on the bioremediation of a hydrocarbon contaminated Antarctic soil [J]. International Biodeterioration & Biodegradation, 2003, 52(2):115-125.
[26] Ivanovo A E, Borzenkov I A, Strelkova E A, et al. Taxonomic diversity of aerobic organotrophic bacteria from clean Vietnamese soils and their capacity for oxidation of petroleum hydrocarbons[J]. Microbiology, 2012, 81(2):233-243.
[27] Wolińska Agnieszka, Kuźniar Agnieszka, and Szafranek-Nakonieczna Anna, et al. Biological Activity of Autochthonic Bacterial Community in Oil-Contaminated Soil: [J]. Water Air & Soil Pollution, 2016, 227(5):130, 1-12.
[28] Ameen F, Hadi S, Moslem M, et al. Biodegradation of engine oil by fungi from mangrove habitat [J]. Journal of General & Applied Microbiology, 2015, 61(5):185-192.
[29] Vanishree M. Biodegradation of Petrol Using Aspergillus sp [J]. Annual Review & Research in Biology, 2014, 4(6):914-923.
[30] Aitkin M D, String fellow W T, Nagel R D, et al. Characteristics of phenanthrene-degrading bacteria isolated from soils contaminated with polycyclic aromatic hydrocarbons. [J]. Canadian Journal of Microbiology, 1998, 44(8):743-752.
[31] Daane L L, Harjono I, Zylstra G J, et al. Isolation and Characterization of Polycyclic Aromatic Hydrocarbon-Degrading Bacteria Associated with, the Rhizosphere of Salt Marsh Plants[J]. Applied & Environmental Microbiology, 2001, 67(6):2683-2691.
[32] Daane L L, Harjono I, Barns S M, et al. PAH-degradation by Paenibacillus spp. and description of Paenibacillus naphthalenovorans sp. nov. A naphthalene-degrading bacterium from the rhizosphere of salt marsh plants [J]. International Journal of Systematic & Evolutionary Microbiology, 2002, 52(1):131-139.
[33] Díez S, Sabatté J, Viñas M, et al. The prestige oil spill. I. Biodegradation of a heavy fuel oil under simulated conditions. [J]. Environmental Toxicology & Chemistry, 2010, 24(9):2203-2217.
[34] Obire O, Nwaubeta O. Biodegradation of refined petroleum hydrocarbons in soil. [J]. Journal of Applied Sciences & Environmental Management, 2001, 5(1):43-46.
[35] Kharusi S A, Abed R M M, Dobretsov S. EDTA addition enhances bacterial respiration activities and hydrocarbon degradation in bioaugmented and non-bioaugmented oil-contaminated desert soils [J]. Chemosphere, 2016, 147:279-286.
[36] Qu Li-na, Jia Hong-bai, WangQiu-yu. Isolation, Identification and Bioaugmentation of Oil-degradation Bacteria from the Petroleum-contaminated Soil [J]. Anhui Agriculture Science Bulletin, 2015, 21(11):17-20.
[37] CHEN Lihua, ZHOU Lihui, LUO Xiaofang, et al. Bioremediation of oil-contaminated soil by microbial agents and ice grass [J]. Journal of Central South University (Science and Technology), 2015(11):4377-4383.
[38] Zyakun AN M, Brodskii E S, Baskunov B P, et al. Bioremediation of oil-polluted soils: Using the [13 C]/ [12 C] ratio to characterize microbial products of oil hydrocarbon biodegradation [J]. Applied Biochemistry & Microbiology, 2014, 50(5):483-491.
[39] Martínez Álvarez L M, Ruberto L, Lo B A, et al. Bioremediation of hydrocarbon-contaminated soils in cold regions: Development of a pre-optimized biostimulation biopile-scale field assay in Antarctica [J]. Science of the Total Environment, 2017:194-203.

# Biodegradation of Phenolic Contaminants: Current Status and Perspectives

**Lin Zhao**[1,a], **Qi Wu**[1], **Aijin Ma**[1,b*]

[1]Institute of Food and Agriculture Standardization, China National Institute of Standardization, Beijing, China
E-mail: [a]zhaolin@cnis.gov.cn, [b*]maaj@cnis.gov.cn

**Abstract.** Phenolic compounds, a class of toxic pollutants in water, come mainly from a variety of industrial processes. The industrial application for biodegradation has become an important topic in recent years. In this review, we discuss the present situation, properties, and pollution characteristics of phenolic contaminants, factors affecting the degradation of phenols, microbial species and biodegradation methods. The challenges and opportunities in developing biodegradation processes of phenolic contaminants are also discussed.

## 1. Introduction

In soil and water pollution, phenolic substances are the most representative environmental endocrine disruptors, and it has toxic effects on almost all organisms. Phenolic contaminants mainly include bisphenol compounds, phenols, alkyl phenols and natural steroid hormones, which are widely, used in industry, such as pesticides, pharmaceuticals, preservatives, fungicides and so on. The annual production of phenol pollutant is very large, for example, the annual output of biphenyl A is up to 4.69 million tons [1]. Thus, the treatment of phenol pollutant is urgent.

Phenol is a kind of compound substituted that hydrogen on aromatic hydrocarbons are replaced by OH radicals. The unique properties of phenolic compounds determine its wide application in industry and great difficulty to degrade naturally. Phenolic-wasted water mainly comes from coking, gas, oil refining and other chemical and pharmaceutical processes [2]. Phenolic compounds, a kind of archetypal poison, can be inhaled into the organism through the mouth or contact with the skin and mucous membranes, resulting in the formation of insoluble proteins and the loss of cell activity, especially in the nervous system. Phenol-containing wastewater has serious effects on water supply and aquatic life. In addition, the toxicity of phenol can inhibit the growth rate of microorganisms in water and affect the ecological balance. The low concentration of phenol-containing wastewater in irrigated farmland will also make the crops contain phenols and affect edible. The high concentration of phenol-containing wastewater directly leads to crop death [3]. Therefore, the harmful effects of phenolic pollutants on the environment are enormous, and how to effectively degrade the phenol pollutants in water has become a problem that needs to be solved in the world.

## 2. Biodegradation

### *2.1. Microbial species*

*2.1.1. Bacteria.* The bacteria degrading phenols are divided into two types: the first type has the ability to use phenols as the sole source of carbon. These bacteria contain enzyme systems degrading phenols, which often live in a bad environment with phenol contamination. This type of bacteria mainly include: staphylococcus, micrococcus, corynebacterium, arthrobacter, acinetobacter and alcaligenes. The second type must rely on other carbon sources to degrade phenols. This type of bacteria mainly decomposes phenols through co-metabolism pattern, which usually needs two or more bacteria working together [4].

The phenol-degrading bacteria generally have their unique properties, inducement of phenol degradation ability, diversity of the matrix degradation and synergistic of phenol degradation. The inducement of phenol degradation ability suggests that the degradation ability of phenol can be improved greatly after adaptive mutation in bacteria. There are two ways to mutate phenol-degrading bacteria: increasing the concentration of phenol and increasing the dosage one time. The results show that the strains which are screened by gradually increasing the concentration of phenol had the strongest degradation ability of phenol. Adaptive mutation of phenol-degrading bacteria is a result of inducing and selecting. In the process of adaptive mutation, the phenol-resistant strains are screened out by increasing the concentration of phenols, which greatly shortens the adaptation period of the bacteria and improves the degradation efficiency [5]. The diversity of matrix degradation means that isolation and screening of phenol-degrading bacteria with a single phenolic substance as carbon source can degrade other phenolic substances and refractory organics. For example, Bacillus coagulans, separated from single carbon source, can also grow with diphenyl dichloride and naphthalene as carbon source. The synergistic of phenol degradation indicates that the degradation ability of mixed strains is significantly higher than that of single strain. The mixed strains even have strong degradation ability when the phenol concentration is high [6].

*2.1.2. Fungi.* The main fungus of phenolic pollutants degradation are filamentous fungi, including fusarium, white-rot fungi, penicillium and so on. In the case of white-rot fungi, it can degrade chlorophenol contaminants through a series of enzymatic reactions. The degradation and mineralization of the chlorophenol contaminants by white rot fungi depend mainly on the system of their extracellular enzyme and lignin-degrading enzyme. Lignin-degrading enzymes include peroxidase and laccase, and the most important peroxidase is lignin peroxidase and manganese peroxidase. Even if the molecular structure change, the white-rot fungi also has the detoxification effect to the chlorophenol and little harm to the sensitive substance. [7]

Yeast mainly includes candida albicans, candida maltosa, trichosporon cutaneum, and oidium and so on. The research shows that the yeasts have a strong ability to degrade phenols and tolerate phenols with high initial concentration. Furthermore, large volume and good sedimentation performance are favorable for recycling and utilization. [8].

*2.1.3. Algae.* Some algae can also be involved in the degradation of phenolic contaminants. The study found that tolypothrix can degrade alkylphenol and the degradation rate is determined by the growth rate of algae and the initial concentration of Phenolic substances [9].

### *2.2. Degradation factors*

The chemical composition and structure of phenolic compounds have an important role on the biodegradation performance. Taking Chlorophenol as an example, those with more halogen substituents are more susceptible to biodegradation [10]. At the same time, the position of substituent has a great effect on biodegradation, such as the degradation sequence of chlorophenol is as follows: Adjacent position > Separated position > Opposing positions. Natural microorganisms often have poor degradation ability of phenolic compounds, but the microbial degradation ability of phenolic compounds can be greatly improved by mutagenesis. The changes of environmental conditions such as temperature, pH value, dissolved oxygen, nutrients and toxic substances, affect the metabolic activity of microorganisms and the physicochemical properties of phenols, thus affecting the biodegradation performance of phenolic compounds.

## 3. Method

The degradation of phenol pollutant is finally going into the sewage treatment stage, thus the selection of sewage treatment method is very important for the degradation of phenol pollutant. At present, the main biodegradation methods include activated Sludge, biofilm reactor, biological contact oxidation, biological fluidized bed and immobilization technology (Table 3-1).

**Table.3-1.** Principe and properties of five main biodegradation methods

| Method | Principle | Properties |
|---|---|---|
| **Activated Sludge** | Adsorption and oxidative decomposition of phenols by aerobic bacteria and other protozoan in activated sludge | Simple equipment,<br>Good treatment effect,<br>High preprocessing requirements,<br>Unsuitable for large-scale sewage treatment |
| **Biofilm reactor** | Using biofilm to stabilize and clarify waste water | Stable water quality,<br>Simple operation,<br>Degradation of Refractory Organic,<br>Finite reaction unit volume,<br>Difficult to control microbial biomass |
| **Biological contact oxidation** | Microorganisms attach to the surface of the filler in the form of a fixed biofilm and are in contact with the required purified sewage | Combination of biofilm and activated sludge |
| **Biological fluidized bed** | Fluidization state of the carrier for attaching microorganism | Combination of membrane separation technology and bioreactor |
| **Immobilization** | To locate Microorganisms or enzymes in limited-space area by chemical or physical means | High microbial density,<br>Fast reaction speed,<br>Low microbial loss,<br>Easy separation of products,<br>Miniaturization of processing equipment |

*3.1.1. Immobilization.* The application of immobilized technology in wastewater treatment is the most extensive, and it also has obvious effect in the degradation of phenol pollutant. Immobilization techniques include immobilized enzyme technology and immobilized microorganism technology.

*3.1.2. Immobilized enzyme.* Laccase is a metal-containing oxidase, which plays an important role in biodegradation of phenolic contaminants. However, the catalytic reaction of laccase is not stable enough under certain operating conditions, the laccase cannot be reused because it is water-soluble molecule and cannot be separated from the substrate and product in the water [11]. The best solution is the immobilization of laccase. The stability of laccase could be improved, and the continuous reaction could be achieved after immobilization [12].

*3.1.3. Immobilized microorganism.* Immobilized microorganism technology is developed from immobilized enzyme technology. Immobilized microorganism technology can increase microbial biomass and keep it high bioactivity. The main advantages of this method as follows: the activity of enzymes will not be reduced without extracting enzymes from cells; microbial density is high; the reaction speed is fast; toxicity tolerance is strong; the product is easy to separate; the equipment is miniaturized. The effect of immobilization is only determined by the nature of the microorganism and carrier and the environmental characteristics [13].

*3.1.4. Immobilization method.* The commonly immobilization methods can be divided into three categories: adsorption, embedding and covalent bonding method (Table.3-2). The embedding method has good comprehensive performance and is the most widely used.

**Table.3-2.** Principe and properties of three main immobilization method

| Method | Principle | Properties |
|---|---|---|
| **Adsorption** | Enzymes or microorganisms adsorbed on the surface of the carrier or electrode by the physical action of hydrophobic interaction and salt bonding | Simple process,<br>More types of carriers,<br>Combination instability |
| **Embedding** | Enzymes or microorganisms are wrapped in the carrier, substrates and products can enter and leave the carrier | Mild reaction conditions<br>High immobilization efficiency<br>Mass transfer resistance. |
| **Covalent binding** | Covalent reaction of the groups on the carrier with the enzyme molecule or microbial surface groups | Combination stability,<br>Complex operation,<br>Activity inhibition. |

## 4. Conclusions and perspectives

Over the past decades, owing to the economic pressure and public concern about environmental pollution, it would be a great chance for researchers to explore new applications and technologies in biodegradation. Current trend based on the focused-directed evolution will continue and even accelerate. As outlined above, the application of Immobilization Technology in wastewater treatment has been more and more extensive, and has also achieved remarkable results in the aspect of phenolic contaminants degradation particularly. However, because of the huge annual production of phenolic contaminants, many phenol-degrading microorganisms have not been able to expand the scale of phenolic wastewater treatment. The problem is that the removal rate of phenol in actual production cannot all reach the national standard, so that may cause secondary pollution. Therefore, we must continually pay attention to the degradation of phenol contaminants, and develop more solutions for industrial applications.

## 5. Acknowledgment

This work was supported by National Key Research and Development Program (No. 2016YYF0202300).

## 6. References

[1] Y. Q. Huang, C. K. Wong, J. S. Zheng, et al, "Bisphenol a (BPA) in China: a review of sources, environment levels, and potential human health impacts," Environment international, vol. 42, pp. 91-99, July 2012.

[2] S. Y. Li, Q. P. Wei, S. B. Xie, T. T. Zeng and L. S. Rong, "Phenolic wastewater treatment via catalytic gasification," Polish Journal of Environmental Studies, vol. 24 (3), pp. 1147-1151, January 2015.

[3] A. Ertani, M. Schiavon, A. Altissimo, C. Franceschi and S. Nardi, "Phenol-containing organic substances stimulate phenylpropanoid metabolism in Zea mays," Journal of Plant Nutrition & Soil Science, vol. 174 (3), pp. 496-503, May 2011.

[4] I. P. Solyanikova and L. A. Golovleva, "Bacterial degradation of chlorophenols: pathways, biochemica, and genetic aspects," J Environ Sci Health B, vol. 39 (3), pp. 333-351, May 2004.

[5] V. Bucci, S. Hoover and F. L. Hellweger, "Modeling Adaptive Mutation of Enteric Bacteria in Surface Water Using Agent-Based Methods," Water Air & Soil Pollution, vol. 223 (5), pp. 2035-2049, June 2012.

[6] J. Rao, B. Nair, A. Rathinam, A. Giridhar and N. Nagiah, "Synergistic effect of pseudomonas

aeruginosa and escherichia coli in the biodegradation of phenolic compounds," Journal-American Leather Chemists Association, vol. 103 (7), pp. 222-226, July 2008.
[7] L. Valentín, H. Oesch-Kuisma, K. T. Steffen, et al, "Mycoremediation of wood and soil from an old sawmill area contaminated for decades," Journal of Hazardous Materials, vol. 260 (6), pp. 668-675, September 2013.
[8] V. L. Santos and V. R. Linardi, "Phenol degradation by yeasts isolated from industrial effluents," Journal of General & Applied Microbiology, vol. 47 (4), pp. 213-221, August 2001.
[9] V. Klekner and N.Kosaric, "Degradation of phenols by algae," Environmental Technology, vol. 13 (5), pp. 493-501, January 1992.
[10] M. Pera-Titus, V. GarcıA-Molina, M. A. Baños, J. Giménez and S. Esplugas, "Degradation of chlorophenols by means of advanced oxidation processes: a general review," Applied Catalysis B Environmental, vol. 47 (4), pp. 219-256, February 2004.
[11] F. Wang, Y. Hu, C. Guo, W. Huang and C. Z. Liu, "Enhanced phenol degradation in coking wastewater by immobilized laccase on magnetic mesoporous silica nanoparticles in a magnetically stabilized fluidized bed," Bioresource Technology, vol. 110 (4), pp. 120-124, April 2012.
[12] F. F. Maria, S. M. Ángeles and M. Diego, "Recent developments and applications of immobilized lactase," Biotechnology Advances, vol. 31 (8), pp. 1808-1825, December 2013.
[13] A. Kapoor, R. Kumar, A. Kumar, A. Sharma and S. Prasad, "Application of immobilized mixed bacterial culture for the degradation of phenol present in oil refinery effluent," Journal of Environmental Science & Health Part A, vol. 33 (6), pp. 1009-1021, December 1998.

# Analysis on Influence Factors of $CO_2$ Emission in Hebei

**Wei Li**[1], **Shu Zhang**[1, a*], **Hongzhi Zhang**[1].

[1] School of Economics and Management, North China Electric Power University, No. 619 Yonghua Street, Baoding, Hebei 071003, China.
E-mail:[a*] ncepuzs2017@163.com.

**Abstract.** China has become the world's largest carbon emitter in the world, while Hebei province, dominated by heavy industry, contributed a mass of carbon emissions for the whole country. It was necessary for Hebei to adjust its industrial and energy structure to reduce carbon emissions. This paper analysed the total energy consumption and $CO_2$ emissions in Hebei from 2000 to 2015, and utilized the Logarithmic Mean Di index Method to decompose $CO_2$ emissions emitted by energy consumption. Six factors were decomposed from the total emissions and the Lagrange descriptions as well as the Eulerian description were both employed to analyse these factors` trajectory of change. The results discovered that the $CO_2$ emissions by energy consumption in Hebei were mainly stimulated by economic scale expansion, especially the production growth of secondary industry. While the energy intensity become the major restraint for causing $CO_2$ emissions declining. The impacts of other factors were not very remarkable. Finally, some macroscopic advices were offered to control carbon emissions in Hebei.

## 1. Introduction

As an industrial region, the development of industrial system began to take shape in Hebei. Conditioned by natural resources and location, the development patterns of relying on resources did not change a lot and the heavy industries like steel, coal, chemical, still pillared the economic growing of Hebei province. And this pattern left $CO_2$ emissions emitted by production activities on the rise.

Factors decomposition analysis was an efficient way to figure out influence factors on carbon emissions growing. And this study utilized Log-Mean Division Index (LMDI) method to study $CO_2$ emission in Hebei and figure out essential factors in it.

There has been much research on carbon emissions through LMDI method, such as Zhou and Ang (2008) [1], Wei et al. (2008) [2], Hammond and Norman (2012) [4]. Roinioti and Koroneos (2017) [5] decomposed $CO_2$ emissions from energy use in the context of the economic crisis in Greece. Lin and Tan (2017) [6] explored the main factors influencing $CO_2$ emissions in China and further estimated the potential carbon emissions in energy intensive industries in China for the future. In regard to the factors decomposed, there were usually three factors [7, 8], four factors [9, 10] and five factors [11-13]. For instance, research by Jiang et al. (2017) [14] discussed the underlying drivers of economic carbon emissions through decomposing four factors. The results indicated that economic expansion were responsible for $CO_2$ emissions growing while the sectoral energy intensity and structure changes caused $CO_2$ emissions declining.

In this paper, it was believed that the change of $CO_2$ emissions in Hebei province was decomposed six factors: the economic scale, population quantity, energy structure, energy intensity, industrial structure and carbon emission coefficient. Compared to other literatures, population was taken into consideration and per capita GRP (Gross Regional Production) was on behalf of the economic factor to eliminate the impact of population change on carbon emissions in Hebei province.

## 2. Material and methodology

### *2.1 Material*

The employed data was taken from the China Energy Statistics Yearbook and Hebei Economic Yearbook (2000-2015). And the gross regional production (GRP) in Hebei was expressed in constant prices using 2000 as benchmark year.

The industry in Hebei was divided into four parts: a) the first industry, namely agriculture; b) modern industry; c) construction industry and d) the third industry, namely modern service industry. As respect to classification of energy resources, primary energy was divided into four parts: a) coal; b) crude oil; c) natural gas; d) primary electricity.

Total emissions of $CO_2$ in was calculated coefficients of standard coal employed the findings of NDRC Energy Research Institute and coefficients of $CO_2$ emissions was cited from GB-T25892008.

$$C = \sum_{j=1}^{n} E_j \bullet \sigma_j \bullet \omega_j \tag{1}$$

### *2.2 Methodology*

The process and formula of $CO_2$ emission decomposition were as follows. This change can be composed into six effects: $\Delta C_U$ represents the change caused by variation of carbon emission intensity, defined as effect of $CO_2$ intensity; $\Delta C_M$ represents the change caused by variation of Energy-resource structure, defined as effect of energy structure; $\Delta C_I$ represents the change caused by variation of Energy intensity of each sector, defined as effect of energy intensity; $\Delta C_S$ represents the change caused by variation of Industrial structure, defined as effect of industrial structure; $\Delta C_{AY}$ represents the change caused by variation of GDP per capita, defined as effect of economy; $\Delta C_P$ represents the change caused by variation of population scale, defined as effect of population.

$$\begin{aligned} \Delta C &= C^T - C^0 \\ &= \sum_{i=1}^{m} \sum_{j=1}^{n} U_{ij}^T \bullet M_j^T \bullet I_i^T \bullet S_i^T \bullet AY^T \bullet P^T \\ &- \sum_{i=1}^{m} \sum_{j=1}^{n} U_{ij}^0 \bullet M_j^0 \bullet I_i^0 \bullet S_i^0 \bullet AY^0 \bullet P^0 \\ &= \Delta C_U + \Delta C_M + \Delta C_I + \Delta C_S + \Delta C_{AY} + \Delta C_P \end{aligned} \tag{2}$$

Further decomposition of $CO_2$ intensity effect was conducted with Eq (4) and Eq (5). And other effect also can ben decomposed in a similar way. However, it was noted that according to IPCC's hypothesis, carbon emission coefficient of a certain energy source can be considered unchanged, namely $\Delta C_U = 0$.

$$\Delta \sum_{i=1}^{m} \sum_{j=1}^{n} B_{ij} \ln\left(\frac{\quad}{U_{ij}^0}\right) \tag{3}$$

$$B_{ij} = \frac{C_{ij}^T - C_{ij}^0}{InC_{ij}^T - InC_{ij}^0} \tag{4}$$

## 3. Energy Consumption and $CO_2$ Emissions in Hebei Province

### *3.1 Energy Consumption*

After entering twenty-first Century, economy of Hebei province has entered a stage of rapid development with regional GDP increasing exponentially. In the process of urbanization and industrialization, a great deal of energy was demanded. As can be seen from Figure.3-1, the total energy consumption and GRP growth trend was about the same: real GRP in Hebei rose from 504.40 billion RMB to 713.29 billion RMB with an average annual growth rate of 2.43%; while total energy consumption rose to 75.62 million tons from 36.22 million tons with an average annual growth rate of 5.38%, which exceeded the average speed of GRP growing. However, total energy consumption has gradually declined since 2012.

Like the whole country, Hebei maintained energy consumption structure with coal dominating. Hebei also was an industrial province in China. The industrial output, especially the heavy industry, promoted provincial economic growth overwhelmingly. Before 2009, energy consumption of industrial sector accounted for over 80%, and dropped to about 75% of the total consumption after 2010. While the share of modern service gradually increased to more than 17%, even 19% since 2010.According to statistics, total energy consumption, total coal consumption and industrial energy consumption in Hebei ranked the second, the fourth and the third respectively in 34 provinces of China.

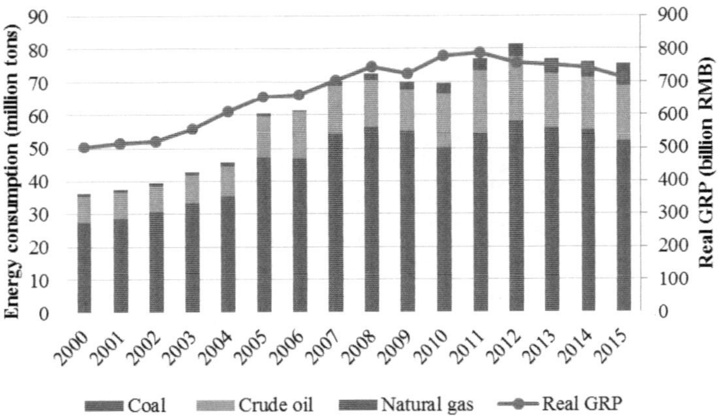


**Figure.3-1.** Hebei`s GRP and energy consumption in 2000-2015.

### *3.2 $CO_2$ Emissions*

$CO_2$ emissions from three primary energy resources in Hebei calculating by Eq.6 were shown in Figure.3-2. Two periods was divided into on the basis of variation: a) the growing stage: 2000-2008 b) the undulate stage: 2009-2015. In the growing stage, total CO2 emissions increased with 9.51% per year. And in the undulate stage, $CO_2$ emissions kept around 180 million tons in total with fluctuation, and reached a maximum of 207.17 million tons in 2012. According to You (Per capita carbon emission of Hebei Province

From an industrial point of view, industry emitted the largest part of $CO_2$ emissions, climbing 9.30% per year and occupying over 80% in the growing stage. Lined with energy consumption, emissions of modern service sector markedly increased and was more than 15% in the undulate stage, where the proportion of industrial $CO_2$ emissions were less than 80%.

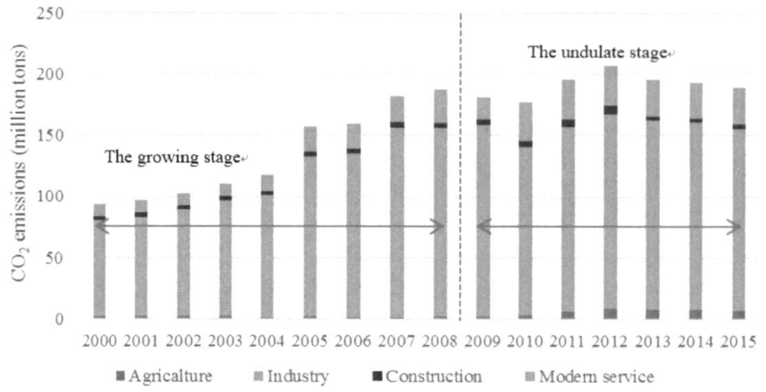


**Figure.3-2.** Hebei`s $CO_2$ emissions in 2000-2015.

## 4. Results

A LMDI model was utilized to decompose six factors responsible for Hebei`s $CO_2$ emissions from 2000 to 2015. On the basis of *2.2*, the effect of $CO_2$ emissions intensity equalled to zero. And the concrete calculation process has been completed in *2.2*. There were two choices of base year in this paper: a) Lagrange description method: unfixed base period, which meat the previous year and the total effect of $CO_2$ emissions in Fig.4-1 was an increment from the previous year; b) Eulerian description method: fixed base period, which meant 2000 was the base year, meaning the total effect in Fig.4-2was an accumulated increment.

### *4.1 Results analysis with Lagrange description*

As shown in Figure.4-1, the economic scale factor was the factor mainly responsible for $CO_2$ increasing in the whole process. From 2001 to 2011, Hebei`s the growth of $CO_2$ emissions has been strongly driven by provincial economic development. And in 2011, the economic effect reached the maximum with contributing 31.58 million tons $CO_2$ in total. Then, contribution of this positive factor began to decline and in 2015 this positive factor only contributed about 1.29 million tons $CO_2$.

The energy intensity effect was the main factor causing the declining in $CO_2$ emissions. As with efficiency of energy utilization, energy intensity reflected regional comprehensive technical level: the more technology advanced, the bigger energy efficiency, the smaller energy intensity namely the less energy consumption per unit of GDP and the less carbon emissions. Except for 2007, the effect of energy intensity was negative, and achieved negative maximum in 2010. In 2009-2010 and 2012-2015, the effect of energy intensity exceeded the economic, which led to a decline in $CO_2$ emissions in these years.

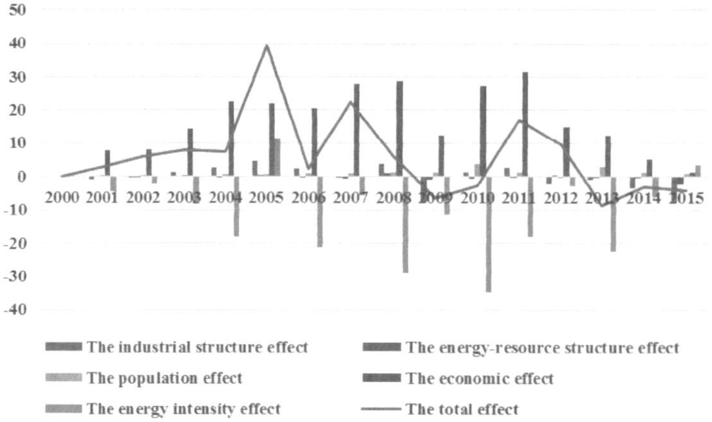


**Figure.4-1.** Decomposition Analysis of $CO_2$ emissions with unfixed base year.

Other effects were smaller than the economic and the energy intensity. Contribution of the industrial structure effect varied in the whole process and kept causing $CO_2$ emissions declined 14.39 million tons since 2012. And the energy-resource structure was negative factor while the population influenced $CO_2$ emissions positively. But both of them impacted on carbon emissions relatively lightly when comparing with the effect of economic and energy intensity.

*4.2 Results analysis with Eulerian description*

According to *3.2*, two stages of $CO_2$ emissions were divided: the growing stage (2000-2008) and the undulate stage (2009-2015). In the growing stage, all factors, except for energy intensity, influenced carbon emissions positively. And the total effect growing rapidly for the reason that the economic effect was obviously bigger than that of energy intensity. To a certain extend changes in industrial structure enhanced the total effect in this stage. The most distinctive feature in undulate stage was that energy intensity strengthened its negative influence. This huge inhibition and no increasing economic effect lowered the growing speed of $CO_2$ emissions and controlled the total effect keeping under 110 million tons since 2012. Compared with the previous phase, the effect of industrial structure generally weakened and even was responsible for -1.85 million tons in 2015. The population effect strengthened in the whole period although its contribution for the total effect was less than 15 million tons. And because of few changes of energy-resource structure the effect of it was unobvious in 2000-2015.

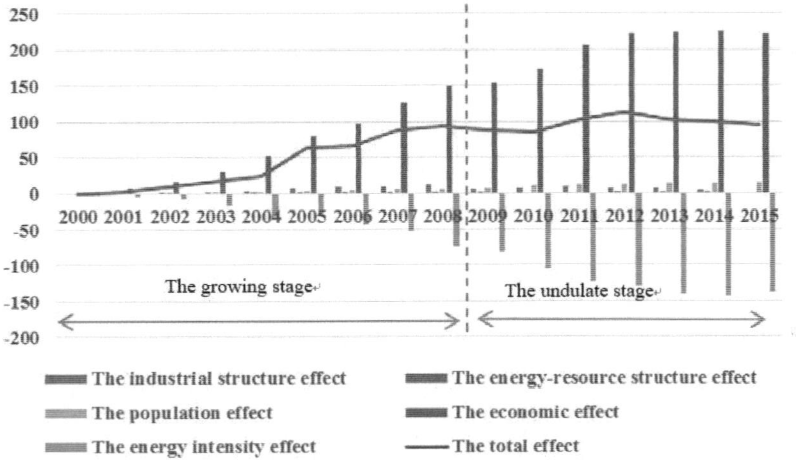


**Figure.4-2.** Decomposition Analysis of $CO_2$ emissions with fixed base year.

## 5. Discussion

On the foundation of previous studies, a LMDI model was employed to analyse factors of $CO_2$ emissions in Hebei province during 2000 -2015. Six factors were decomposed and impact by these factors was quantified. Effects of these factors was analysed with Lagrange description and Eulerian description method. The results can be concluded into four points.

a) Both of Lagrange and Euler description method showed that compared with other positive factors, the economic effect represented by per capita GRP was the maximum factors, which meant economic growth was a powerful driving force for the aggrandizement of $CO_2$ emissions in Hebei province.

b) The decline of energy intensity mainly leading $CO_2$ emissions to inhibit in Hebei province. During 2000 -2015, the energy intensity in Hebei dropped by 0.46.This change had saved over 100 million tons $CO_2$ emissions each year since 2010.

c) The factor of population was responsible for part of carbon emissions increment. Impacts of industrial structure varied from positive to negative ceaselessly but began to lowered $CO_2$ emissions after 2012. This change was in accordance with the reform of industrial structure in Hebei, which required the proportion of the modern service sector raised and the industry sectors declined.

d) Compared with other factors, the energy-resource structure had the least influence on $CO_2$

emissions. The reason was that the proportion of coal, oil and natural gas in primary energy consumption did not change much during 2000-2015.

Some suggestion can be made to decrease $CO_2$ emissions in Hebei province. First of all, energy efficiency should be improved through technological research and application, which means reduce energy intensity and intensify negative influence on $CO_2$ emissions. Secondly, reform on industrial structure adjustment should be continued. Policy measures of increasing the production proportion of the modern service sector ought to be implemented. Last not the least; popularize usage of natural gas and renewable energy to shrink percentages of coal and oil in energy consumption.

## 6. References

[1] Zhou P and Ang BW 2008. Decomposition of aggregate CO2emissions: a production-theoretical approach. *Energy Economics.* **30** 1054-1067

[2] Yiming W, Lanzhi L, Ying F and Gang W 2008. *China Energy Report (2008): $CO_2$ Emissions Research* (Beijing: Science Press)

[3] Hammond GP and Norman JB 2014. Decomposition analysis of energy-related carbon emissions from UK manufacturing. *Energy* **41** 220–227

[4] Argiro R and Christopher K 2017. The decomposition of $CO_2$ emissions from energy use in Greece before and during the economic crisis and their decoupling from economic growth. *Renewable and Sustainable Energy Reviews* **76** 448-459

[5] Lin B and Tan R 2017. Sustainable development of China's energy intensive industries: From the aspect of carbon dioxide emissions reduction. *Renewable and Sustainable Energy Reviews* **77** 386-394

[6] Malla S 2009. $CO_2$ emissions from electricity generation in seven Asia-Pacific and North American countries: a decomposition analysis. *Energy Policy* **37** 1–9

[7] Zhang W, Li K, Zhou D, Zhang W and Gao H 2016. Decomposition of intensity of energy-related CO2 emission in Chinese provinces using the LMDI method. *Energy Policy* **92** 369–81

[8] Guo Y, Lin F, Bai Y and Wu M 2009. The decomposition research on energy-related carbon emissions of Shanghai. *Environmental Pollution & Control* **31**(09) 68-72

[9] Zhao X and Long R 2010. The Status and Decomposition Model of Carbon Emissions for Jiangsu Province. China Population. *Resources and Environment* **20**(7) 25-30

[10] Lin B and Ouyang X 2014. Analysis of energy-related $CO_2$ (carbon dioxide) emissions and reduction potential in the Chinese non-metallic mineral products industry. *Energy* **68** 688–697

[11] Liu C, Chen C and Chen Q 2010. Evolution Mechanisms of $CO_2$ Emissions in Beijing City during the Period 1997-2007. *Resources Science* **32**(2) 235-241

[12] Qin Z, Xizhe P, Zhiming C and Kaiya W 2009. Factors Decomposition and Empirical Analysis of Variations in Energy Carbon Emission in China. *Resources Science* **31**(12) 2072-2079

[13] Jiang, Ye B, Xie D, Li J, Miao L and Yang P 2017. Sector decomposition of China's national economic carbon emissions and its policy implication for national ETS development. *Renewable and Sustainable Energy* **75** 855-867

IOP Conf. Series: Earth and Environmental Science **111** (2018) 012026 doi:10.1088/1755-1315/111/1/012026

# Decomposition of $CO_2$ Emission Factors in Baoding

**Wei Li [1], xuyang Wang [1, a*], Hongzhi Zhang [1]**

[1]School of Economics and Management, North China Electric Power University, No. 619 Yonghua Street, Baoding, Hebei 071003, China.
E-mail:[a]* ncepuwxy@126.com.

**Abstract.** Baoding, as one of the first "five provinces and eight cities" low carbon pilot cities, undertakes an important task and mission. The urgent task is to explore a peak route and emission reduction path suitable for Baoding's own development, so as to provide reference for the construction of low-carbon pilot cities. At present, the carbon emissions of Baoding city and its subordinate districts and counties are not clear, and the carbon emissions, change trends and emission characteristics of various industries have not been systematically studied. This lead researcherscan not carry out further attribution analysis, the prediction of future emissions trends and put forward specific measures to reduce emissions are impossible.If the government can not accurately and comprehensively understand the problems faced in the construction and development of low-carbon cities, it is difficult to fundamentally put forward effective emission reduction policies and measures.

## 1. Introduction

The carbon emissions were taken as the research object. Angapplythe new proposed Division indexdecomposition method to study Chinese carbon dioxide and the energy consumption of the industrial sector to solve the remaining problems and data decomposition in 0 value problems. This decomposition covers 4 kinds of fuel and the 8 sectors, the results of the study showed that 1985-1990 the industrial sector carbon emissions have a positive effect on the change in total output, and the change of industrial energy consumption intensity play a an inhibitor effect on the sector's carbon emissions [1].Wang used logarithmic Division index decomposition method to discuss Chinese carbon dioxide emissions in 1957-2000, decomposition results showedthatenergy consumption intensity is the strongest factor to reduce carbon emissions, energy structure effect restrain carbon emissions growth, economic growth factors play a role to increase carbon emissions[2].Taking carbon emission intensity as the object of study. Shrestha and Timilsina address Division index decomposition method to study the 12 Asian country electric power industry carbon dioxide emissions intensity changes, including China, the results showed that the effect of carbon dioxide emissions from 1980-1990 Asia power industry are the change of fuel intensity[3]. Fan adopted adaptive weighted Divisia decomposition method devoted to China carbon emission intensity in 1980-2003, the $CO_2$ emissions trend in Chinais rising, but the intensity of carbon emissions continued to decline, and by distinguishing a carbon intensity of primary energy consumption intensity and terminal energy consumption factors. The primary energy carbon emission intensity has a significant influence on the carbon emissions intensity change, the energy structure factors, which have an important influence on the carbon emission intensity [4].Constructing linear relationship model to forecast carbon emission scenarios. Lesterbased on the combination of 3 kinds of Kaya equation set the situation Waxman-Markey and Obama policy target to forecast U.S. carbon emissions. [5].

## 2. Material and methods

Based on the six traditional industries, as well as 17 kinds conventional energy, to decompose the Baoding carbon dioxide emissions factor, which can be decomposed into six factors: population effect, economic output effect, industrial structure effect, energy intensity effect, energy structure effect and carbon dioxide emission intensity effect, according to the energy consumption of carbon emissions equation:

$$C = \sum_{i,j} C_{i,j} = \sum_{i,j} \frac{C_{i,j}}{E_{i,j}} \times \frac{E_{i,j}}{E_i} \times \frac{E_i}{Q_i} \times \frac{Q_i}{Q} \times \frac{Q}{P} \times P \quad (1)$$

Where C represents the CO2 emissions in Baoding andCij indicates that the amount of CO2 emissions consumed by the energy j in sector i,Eijexpressed the energy consumption j in sector i, Ei represents the amount of energy consumed by the sector i, Qirepresent the economic output of the sector i, Q delegate the city's GDP, P indicate the population.

$$C = \sum_{i,j} C_{i,j} = \sum_{i,j} F_{i,j} \times M_{i,j} \times R_i \times H \times Y \times P \quad (2)$$

In the equation, $F_{i,j} = \frac{C_{i,j}}{E_{i,j}}$ represents the intensity of $CO_2$ emissions from the energy consumed by the i sector, and the changes in carbon emissions, due to the change of $CO_2$ emissions is called the $CO_2$ emission intensity effect. $M_{i,j} = \frac{E_{i,j}}{E_i}$ Represent in the sector i, the consumption of j energy accounted for the proportion of total energy consumption, due to the changes in the energy structure of the various sectors caused by changes in the total amount of carbon emissions is known as the energy structure effect. $R_i = \frac{E_i}{Q_i}$ indicate the energy intensity of the sector i, the impact of its changes on carbon emissions is called energy intensity effects; $H = \frac{Q_i}{Q}$ indicate the economic output of the sector i accounted for the proportion of the city's GDP, industrial adjustment to influence on the carbon emissions is called the industrial structure effect. $Y = \frac{Q}{P}$ represent per capita GDP, its changes reflect the impact of economic growth on Baoding carbon emissions is called the economic output effect's represents the population, and the impact of demographic changes on the carbon emissions of Baoding is called the population effect.

$$\Delta C = C^T - C^0 = \Delta C_F + \Delta C_M + \Delta C_R + \Delta C_H + \Delta C_Y + \Delta C_P \quad (3)$$

$\Delta C$ Indicate the total $CO_2$ emissions changes, $\Delta C_F$ indicate the $CO_2$ emission intensity effect, $\Delta C_M$ represent the energy structure effect, $\Delta C_R$ indicate the energy intensity effect, $\Delta C_H$ represent the industrial structure effect, $\Delta C_Y$ indicate the economic output effect, $\Delta C_P$ and represent the population effect. 0 and T represent the base year and the target year respectively. In this study, the LMDI method is used to decompose in the equation.

$$\Delta C_F = \sum_{i=1}^{6} \sum_{j=1}^{12} \left[ L(C_{i,j}^T, C_{i,j}^0) \times \ln \frac{F_{i,j}^T}{F_{i,j}^0} \right] \quad (4)$$

$$\Delta C_M = \sum_{i=1}^{6}\sum_{j=1}^{12}\left[ L(C_{i,j}^T, C_{i,j}^0) \times \ln \frac{M_{i,j}^T}{M_{i,j}^0} \right] \tag{5}$$

$$\Delta C_R = \sum_{i=1}^{6}\sum_{j=1}^{12}\left[ L(C_{i,j}^T, C_{i,j}^0) \times \ln \frac{R_i^T}{R_i^0} \right] \tag{6}$$

$$\Delta C_H = \sum_{i=1}^{6}\sum_{j=1}^{12}\left[ L(C_{,i,j}^T, C_{i,j}^0) \times \ln \frac{H^T}{H^0} \right] \tag{7}$$

$$\Delta C_Y = \sum_{i=1}^{6}\sum_{j=1}^{12}\left[ L(C_{i,j}^T, C_{i,j}^0) \times \ln \frac{Y^T}{Y^0} \right] \tag{8}$$

$$\Delta C_P = \sum_{i=1}^{6}\sum_{j=1}^{12}\left[ L(C_{i,j}^T, C_{i,j}^0) \times \ln \frac{P^T}{P^0} \right] \tag{9}$$

$$L(C_{i,j}^T, C_{i,j}^0) = \begin{cases} \dfrac{C_{i,j}^T - C_{i,j}^0}{\ln C_{i,j}^T - \ln C_{i,j}^0} & C_{i,j}^T \neq C_{i,j}^0 \\ C_{i,j}^T & C_{i,j}^T = C_{i,j}^0 \end{cases} \tag{10}$$

Analyze the decomposition process, It can be seen that the $CO_2$ emissions energy intensity of all kinds energy fixed constant so that $\Delta C_F = 0$

## 3. Results and discussion

Based on the LMDI model, the influencing factors of $CO_2$ emission in Baoding City from 2004 to 2016 were analyzed. On the basis of the analysis, the influence of these factors on the $CO_2$ emissions in Baoding was analyzed quantitatively. The results of the analysis can be concluded that the economic output effect is the most positive factor for the change of $CO_2$ emissions in Baoding. The economic output effect makes the city's carbon footprint increase by 35.24 million tons in 2004-2016, and the industrial structure effect also caused Baoding $CO_2$ emissions in the research period increased by 2.72 million tons, the population effect of Baoding $CO_2$ emissions increased by2.52 million tons. Energy structure effect in the study period to suppress the Baoding $CO_2$ emissions increased by 4.04 million tons, the energy intensity effect is the strongest factor to suppress the growth of carbon emissions. In 2004-2016 total inhibition of the city's $CO_2$ emissions is 31.19 million tons, under the combined effect of various effects, $CO_2$ emissions in Baoding City increased by 1.99 million tons throughout the study period.

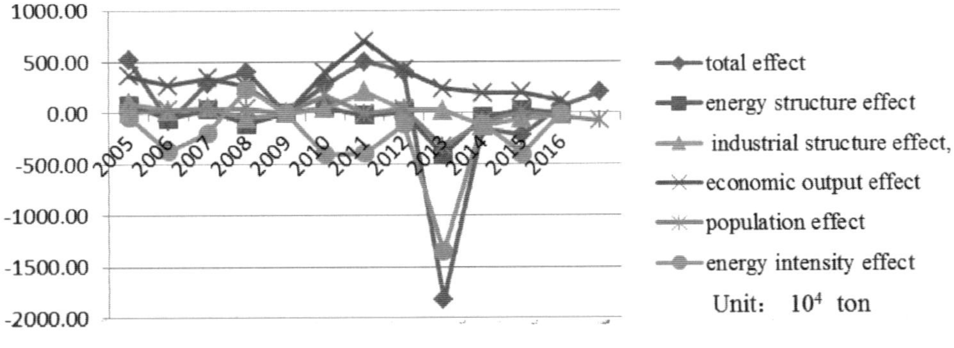


**Figure.3-1.** LMDI Decomposition Results in 2004-2012.

Economic output effect is the main factor affect the carbon emissions growth, of Baoding City, industrial structure effect and population effect to promote the Baoding City carbon emissions increase, economic development and population growth of the carbon emissions associated with the other areas of China in the $CO_2$ emission growth, so in the future, The government should pay attention to balance the relationship between economic development and population growth and energy consumption; in the processofdevelopment, Baoding city should develop actual policy according to their own status, and grasp the integration of Beijing Tianjin Hebei; along with Beijing and Tianjin industrial transfer opportunities, accelerate the optimization and upgrade its industrial structure, build the new energy industries, achieved the transformation to promote the city's carbon emissions growth factors to inhibit the growth of carbon emissions index to the industrial structure effect, On the basis, it is also seriously to follow the impact of low-carbon pilot city construction in Baoding on the concept of low carbon life of residents, which is one of the influencing factors to reduce carbon emissions.
The energy intensity is the strongest inhibition effect on Baoding $CO_2$ emissions growth, which indicates that the energy efficiency of Baoding increased in the past period, although the amount of energy consumption of unit GDP in 2016 decreased significantly, compared with 2004.Even so, in the "13th Five-Year" period, Baoding should also focus on research and promote energy-saving technology to reduce energy intensity and energy consumption, promote low-carbon and clean energy used, change the mode of economic growth, and strive to achieve the city's carbon emissions peak.
The energy structure effect in the 2004-2016 study period is one of the driving factors of Baoding City, inhibited the growth of $CO_2$emissions. In recent ten years the energy consumption structure of Baoding, the proportion of low carbon energy increased, clean energy application and project construction work to further promote, but it cannot ignored that in the energy structure. Coal consumption is still accounted for more than 50%, thus promote the use of clean and low carbon energy and non fossil energy, fully tap the potential of emission reduction; still the current emission reduction in Baoding, from the long-term development, the optimization of energy structure will be the carbon emission reduction potential. Draw the lessons of home and abroad, the energy structure effect has become one of the important factors inhibited the growth of carbon emissions.

## 4. References

[1] Ang B W, Choi K 1997.Decomposition of Aggregate C and Gas Emission Intersities for Industry: A Refined Divisia Index Method. *The Energy Journal*, **3** 59-73.

[2]Wang C, Chen J and Zou J 2005. Decomposition of Energy-Related CO2 Emission in China 1957-2000. *Energy*, **30**, 77-83.

[3]Shretha R M, Timisina G R 1996. Factors Affecting CO2 Inttensities of Power Sector in Asia: A Division Decomposition Analysis. *The Energy Economics*, **18**(4) 283-293.

[4]Fan, Y, Liu, Li, CWu, G., Tsai.H.T. And Wei Y 2007. Changes in Carbon Intensity in China. Empirical Findings from 1980-2003. *Ecological Economics* **62** 15-30.

[5].Lester R K, Finan A 2009. Quantifying the impact of proposed carbon Emissions on reductions on the U S energy infasture. MIT-IPC- Energy-InnocationWoeking Paper,

IOP Conf. Series: Earth and Environmental Science **111** (2018) 012027 doi:10.1088/1755-1315/111/1/012027

# Non-invasive Continuous Monitoring of Cerebral Edema Using Portable Microwave Based System

**Yuhao Jiang[1], Minji Zhao[2], Huiqian Wang[1], Guoquan Li[1, a*]**

[1] Chongqing University of Posts and Communications, Chongqing 400065, China
[2] Department of Mechanical Science and Engineering, Graduate School of Engineering, Nagoya University, Nagoya 464-8603, Japan
E-mail: [a]ligq@cqupt.edu.cn

**Abstract**. A portable non-invasive head detecting system based on microwave technology was developed for evaluation of cerebral edema change inside human brain. Real-time monitoring of cerebral edema in the brain helps the clinician to assess medical condition and treatment. In this work, a microwave signal was transmitted and coupled into an open-end circular waveguide sensor, incident on a 3D printed head phantom, and reflected back to receiver. Theoretically, the operation of this instrument depends on the conductivity contrast between cerebral edema and healthy brain tissues. The efficacy of the proposed detecting system is verified using 3D printed anatomically and dielectrically realistic human head phantoms with simulated cerebral edema targets with different size. Changes in the amplitude of time domain result were shown to be induced by the expansion or decrease of the edema volume. The eventual goal of this proposed head evaluating system is use in the hospital as an effective real-time monitoring tool.

## 1. Introduction

Cerebral edema is responsible for a huge number death and disabled around the world. For the reason that cerebral edema growth is a determinant of mortality and poor outcomes, and detecting edema expansion is an important prognostic indicator for treatment [1]. Continuous accurate detection of cerebral edema is the vital of timely adequate medication.

Because the general recommended imaging techniques, MRI or CT scan, are impossible to be monitoring tools, diagnosing the change of edema and evaluating the therapeutic effects only depend on the clinician's experience in the rest of time. Thus, it is obviously that there is a clinical demand for a non-invasive, easy–to-operation, portable instrument, which is used for bed-side monitoring to observe the change of edema.

Microwave based detection has attracted much attention from the researchers as a non-invasive portable complementary diagnostic tool, due to its non-ionizing radiation, low power and low cost features [2-7]. Moreover, when brain edema occurs, there is accumulation of fluid and a change in the water composition of the tissue, which is bound to change the conductivity of edema region [8]. In this work, a novel microwave based head detection system for continuous assessment of edema is proposed. Moreover, the proposed method is verified in experiments using 3D printed anatomically and dielectrically realistic human phantom.

IOP Conf. Series: Earth and Environmental Science **111** (2018) 012027 doi:10.1088/1755-1315/111/1/012027

## 2. Methods and material

### *2.1. Head detection system*

The prototype system is consist of a circular waveguide which is employed as detecting sensor, a coaxial cable to transmit and receive microwave signals, and a Vector Network Analyzer which acts like a data storing and processing unit. The inner cable should be long enough to produce strong signals in the circular waveguide, and the circular waveguide used in this study has the internal radius a of 90.0 mm and the length d of 115.0 mm as shown in figure.2-1. A soft plastic bag filled with matching medium is placed between head and sensor. The use of the matching medium ensures a good electromagnetic coupling between the sensor and the head tissue.

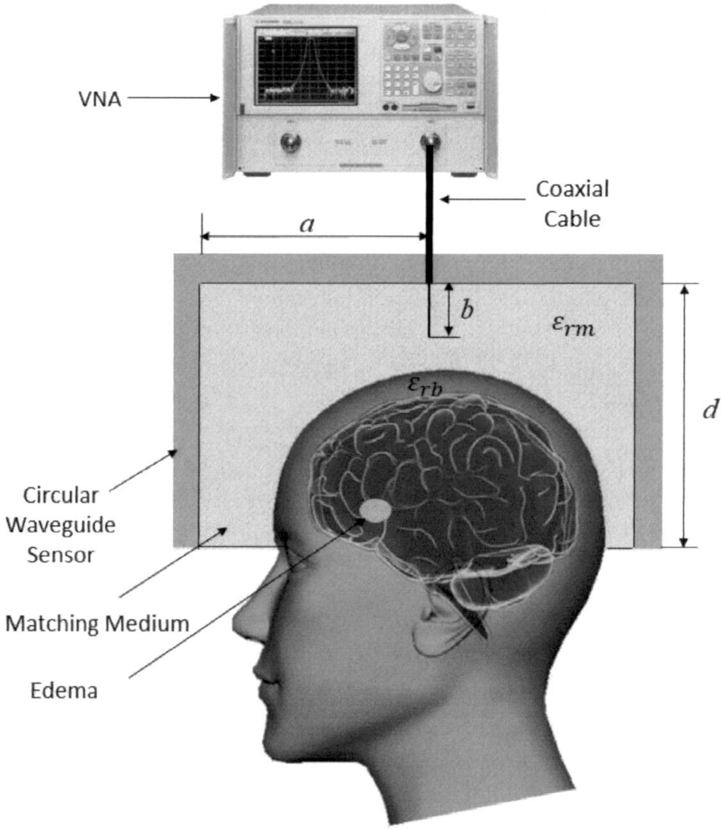


**Figure.2-1.** The schematic diagram of the head detecting system

### *2.2. Head mimicking phantom*

In this work, an anatomically and dielectrically realistic head phantom is manufactured using the stereo-lithographic machine (Eden250, Stratasys Inc.). The basic design of the 3D voxel phantom is derived from MRI scan data of head. And the head model is divided into skin, skull, gray matter, white matter and cerebral edema portions. The model has separate compartments, so that the internal cavities can be filled with corresponding tissue equivalent materials, which are composed of gelatin, water, flour, sodium chloride with different recipes [9,10]. Figure.2-2 illustrates the 3D geometry of the brain phantom. In addition, the cerebral edema was simulated using mimicking material with conductivity of 0.67 S/m and the volumes (0, 5, 10, 15, 20 ml) of edema had been tested in this study.

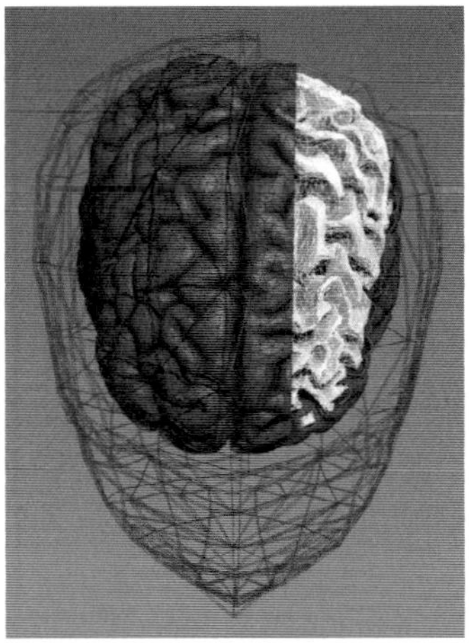

**Figure.2-2.** Three-dimensional geometry of the head phantom.

## 3. Theory and calculation

Because the microwave propagates in waveguide incident from the dielectric to the air at the termination interfaces of waveguide sensor give rise to high coefficient reflection, interior reflections are produced and most of energy is confined in the head and medium.

In this work, loss due to conductivity $\sigma$ should be taken into account, so the complex permittivity $\varepsilon$ can be written as

$$\varepsilon=\varepsilon_0\left(\varepsilon_r - j\frac{\sigma}{\omega\varepsilon_0}\right) \tag{1}$$

Where $\omega$ , $\varepsilon_0$ , $\varepsilon_r$ denote angular frequency, permittivity of free space, dielectric constant, respectively.

Due to the electromagnetic wave couples to the circular waveguide and then transmits along longitudinal direction, the complex propagating constant $\gamma$ in circular waveguide is shown as

$$\gamma= \alpha + j\beta= \left(k_c^2 - \omega^2\mu_0\varepsilon_0\varepsilon_r + j\omega^2\mu_0\varepsilon''\right)^{1/2} \tag{2}$$

Where $\alpha$ and $\beta$ denote attenuation constant due to conductivity loss and propagating constant, respectively. Then and can be expressed as respectively

$$\alpha=\left(\frac{1}{2}\left(\left(k'^2 - k_c^2\right)^2 + k''^4\right)^{1/2} - \frac{1}{2}(k'^2 - k_c^2)\right)^{1/2} \tag{3}$$

$$\beta=\left(\frac{1}{2}\left(\left(k'^2 - k_c^2\right)^2 + k''^4\right)^{1/2} + \frac{1}{2}(k'^2 - k_c^2)\right)^{1/2} \tag{4}$$

Where $k' = \omega\left(\mu_0\varepsilon_0\varepsilon_r\right)^{1/2}$, $k'' = \omega\left(\mu_0\varepsilon''\right)^{1/2}$, $\varepsilon''$ and denotes the imaginary part of $\varepsilon$ . As a result, the change of edema volume can be evaluated in light of the relationship between attenuation constant $\alpha$ and the conductivity $\sigma$ in the waveguide.

## 4. Results

As mentioned in section 2, head phantoms with different volume of cerebral edema (0, 5, 10, 15, 20 ml) had been detected in this work. In addition, when a range of frequencies are swept in the frequency domain, the wave package is consist of such multiple frequencies. In this work, the experimental frequency range of wave group is from 150 to 450 MHz. The time domain experimental results of various cerebral edema models are shown in figure 4-1. These results are obtained from frequency domain results by Inverse Fast Fourier transform. While the zero time reference should be set as the end point of the VNA cable, the first significant peak relates to the connecting point between coaxial cable and waveguide sensor, and the second significant peak corresponds to the waveguide open-ended end reflection. Moreover, the amplitude difference between these two significant peaks corresponds to the power attenuation due to conductivity loss. Although there is little difference in conductivity in these cerebral edema models, the amplitude still can be clearly observed if enlarge the results, i.e., enlarged the second significant peak of the results over time range of 2.5 to 6 ns as shown in figure.4-2. Seen from figure.4-1, it is interesting to see that the magnitude shifts in different cerebral edema models, indicating that the amplitude vary with change in the conductivity of the head phantom. To illustrate this, the conductivity of edema tissue is greater than other tissues, edema growth will give rise to overall conductivity of brain increasing. In this experiment, the volume of cerebral edema expand from 0 to 20 ml, the amplitude have decreased 0.858 dB.

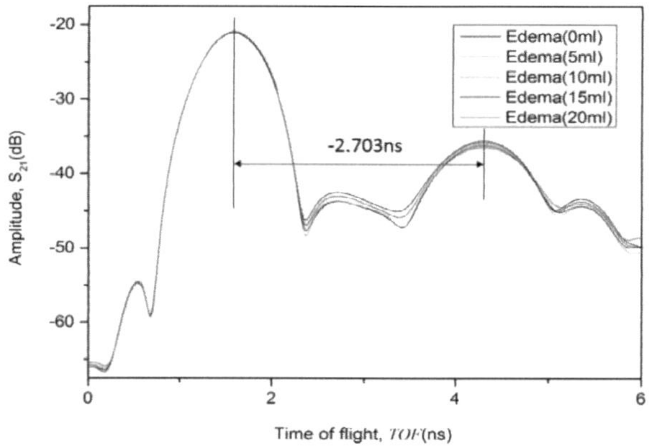


**Figure.4-1.** Time domain experimental results of various edema models

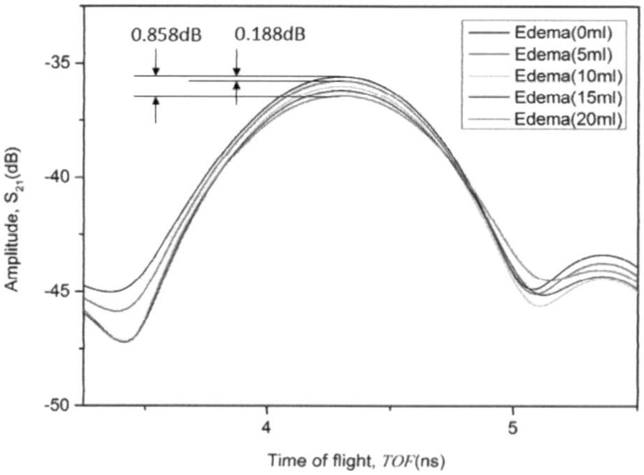


**Figure.4-2.** Enlarged the second significant peak of the experimental results shown in figure.4-1.

## 5. Discussion and conclusion

The experimental time domain performance shows that the power of microwave signal clearly decreased due to conductivity loss. Moreover, for the reason that the microwave instrument has the advantage of extremely high detecting resolution, this proposed method has a quite high resolving power for evaluating the conductivity of brain.

Since the left significant peak relates to the connecting point between the coaxial cable and the waveguide sensor, and the right significant peak corresponds to the waveguide open-ended end reflection, the wave group velocity can be calculated by $V_g = 2L/t$, where $L$ denotes the length of wave group propagated, and $t$ denotes the time of flight. The proposed system is capable of not only measuring the conductivity of brain but also evaluating the relative permittivity of brain theoretically.
In summary, this work demonstrated that a non-invasive system for the continuous monitoring of cerebral edema and verified it using a realistic 3-dimensional head phantom to illustrate the performance in term of sensitivity and accuracy. By using the proposed system to measure the magnitude of microwave signal, the conductivity of brain can be evaluated. As a result, there is a very satisfactory accordance between the measured values and the nominal ones, which shows that this proposed system has the potential to real-time monitor the rate and size of edema expansion in clinical environment.

## 6. Acknowledgement

This work was supported in part by Scientific and Technological Research Program of Chongqing Municipal Education Commission (Grant No. KJ1704102); in part by Doctoral Foundation of Chongqing University of Posts and Telecommunications (Grant No. A2016-77).

## 7. References

[1] Ninds I C H 2005 *Stroke* Priorities for clinical research in intracerebral hemorrhage: report from a National Institute of Neurological Disorders and Stroke workshop **36**(3) e23.
[2] Mobashsher A T, Bialkowski K S, Abbosh A M and Crozier S 2016 *Plos one* Design and experimental evaluation of a non-invasive microwave head imaging system for intracranial haemorrhage detection **11**(4) e0152351.
[3] Mobashsher A T, Abbosh A M 2016 *Sci. Rep.* On-site Rapid Diagnosis of Intracranial Hematoma using Portable Multi-slice Microwave Imaging System **6** 37620.
[4] Korfhagen J J, Kandadai M A, Clark J F, Adeoye O and Shaw G J 2013 *J. Neurosci. Meth.* A prototype device for non-invasive continuous monitoring of intracerebral hemorrhage **213**(1) 132-137.
[5] Wang X X, Jin L, Li J, Ran Y and Guan B O 2014 *Opt. Express.* Microfiber interferometric acoustic transducers **22** 8126-8135.
[6] Kandadai M A, Korfhagen J J, Beiler S, Beiler C, Wagner K, Adeoye O M and Shaw G J 2014 *J. Neurosci. Meth.* In vivo testing of a non-invasive prototype device for the continuous monitoring of intracerebral hemorrhage **235** 117-122.
[7] Chandra R, Zhou H, Balasingham I and Narayanan R M 2015 *IEEE T. Bio-med. Eng.* On the opportunities and challenges in microwave medical sensing and imaging **62**(7) 1667-1682.
[8] González C A, Rubinsky B 2006 *Physiol. Meas.* The detection of brain oedema with frequency-dependent phase shift electromagnetic induction **27**(6) 539.
[9] Kandadai M A, Raymond J L and Shaw G J 2012 *Mat. Sci. Eng. C* Comparison of electrical conductivities of various brain phantom gels: Developing a 'brain gel model' **32**(8) 2664-2667.
[10] Mobashsher A T, Abbosh A M 2015 *IEEE Microw. Mag.* Artificial human phantoms: human proxy in testing microwave apparatuses that have electromagnetic interaction with the human body **16**(6) 42-62.

IOP Conf. Series: Earth and Environmental Science **111** (2018) 012028 doi:10.1088/1755-1315/111/1/012028

# Integration of Energy Consumption and $CO_2$ Emissions into the DES Tool with Lean Thinking

**Reda Nujoom and Qian Wang**

School of Engineering, University of Portsmouth, Portsmouth P01 3DJ, UK
Email: qian.wang@port.ac.uk

**Abstract**. Products are often made by accomplishing a number of manufacturing processes on a sequential flow line which is also known as manufacturing systems. In a traditional way, design or evaluation of a manufacturing system involves a determination or an analysis of the system performance by adjusting system parameters relating to such as system capacity, material processing time, material-handling and transportation and shop-floor layout. Environment related parameters, however, are not considered or considered as separate issues. In the past decade, there has been a growing concern about the environmental protection and governments almost in all over the world enforced certain rules and regulation to promote energy saving and reduce carbon dioxide ($CO_2$) emissions in manufacturing industry. To date, development of a sustainable manufacturing system requires designers who need not merely to apply traditional methods of improving system efficiency and productivity but also examine the environmental issues in production of the developed manufacturing system. Most researchers, however, focused on operational systems, which do not incorporate the effect of environmental factors that may also affect the system performance. This paper presents a research work aiming to addresses these issues in design and evaluation of sustainable manufacturing systems incorporating parameters of energy consumption and $CO_2$ emissions into a DES (discrete event simulation) tool.

## 1. Introduction

Lean concepts are widely adopted by many manufacturing plants as a popular model when designing, implementing, operating or managing a manufacturing system. This is because it is a cost-effective approach promoting a sustainable manufacturing system by consistently seeking and removing any non-value added wastes in production without much additional investments. Lean thinking can also be described as an enterprise culture in which any creation of ideas or methods, which can be used effectively and economically for improving efficiency and productivity of a manufacturing system and reducing production costs, should be encouraged and embedded in any manufacturing-related activities or organizations. Manufacturing sustainability may be defined as the creation of manufactured products or systems by reducing the environmental impact on manufacturing systems due to usage of energy or natural resources. The conventional lean manufacturing method, however, does not consider environmental wastes of such as energy consumption and $CO_2$ emissions which are also considered as wastes of non-values added on manufactured products. Thus, to develop a sustainable manufacturing system, system designers need not merely to apply traditional methods to improve efficiency and productivity of a manufacturing system but also examine the environmental impact on the developed manufacturing system. This is also because governments in many countries have been enforcing ever-stricter environmental policies and regulations by promoting energy saving and low $CO_2$ emissions in manufacturing industry.

There are a few studies in considering environmental aspects related to developments of sustainable manufacturing systems using the modelling simulation methods. Heilala et al argued that manufacturing system designers need to not merely rely on traditional methods in improvements of system efficiency and productivity but also incorporate environmental considerations into design and operation of the developed manufacturing processes or systems [1]. Koc and Kaplan presented an investigation into energy consumption for a particular ring type yarn manufacturing system [2]. Lind et al developed a production simulation tool that was used for making a trade-off decision based on considerations of ergonomic constraints, levels of automation and environmental impacts [3]. Wang et al proposed the process integration (PI) method that was used for evaluating $CO_2$ emissions for a steel industry [4]. Gutowski et al conducted a thermodynamic analysis by examining the resources used in manufacturing processes [5]. Branham et al used the quantitative thermodynamic analysis for quantifying energy in different categories applied into manufacturing processes or systems [6]. Moreover, current modelling simulation tools in the market for manufacturing systems evaluation do not provide facilities that allow system designers to combine parameters of energy consumption and $CO_2$ emissions into an investigation of the overall manufacturing system performance. This is because, for example, in a discrete event simulation (DES) model, facilities are defined and treated as resources the same as parts, machines, conveyors and so on. The application of DES models is therefore restricted to predicting only such variables as the required number of workers, their shift patterns and routes [7]. This paper presents a study in mapping parameters relating to energy consumption and $CO_2$ emissions for manufacturing systems design and evaluation by proposing a methodology that can incorporate these parameters into the existing DES tools.

## 2. Energy Consumption and $CO_2$ Emissions for Manufacturing Systems Design and Evaluation

In a manufacturing system, energy is used for operating machines, illumination systems, air conditioning systems and other relevant supporting equipment such as compressors which supply compressed air to some of these machines. To describe amounts of energy consumption and $CO_2$ emissions mathematically, the following notations are used:

$m$: number of processes in a manufacturing system
$n_i$: number of machines involved in process $i$, where $i \in \{1, 2, ...., m\}$
$E_i$ (kWh): energy consumption for a machine involved in process $i$
$E_i^{cond}$ (kWh): energy consumption of an air conditioning system
$E_i^{illum}$ (kWh): energy consumption of an illumination system
$E_i^{air\ comp}$ (kWh): energy consumption of compressed air needed for a machine involved in process $i$
$TE$ (kWh): total energy consumption of a manufacturing system
$N_i$ (kw): installed power for a machine involved in process $i$
$R_i$ (kg/h): manufacturing rate for a machine involved in process $i$
$\tau_i$ (hr): operating time for a machine involved in process $i$
$\mu_i$ (%): efficiency for a machine involved in process $i$
$\partial_i$ (kg): mass of materials transferred from a machine involved in process $i$
$G_i$ (kg): mass production per month
$¥_i$ (%): waste ratio for a machine involved in process $i$
$Ę_i$ (kWh): energy consumption of air conditioning per month
$Ě_i$ (kWh): energy consumption of illumination per month
$\varsigma_i^{air\ comp}$ (kWh/m$^3$): energy consumption per cubic meter of a compressor
$\upsilon_i$ (m$^3$/h): compressed air used for a machine involved in process $i$ per hour
$\rho_i^{air\ comp}$ (m$^3$/h): capacity of compressed air in cubic meter per hour of a compressor
$N_i^{air\ comp}$ (kWh): installed power for a compressor
$e_i$ (kg/kWh): amount of $CO_2$ emissions per kWh released from a machine involved in process $i$
$Te_i$ (kg/kWh): amount of $CO_2$ emissions per KWh of a machine, an air conditioning system and anillumination system involved in process $i$

$\omega$ : $CO_2$ emission factor using different energy sources
*Te* (kg/kWh): total amount of $CO_2$ emissions released from a manufacturing system
$q_i$ (kg): mass of materials involved in process *i*

### 2.1. Energy Consumption

The energy consumption $E_i$ for a machine involved in process *i* is given by:

$$E_i = \tau_i \times N_i \times n_i \tag{1}$$

The operating time $\tau_i$ for a machine involved in process *i* is calculated by:

$$\tau_i = \frac{q_i}{R_i \times \mu_i} \tag{2}$$

Mass of materials $q_i$ transferred from a machine involved in process *i* is obtained by:

$$q_i = \partial_i \times (1 + ¥_i) \tag{3}$$

The energy consumption for air conditioning $E_i^{cond}$ in a manufacturing system is given by:

$$E_i^{cond} = Ę_i \times \frac{\partial_i}{G_i} \tag{4}$$

The energy used for an illumination system $E_i^{illum}$ is calculated by:

$$E_i^{illum} = Ě_i \times \frac{\partial_i}{G_i} \tag{5}$$

The energy consumption of a compressed air needed for a machine involved in process *i*, thus $E_i^{air\ comp}$ is calculated by:

$$E_i^{air\ comp} = \tau_i \times \zeta_i^{air\ comp} \times \upsilon_i \times n_i \tag{6}$$

Where $\zeta_i^{air\ comp}$ can be determined by:

$$\zeta_i^{air\ comp} = \frac{N_i^{air\ comp}}{\rho_i^{air\ comp}} \tag{7}$$

Thus, the total energy consumption TE for a manufacturing system is given below:

$$TE = \sum_{i=1}^{m} (E_i + E_i^{air\ comp} + E_i^{cond} + E_i^{illum})$$

Where $i \in \{1, 2, ...., \mathrm{m}\}$ (8)

Hence, equation 8 can be given below:

$$TE = \sum_{i=1}^{m} \left(\frac{\partial_i \times (1 + ¥_i)}{R_i \times \mu_i} \times N_i \times n_i + \tau_i \times \zeta_i^{air\ comp} \times \upsilon_i \times n_i + Ę_i \times \frac{\partial_i}{G_i} + Ě_i \times \frac{\partial_i}{G_i}\right) \tag{9}$$

### 2.2. $CO_2$ Emissions

The amount of $CO_2$ emissions $e_i$ released from a machine involved in process *i* is calculated by

$$e_i = \omega \times E_i \tag{10}$$

IOP Conf. Series: Earth and Environmental Science **111** (2018) 012028 doi:10.1088/1755-1315/111/1/012028

Where, the value of $CO_2$ emission factor $\omega$ is given in Table 1 [8].

**Table 1.** $CO_2$ emission factor $\omega$ (per kWh) using different energy sources

| Energy source | $\omega$ (kg/kWh) |
|---|---|
| Oil as direct energy source to generate thermal energy | 0.5 |
| Oil as indirect energy source to generate electricity | 0.6895 |
| Solar as indirect energy source to generate electricity | 0.05 |

The total amount of $CO_2$ emissions $Te$ can be calculated as follows [9]:

$$Te = \sum_{i=1}^{m} (e_i \times q_i + 0.6895 \times (E_i^{air\ comp} + E_i^{cond} + E_i^{illum}))$$

Where $$i \in \{1, 2, ...., m\} \tag{11}$$

Thus, Te can be expressed below:

$$Te = \sum_{i=1}^{m} (\omega \times \tau_i \times N_i \times n_i \times \partial_i \times (1 + ¥_i) + 0.6895 \times (\tau_i \times \zeta_i^{air\ comp} \times \upsilon_i \times n_i + Ȩ_i \times \frac{\partial_i}{G_i} + \check{E}_i \times \frac{\partial_i}{G_i}) \tag{12}$$

## 3. Implementation of Energy Parameters into a DES Tool

Until recently, there has been a lack of feasible and practical modeling simulation tools which permit system designers, at the early design stage, to assess the performance of manufacturing operations by taking the environmental aspect into account within manufacturing systems. For instance, discrete event simulation (DES) tools are widely used for manufacturing systems design and evaluation, and these modeling simulation packages have now evolved from 2D to 3D. Nevertheless, these tools do not have built-in functionalities that allow system designers to evaluate the performance of manufacturing systems by incorporating considerations of such as energy consumption and $CO_2$ emissions relating to manufacturing activities within a manufacturing system. In order to resolve this, parameters of energy consumption and $CO_2$ emissions can be modeled into either external MS Excel worksheets that can be linked into a DES tool or internal program using the DES simulation language; these parameters interact with parameters of physical elements (built in the DES tool) of machines and conveyors etc., together with logical interrelationship for operational activities in a manufacturing system. With this approach, by altering a value of each parameter of energy consumption and $CO_2$ emissions, this enables system designers to evaluate the overall performance of a manufacturing system considering not just operational activities bus also amounts of energy consumption and $CO_2$ emissions that are generated for running manufacturing facilities and processes. This integrated modeling simulation methodology also permits users to determine the relevant impact on logical interactions and interrelationships between parameters in manufacturing operations, energy consumption and $CO_2$ emissions within a manufacturing system. With the integrated approach, a manufacturing system can be modeled, simulated and examined concurrently to gain a comprehensive preview of the system performance and provide the performance prediction that allows designers to build a system based on an optimal solution before implementing a real system. Moreover, this technique can be implemented as a strategic planning tool for systems analysts and designers to quickly provide a visible preview of the integrated system performance at an early stage in the design process. Figure 1 illustrates an integrated DES model with which a manufacturing system design can be investigated aimed to minimize unnecessary wastes of materials handing activities by implementing lean management rules and reduce the environment impact of energy consumption and $CO_2$ emissions.

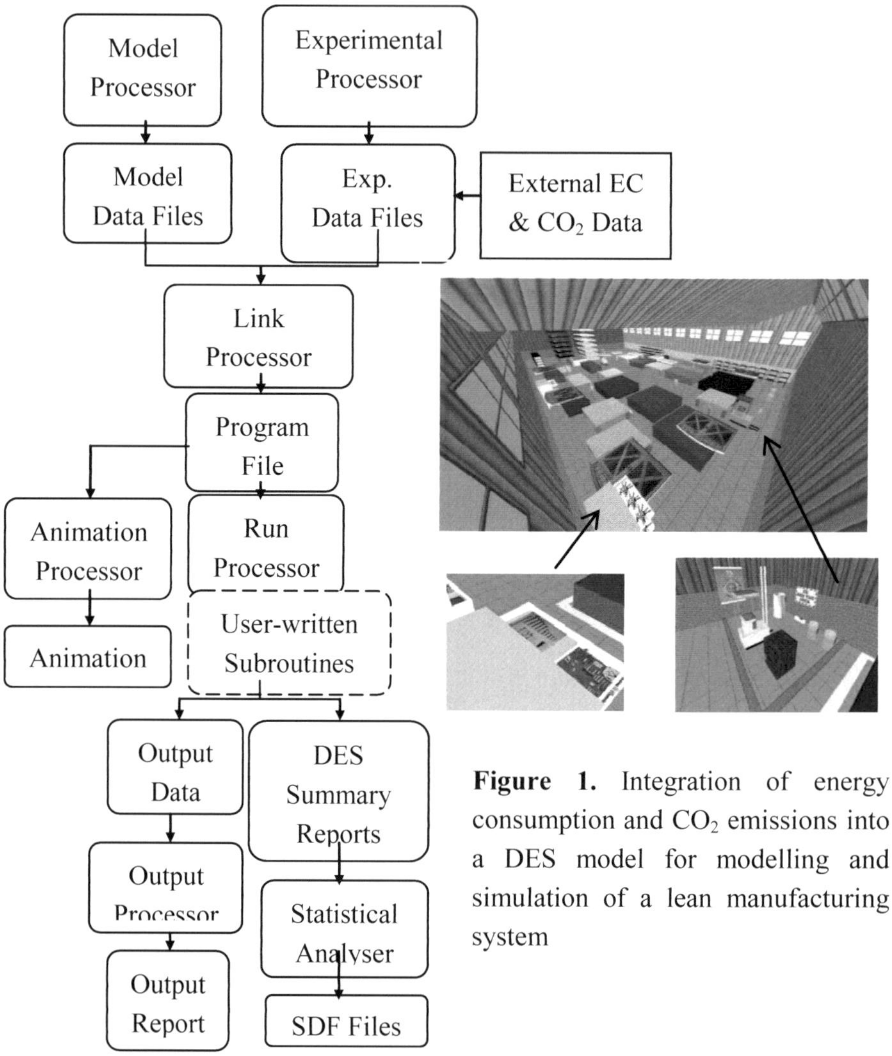


**Figure 1.** Integration of energy consumption and $CO_2$ emissions into a DES model for modelling and simulation of a lean manufacturing system

## 4. Conclusions and Future Work

When designing a manufacturing system, engineers used to focus on key performance indicators in terms of such as system productivity, capacity and other operation-related issues; environmental considerations are often overlooked or considered separately. Moreover, a literature study by authors shows that there is a lack of DES modeling simulation tools for manufacturing systems design and evaluation by combining parameters associated with environmental considerations of such as energy consumption and $CO_2$ emissions. This paper presents a study in proposing an integrated DES model by addressing energy consumption and $CO_2$ emissions relating to manufacturing activities of manufacturing systems. The proposed DES tool has capability of incorporating a number of environmental parameters in terms of energy consumption and $CO_2$ emissions allowing a joint analysis of a manufacturing system performance under environmental constraints. The integrated method can be useful for helping decision-makers to develop a manufacturing system considering amounts of energy consumption and $CO_2$ emissions. In the latest developments of DES tools, although developers have been slowly introducing cloud-based technologies to facilitate the mobility of applications and the interoperability between different users or partners, none of these DES tools has built-in functions incorporating parameters of energy consumption and $CO_2$ emissions for manufacturing systems design and evaluation. Apart from this, current DES tools have a lack of a great variety of functions and data exchange among different domains which cause difficulties in interoperability and collaboration between systems or users. These issues ought to be addressed in the future development with utilization and integration of multi-disciplinary and multi-domain simulation tools. Through the literature studies, there were a few studies in product life-cycle costs relating to environmental issues that were incorporated into DES tools. As a result, simulation tools that assure the multi-level integrations are still under developments. Also, efforts ought to be made to create smart, intelligent and self-learning DETS tools. Research is required in order to develop more intelligent tools that will lead to autonomous and self-adapting systems for automated optimization of system parameters [7].

## 5. References

[1] J. Heilala S, Vatanen H, Tonteri J, Montonen S, Lind B and Stahre J 2008 Simulation-based sustainable manufacturing system design *Proceedings of the 2008 Winter Simulation Conference* pp 1922–1930.

[2] Koç E and Kaplan E 2007 An investigation on energy consumption in yarn production with special reference to ring spinning *J Fibr & Texti in Eas* Eur 4 pp 18–24.

[3] Lind S B, Krassi B, Johansson B, Viitaniemi J, Heilala J, Stahre S, Vatanen A and Berlin C 2008 SIMTER: A production simulation tool for joint assessment of ergonomics, level of automation and environmental impacts *Proceedings of the 18th International Conference on Flexible Automation and Intelligent Manufacturing*.

[4] Wang C, Larsson M, Ryman C, Grip C E, Wikström J O, Johnsson A and Engdahl J 2008 A model on $CO_2$ emission reduction in integrated steelmaking by optimization methods *International Journal of Energy Research* 32(12) pp 1092–1106.

[5] Gutowski T G, Branham M S, Dahmus J B, Jones A J, Thiriez A and Sekulic D P 2009 Thermodynamic analysis of resources used in manufacturing processes *Environmental Science & Technology* 43(5) pp 1584–1590.

[6] Branham M, Gutowski T G, Jones A and Sekulic D P 2008 A thermodynamic framework for analysing and improving manufacturing processes, *Electronics and the Environment* pp. 1–6.

[7] Wang Q and Chatwin C R 2005 Key issues and developments in modelling and simulation-based methodologies for manufacturing systems analysis, design and performance evaluation *International Journal of Advanced Manufacturing Technology* 25(11-12) pp 1254–1265.

[8] EPA *The Lean and Environment Toolkit* US Environmental Protection Agency.

[9] Nurjanni K P, Carvalho M S and da Costa L A A F 2014 Green Supply Chain Design with Multi-Objective Optimization, *Proceedings of the 2014 International Conference on Industrial Engineering and Operations Management* pp 7–9.

IOP Conf. Series: Earth and Environmental Science **111** (2018) 012029 doi:10.1088/1755-1315/111/1/012029

# Cyber Physical System Modelling of Distribution Power Systems for Dynamic Demand Response

**Xiaodong Chu[1,3], Rongxiang Zhang[1], Maosen Tang[1], Haoyi Huang[1] and Lei Zhang[2]**

[1]School of Electrical Engineering, Shandong University, Jinan, China
[2]Electric Power Research Institute of State Grid Shandong Electric Power Company, Jinan, China
E-mail: chuxd@sdu.edu.cn

**Abstract**. Dynamic demand response (DDR) is a package of control methods to enhance power system security. A CPS modelling and simulation platform for DDR in distribution power systems is presented in this paper. CPS modelling requirements of distribution power systems are analyzed. A coupled CPS modelling platform is built for assessing DDR in the distribution power system, which combines seamlessly modelling tools of physical power networks and cyber communication networks. Simulations results of IEEE 13-node test system demonstrate the effectiveness of the modelling and simulation platform.

## 1. Introduction

Dynamic demand response (DDR) is a package of control methods to enhance power system security and/or improve operational efficiency through actively modulating power demand, which has gained an increasing attention in the field [1]. The implementation of DDR is of great challenge since the controllable power demand devices are dispersed and very large in number, e.g., hundreds of thousands or even larger for a power system. A key issue to control such large number of devices in a coordinated way is the communication burden [2].

Typical control architectures of DDR can be categorized into centralized, distributed and decentralized ones [1, 2]. For the centralized control, the communication burden is extremely high since each power demand device transfers its working states to the power dispatch center while the dispatch center sends the control commands to the demand devices. The communication burden reduces much for the distributed control considering that data exchanges are limited in the neighborhood of the demand devices. The decentralized control is of the lowest communication burden among all types of control architectures. Nevertheless, an appropriate model for the communication network of the power system is required to depict the information flow of DDR correctly.

A distribution power system plays the role to bridge the transmission power system that is responsible for long-distance and large-capacity power transfer with the electricity end users. Therefore, the interface layer of DDR is in the distribution power system. A cyber-physical system (CPS) model for the distribution power system is necessary to reflect interactions between the information flow and power flow [3]. CPS modelling has gained much more attention in the fields of computer science, control theory, and communication engineering [4, 5].

A CPS modelling and simulation platform for DDR in distribution power systems is presented in this paper. Firstly, CPS modelling requirements of distribution power systems are elaborated. The heterogeneous nature of the ICT architecture of the distribution power system necessitates the CPS model. Secondly, a coupled CPS modelling platform is built for assessing DDR in the distribution power system, which combines seamlessly modelling tools of physical power networks and cyber communication networks. Finally, simulations are conducted on a test system, which demonstrate the effectiveness of the modelling and simulation platform.

## 2. CPS Modelling Requirements of Distribution Power System

Distribution power systems are undergoing the evolvement from the passive receiving-end power network to the active user-centric power system with multiple types of distributed resources integrated [6]. The conventional data acquisition system for the distribution power system cannot satisfy the information processing requirements of the evolving distribution power systems. Advanced distribution management system (DMS) aims at providing the comprehensive measurement, control, and decision capabilities for the distribution power system [7]. The information and communication (ICT) technology is the core of a DMS.

### *2.1. ICT Architecture for Distribution Power System*

The ICT architecture for a distribution power system is of multiple levels with heterogeneous technologies employed in each level, which is adaptive to the physical structure of the distribution power system. In China, typical voltage levels of a distribution power system include 380 V, 10 kV, 35 kV, and 110 kV. The voltage level of 380 V is the phase to phase voltage grade of residential or commercial users, in which the short-range or medium-range wireless communication technology is a good option to construct a home area network (HAN) or neighborhood area network (NAN). The users are aggregated and connected through 380 V/10 kV transformers to the 10 kV feeder where power line carrier or long-range wireless communication media are available, e.g., WiMAX or cellular networks. The voltage level of 35 kV is suitable for larger industrial users, where a hybrid package of wired or wireless communication technologies can be used. The fiber network is applied to the voltage level of 110 kV and above to ensure the communication quality [8].

The heterogeneous nature of the ICT architecture imposes great challenges to the modelling and management of DMS. It is not viable to assume perfect communication conditions or even neglect the dimension of communication as what the conventional modelling approaches do. The CPS modelling methods are required to depict correctly the impacts of the ICT conditions on the operation of the distribution power system [9].

### 2.2. CPS Model

The CPS model for the distribution power system should reveal appropriately the coupling mechanism between the physical power network and cyber communication network, which is not simply merging the two parts. Dynamics of the physical power network are usually time-driven whereas it is appropriate to depict the cyber communication network using the event-driven model [10].

Dynamical model of the physical power network can be expressed by using a set of differential equations

$$\frac{d\boldsymbol{x}}{dt} = \boldsymbol{f}(\boldsymbol{x}, \boldsymbol{u}, t) \tag{1}$$

where $\boldsymbol{x}$ and $\boldsymbol{u}$ are vectors of state variables and input variables, respectively. The above differential equations can be substituted by corresponding difference equations. The occurrence of events in the cyber communication network is stochastic, which is recorded in the list of events along the time axis as shown in figure 1.

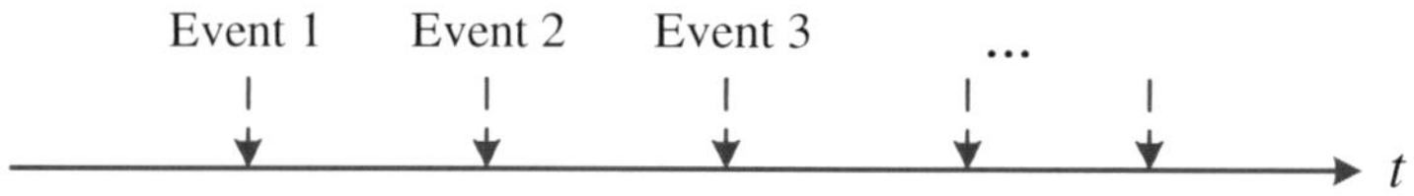


**Figure 1.** Discrete events in cyber communication network.

A key issue of the CPS model for the distribution power system is the synchronization between the physical power network and the cyber communication network. There are mainly three synchronization approaches including time-stepped, master-slave, and global event-driven approaches [10]. Compared with the other two methods, the global event-driven approach is superior in its performance. The synchronization mechanism of the approach can be explained briefly as follows, which is illustrated in figure 2 [10]. Each time step in the difference equations of the physical power network is treated as a discrete event. Through such conversion, both the physical power network and cyber communication network are event-driven. All events are recorded in the event list in accordance with their occurring time. As shown in figure 2, initially event 1 occurs in the communication network followed by event 2. The jump from event 1 to event 2 is denoted as step 1. The model then traces back to the time instant when event 1 occurs (step 2) and drives dynamics in the power network to the time instant of event 2 (step 3). After the data exchange between the power network and communication network is completed, the model returns to the communication network (step 4) for the subsequent process.

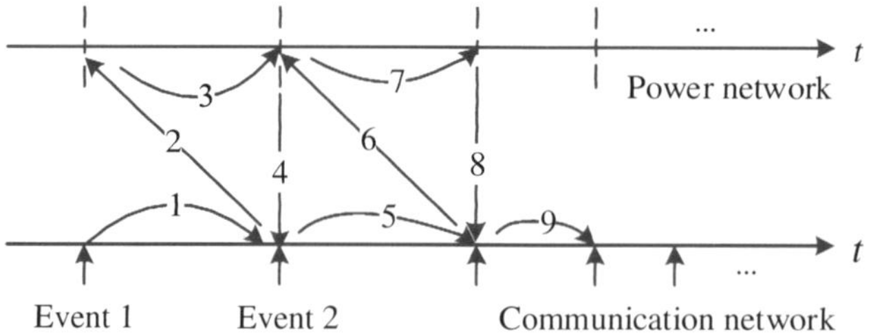


**Figure 2.** Synchronization mechanism between physical power network and cyber communication network using global event-driven approach.

## 3. CPS Modelling for DDR

### *3.1. Modelling Tools*

There are various tools available for modelling and simulation of physical power networks, which include, but are not limited to, PSS/E, PSLF, PowerFactory, Eurostag, and OpenDSS [11]. OpenDSS is a tool more suitable for distribution power systems, which provides versatile modelling functions associated with planning, design, and operation stages of a system. A great advantage of OpenDSS is that it is an open-source tool. Therefore, the modeler can easily add and/or modify components and modules besides the built-in ones. OpenDSS supports a Component Object Model (COM) interface facilitating custom solution modes from an external program, e.g., MATLAB.

The most commonly used tools for modelling and simulation of cyber communication networks are NS-2, NS-3, OPNET, and OMNET++ [10]. NS-2 is preferable considering its several advantages. Firstly, NS-2 is open-source easing further development. Secondly, it supports multiple communication protocols and heterogeneous networks including wired, wireless, and satellite. Thirdly, a good balance can be maintained between modularity and speed since NS-2 is written with C++ and OTcl separately, in which C++ is used for data whereas OTcl is applied to control.

### *3.2. Coupled CPS Modelling for DDR*

A coupled CPS modelling scheme is used in this paper to model appropriately the interactions between the physical power network and the cyber communication network in a distribution power system. OpenDSS and NS-2 are employed respectively in the coupled modelling scheme while MATLAB functions as a bridge between the two tools [12]. The main tasks of MATLAB can be summarized as follows.

Synchronization mechanism: It is through MATLAB to coordinate and synchronize modelling and simulation of the physical power network and the cyber communication network by using OpenDSS and NS-2 respectively.

Data exchange: MATLAB functions as a medium to exchange required data between OpenDSS and NS-2.

DMS emulation: MATLAB emulates DMS in the distribution power system to provide dispatch and control decisions to other entities.

Both of OpenDSS and MATLAB run under the Windows operating system (OS). They communicate with each other via the COM interface. NS-2 does not support the Window OS, for which Cygwin is employed to build the virtual Linux OS. Secure Shell (SSH) protocol and Secure Copy (SCP) protocol are used for data exchanges between NS-2 and MATLAB. The coupled CPS modelling scheme for a distribution power system is illustrated in figure 3 [12].

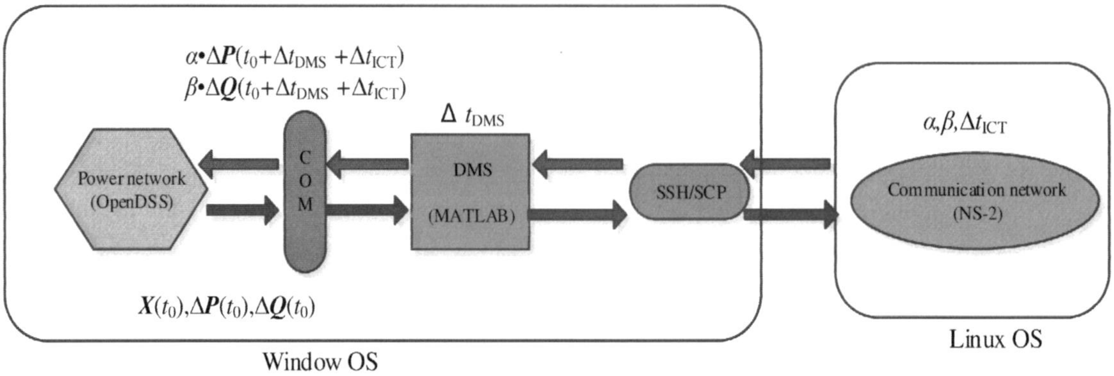


**Figure 3.** Coupled CPS modelling of distribution power system.

A coupled CPS modelling platform is built for assessing DDR in the distribution power system. Since controllable power demand devices applied to DDR are dispersed in the system, the impact of the communication environment cannot be neglected. The amount of aggregated DDR in the system is computed by OpenDSS, which is $\Delta \boldsymbol{P}(t_0)$ and $\Delta \boldsymbol{Q}(t_0)$ (representing active power and reactive power response, respectively) given the system states as $\boldsymbol{X}(t_0)$ at time $t_0$. If it is in a perfect communication environment, the available provision of DDR will be $\Delta \boldsymbol{P}(t_0)$ and $\Delta \boldsymbol{Q}(t_0)$. Nevertheless, the real imperfect communication conditions will cause time delay, bit error and packet loss, which are estimated by NS-2. The symbol $\Delta t_{\mathrm{ICT}}$ represents the time delay weighing data transfer through the communication network and the symbols of $\alpha$ and $\beta$ denote metrics for bit error and packet loss. The decision process of DMS will produce time delay denoted as $\Delta t_{\mathrm{DMS}}$, which is mimicked by MATLAB. Consequently, the available provision of DDR will be $\alpha \bullet \Delta \boldsymbol{P}(t_0+\Delta t_{\mathrm{DMS}}+\Delta t_{\mathrm{ICT}})$ and $\beta \bullet \Delta \boldsymbol{Q}(t_0+\Delta t_{\mathrm{DMS}}+\Delta t_{\mathrm{ICT}})$.

## 4. Simulation Results

### *4.1. Test System*

IEEE 13-node distribution power system is used as the test system [13], which is shown in figure 4. Node 1 is the root node of the distribution power system, which is the connection point to the upstream power system. Controllable power demand devices applied to DDR are air-conditioners, which are assumed to be aggregated at Node 670. The number of air-conditioners is 2000 and the aggregated active power $\Delta P$ is 1000 kW with reactive power omitted, i.e., $\Delta Q$ is zero.

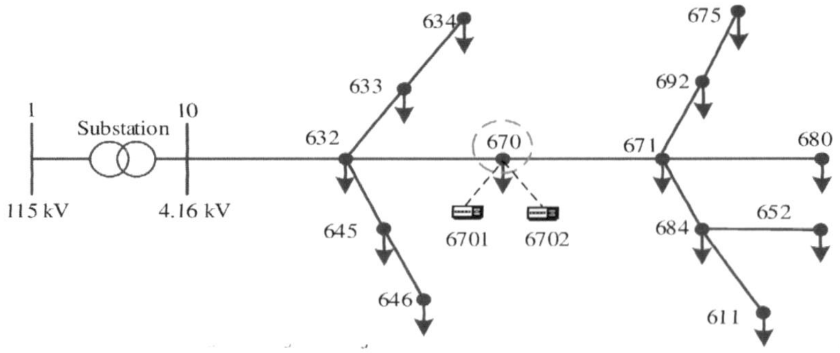


**Figure 4.** IEEE 13-node distribution power system.

### *4.2. Impact of Communication Conditions on DDR*

Communication conditions for the test system can be measured by the data transmission rate of each corresponding communication line. Assuming that there are two DDR vendors providing aggregated DDR at Node 670, which are named as 6701 and 6702, respectively. As shown in figure 5, data transmission rates of communication lines 6701-670, 6702-670, and 670-1 are limited and time variant, which reflects real imperfect conditions of the communication network.

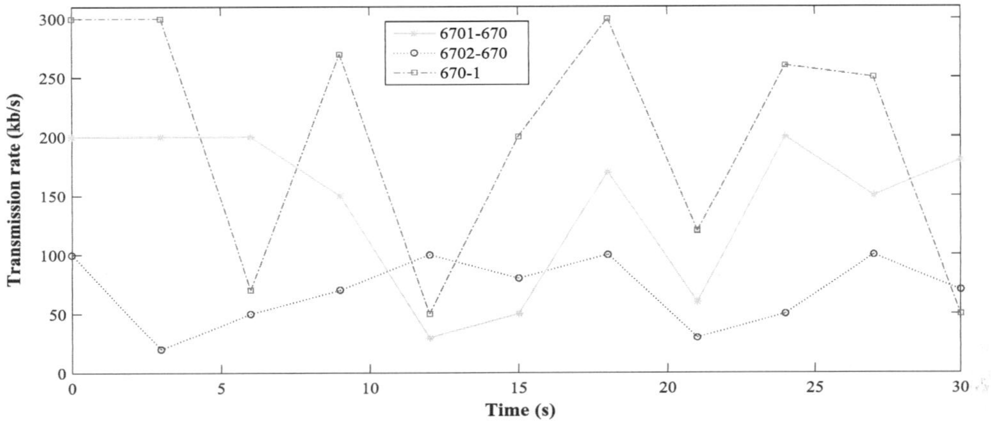


**Figure 5.** Data transmission rates of communication lines.

Assuming that the activation time instant of DDR is 5 s, the voltage value of node 670 will increase after that instant with the release of partial load demand, which simulates a typical scenario that stressed operating states of the distribution power system can be relieved via DDR. Nevertheless, imperfect conditions of the communication network will discount the effect of DDR. As illustrated in figure 6, the response of node 670 will be postponed accounting for the time delay in in the communication network.

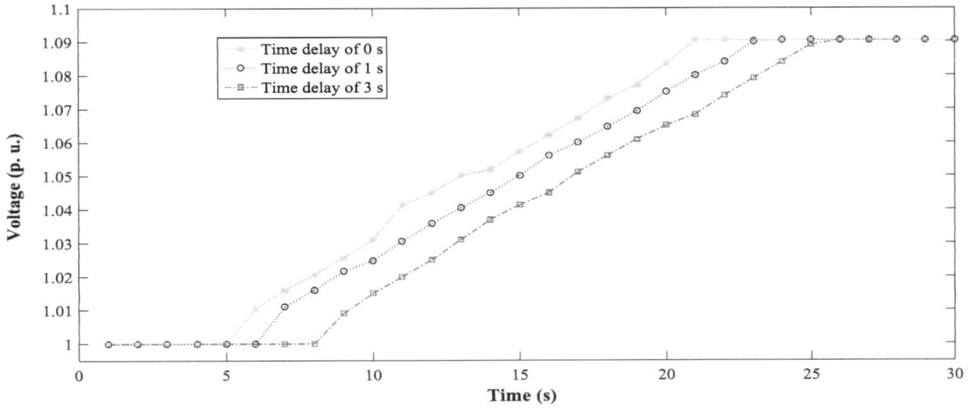


**Figure 6.** Voltage dynamics of node 670 with communication delay.

## 5. Conclusion

DDR is a set of control approaches to enhance power system security and/or improve operational efficiency through actively modulating load demand. In this paper, a CPS modelling and simulation platform is built for DDR in distribution power systems. Firstly, CPS modelling requirements of distribution power systems are analyzed. A coupled CPS modelling platform is then constructed for assessing DDR in the distribution power system, which combines modelling tools of physical power networks and cyber communication networks. Simulations results of IEEE 13-node test system demonstrate the effectiveness of the modelling and simulation platform.

## 6. Acknowledgments

This work was financially supported by Science and Technology Project of State Grid Corporation of China (SGSDDK00KJJS1600061).

## 7. References

[1] Yoon J H, Baldick R and Novoselac A 2014 Dynamic demand response controller based on real time retail price for residential buildings *IEEE Trans. Smart Grid* **5** 121-9

[2] Pourmousavi S A and Nehrir M H 2014 Introducing dynamic demand response in the LFC model *IEEE Trans. Power Syst.* **29** 1562-72

[3] Sridhar S, Hahn A and Govindarasu M 2012 Cyber–physical system security for the electric power grid *Proc. IEEE* **100** 210-24

[4] Derler P, Lee E A and Vincentelli A S 2012 Modeling cyber–physical systems *Proc. IEEE* **100** 13-28

[5] Memon I, Hussain I, Akhtar R and Chen G 2015 Enhanced privacy and authentication: An efficient and secure anonymous communication for location based service using asymmetric cryptography scheme *Wirel. Pers. Commun.* **84** 1487-1508

[6] Martins V F and Borges C L T 2011 Active distribution network integrated planning incorporating distributed generation and load response uncertainties *IEEE Trans. Power Syst.* **26** 2164-72

[7] Gungor V C, Sahin D, Kocak T, Ergut S, Buccella C, Cecati C and Hancke G P 2013 A survey on smart grid potential applications and communication requirements *IEEE Trans. Ind. Inform.* **9** 28-42

[8] Gungor V C, Sahin D, Kocak T, Ergut S, Buccella C, Cecati C and Hancke G P 2011 Smart grid technologies: Communication technologies and standards *IEEE Trans. Ind. Inform.* **7** 529 39

[9] Vallejo A, Zaballos A, Selga J M and Dalmau J 2012 Next-generation QoS control architectures for distribution smart grid communication networks *IEEE Commun. Mag.* **50** 128-34

[10] Li W, Ferdowsi M, Stevic M, Monti A and Ponci F 2014 Cosimulation for smart grid communications *IEEE Trans. Ind. Inform.* **10** 2374-84

[11] Mahmud K and Town G E 2016 A review of computer tools for modeling electric vehicle energy requirements and their impact on power distribution networks *Appl. Energy* **172** 337-59

[12] Celli G, Pegoraro P A, Pilo F, Pisano G and Sulis S 2014 DMS cyber-physical simulation for assessing the impact of state estimation and communication media in smart grid operation *IEEE Trans. Power Syst.* **29** 2436-46

[13] Wang H and Schulz N N 2004 A revised branch current-based distribution system state estimation algorithm and meter placement impact *IEEE Trans. Power Syst.* **19** 207-13

# Study on Calculation of Liquid Level And Storage of Tanks for LNG-fueled Vessels

**Kun Li[1, a], Guoqing Wang[2, b] and Chang Liu[1, c,*]**

[1]China Waterborne Transport Research Institute, Beijing, China
[2]Wuhan Jiao da New Energy Technology Co., Ltd, Wuhan, China
Email: [a] likun@wti.ac.cn, [b] wgq-msp@163.com, [c] liuchang@wti.ac.cn

**Abstract.** As the ongoing development of the application of LNG as a clean energy in waterborne transport industry, the fleet scale of LNG-fueled vessels enlarged and the safety operation has attracted more attention in the industry. Especially the accurate detection of liquid level of LNG tanks is regarded as an important issue to ensure a safe and stable operation of LNG-fueled ships and a key parameter to keep the proper functioning of marine fuel storage system, supply system and safety control system. At present, detection of LNG tank liquid level mainly adopts differential pressure detection method. Liquid level condition could be found from the liquid level reference tables. However in practice, since LNG-fueled vessels are generally not in a stationary state, liquid state within the LNG tanks will constantly change, the detection of storage of tanks only by reference to the tables will cause deviation to some extent. By analyzing the temperature under different pressure, the effects of temperature change on density and volume integration calculation, a method of calculating the liquid level and storage of LNG tanks is put forward making the calculation of liquid level and actual storage of LNG tanks more accurately and providing a more reliable basis for the calculation of energy consumption level and operation economy for LNG-fueled vessels.

## 1.Introductions

Liquefied natural gas (LNG) is a kind of high quality clean energy fuel, and it is widely used in ships. Storage tank is the medium for storing LNG fuel. LNG is an ultra-low temperature liquid. When it is stored, there is a gas-liquid two-phase state, and the volume expansion of LNG is about 600 times after gasification. When the temperature rises, the pressure will increase dramatically and is easy to make the tank in a dangerous state, Accurate measurement of the liquid level in the tank has become a security problem that can not be ignored. At the same time, by measuring the real-time liquid level of the tank, the energy consumption level of the ship can be calculated, so as to provide the basis for the ship to grasp the economic operation of the ship.

The existing research on the LNG tank level detection using differential pressure detection, reference level control table check the liquid level[1-2]. In practice, due to LNG power ship in a non-stationary state, the LNG storage tank of liquid state will change constantly, the storage tank storage only through the reference level table will have a certain degree of error. In this paper, the relationship between LNG tank height and volume, LNG density and temperature, LNG pressure and temperature, and LNG tank level and pressure difference are analyzed. puts forward a method to calculate the liquid level and reserves of LNG dynamic storage tank, and solves this problem well.

Liquefied Natural Gas (LNG) is widely applied in transport industry due to it is clean and of high quality. As the formulation of 'Emission Standards for Pollutants from Shipbuilding Industry', ship emission control zones have been set up in the waters of the Pearl River Delta, Yangtze River Delta and

Bohai Bay to control ship emissions of sulfur oxides, nitrogen oxides and particulate matter and improve environment and air quality of coastal and river area especially in port cities, the advantages of LNG as a shipping fuel are widely accepted and promoted.

## 2. The Storage of LNG for LNG-fueled Vessels

### *2.1. The Structure of LNG Tanks*

The structure of tanks for LNG-fueled vessels is double metal cylindrical tanks, including container, hull and holder, insulating layer, piping and valve, instrument and security accessories[3]. Type-C horizontal standard ellipse head structure is generally adopted. Inside container and its piping are made of austenite stainless steel (S30408) and the design of main pressure parts are strictly compliance with GB24511-2009, the hull is as well as made of austenite stainless steel (S30408) according to 'Rules for Natural Gas Fueled Ships' issued by China Classification Society (CCS) in 2013. The tank settings include upper and lower inflow liquid, outflow liquid, BOG, self-pressurization, overflow, pressure gage, liquid indicator and thermometer, on-site or remote pipeline monitoring. The thermal insulation performance directly affects the container, in other words, good insulation is a key factor to ensure the storage for a long time. Tanks generally use vacuum powder and high vacuum multiple wraps so far now. LNG in tanks is stored in saturation liquid, the temperature inside tank below 163°C and the top pressure is around 11 bar[4-6].

### *2.2. Monitoring System*

Since liquefied natural gas stored in tanks of LNG-fueled vessels is low temperature, flammable and explosive, a monitoring system to ensure normal operation and crew safety is necessary. The system presents status parameter such as liquid level and pressure on the spot. At the same time a monitoring instrument with the function of remote transmission to provide status parameter for chain control of safety control system is required.

Differential pressure method is the most common way to measure the liquid level of low temperature liquid in tanks. Gas derived through the bottom and top of the tank (see A, B tubes in Figure 1, pressure in tube A is for bottom pressure, pressure in tube B is for top pressure), the liquid level height inside the tank is confirmed on the spot through the liquid indicator which measures the pressure difference from tube A and B (the pressure difference is the static pressure made by the liquid column inside the tank). The pressure difference made by the liquid level could also be confirmed through transmitter sending the signal to the central control room participating control interlock of the tank or system.

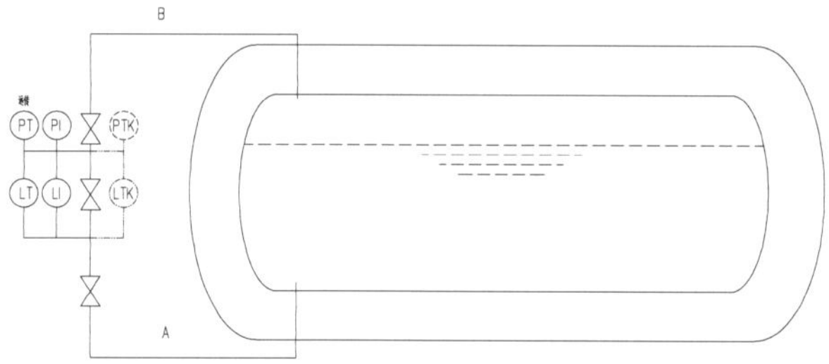


**Figure.1** Liquid level detection of pressure in the tank

## 3. Calculation Parameters of Liquid Storage Capacity for Tanks

### *3.1. The Relationship Between Height and Volume of Liquid Level*

Tank volume could be divided into two parts for easy calculation, see figure 2 the straight pipes and heads, two of which (standard elliptical heads) could be regarded as one ellipsoid[1]. Conditions are as follows:

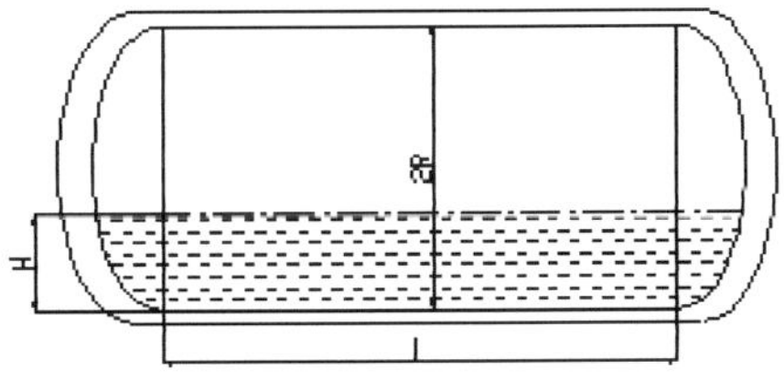


**Figure.2** Standard Elliptical Heads Type-C Horizontal Tank

R——Inside tank radius
L——Inside tank straight pipe length (including the head straight side)
H——Inside tank liquid height

*3.1.1. Calculation the Ellipsoid Liquid Volume*

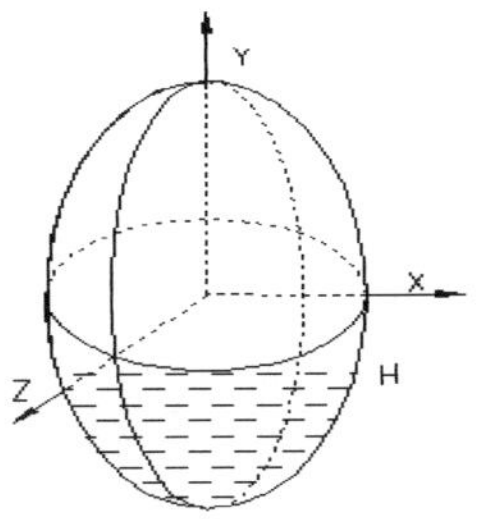


**Figure.3** Ellipsoid 3D Coordinate System

Ellipsoid equation is $\frac{x^2}{a^2}+\frac{y^2}{b^2}+\frac{z^2}{c^2}=1$ of which $a=\frac{R}{2}; b=R; c=R$.

Shadow in XY plane is ellipse model, a cross section coordinate system is established as follows:

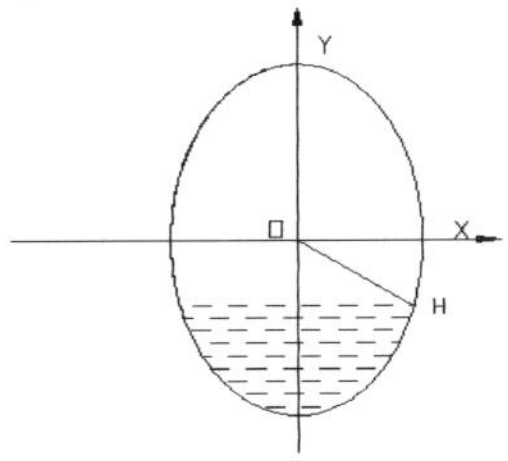


**Figure.4** XY plane coordinate system

Ellipse equation is $\frac{x^2}{a^2}+\frac{y^2}{b^2}=1$ of which $a=\frac{R}{2}; b=R$

When liquid level height is H, $y=H-R$

Plug it into equation to get H point: $x=\frac{\sqrt{R^2-(H-R)^2}}{2}$

Shadow in YZ plane is round model, cross section coordination system is established as follows:

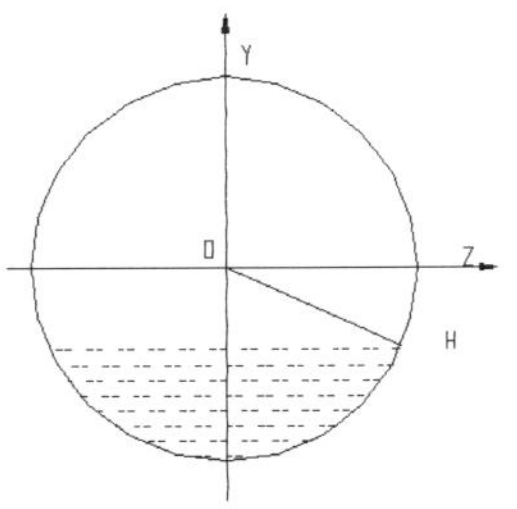


**Figure.5** YZ plane coordinate system

Round equation is $\frac{y^2}{b^2}+\frac{z^2}{c^2}=1$ of which $b=R; c=R$

When liquid level height is H, $y=H-R$

Plug it into equation to get H point: $z=\sqrt{R^2-(H-R)^2}$;

When liquid level in XZ coordinate system is H, shadow of liquid level is ellipse, of which major semi-axis A=z and minor semi-axis B=x.

According to the formula of ellipse area: $S_H=\pi AB$, the area of liquid level where the height is H is: $S_1=\pi\frac{R^2-(H-R)^2}{2}$

Therefore when the liquid level height is H and $y=R-H$, volume of ellipsoid liquid is: $V_1=\int_{-R}^{R-H}S_1dy=\frac{\pi}{2}\left(RH^2-\frac{1}{3}H^2\right)$;

*3.1.2. Calculation the Cylinder Liquid Volume.* When liquid height is H, shadow of liquid level in YZ plane is as figure 5. Shadow in XZ plane is rectangle, of which length is L, width is $z=\sqrt{R^2-(H-R)^2}$.

Area of liquid level is $S_2=L\sqrt{R^2-(H-R)^2}$

When liquid level of cylinder is H, y= R-H, the cylinder volume is: $V_2=\int_{-R}^{R-H}S_2dy=L\left[R^2\cos^{-1}\frac{R-H}{R}-\sqrt{R^2-(R-H)^2}(R-H)\right]$.

*3.1.3. Total Volume of Liquid in Tanks.* When liquid height is H, the total volume of liquid is the sum of the above two parts:

$$V_{liquid}=L\left[R^2\cos^{-1}\frac{R-H}{R}-\sqrt{R^2-(R-H)^2}(R-H)\right]+\frac{\pi}{2}\left(RH^2-\frac{1}{3}H^2\right). \quad (1)$$

The gross capacity of tank is

$$V_{total}=\pi LR^2+2\pi R^3-\frac{2}{3}\pi R^2. \quad (2)$$

Unit: V -- m³; R, H -- m.

The relationship between LNG density and temperature

Density of LNG depends on its constitution, generally is 430~470Kg/m3, the higher the methane content, the lower the density, and the variation gradient is 1.35 Kgm-³/°C. A similar calculation of Low temperature fluid density (kg/m³) and temperature (thermodynamics temperature K) referred in 'Low Temperature Engineering Technology'[7] is:

$$\rho=AB^{-(1-T/T_c)^n}. \quad (3)$$

A, B, n is regression coefficient of compounds

A=0.15998, B=0.2881, n=0.277

Tc -- critical temperature; Tc=190.58

Since the design pressure of LNG is generally 1.2MPA, actually the storage temperature is between 110K and 150K. The saturated vapor pressure will exceed the design pressure if more than 150K. Pressure will discharge through safety valve when overpressure occurs. Therefore the article adopts the temperature range from 110K to 150K for calculation(As shown in Table 1).

IOP Conf. Series: Earth and Environmental Science **111** (2018) 012030 doi:10.1088/1755-1315/111/1/012030

**Table.1.** Table of temperature and density

| T | Calculation Density | Physical Property Density[7] | Difference | T | Calculation Density | Physical Property Density | Difference |
|---|---|---|---|---|---|---|---|
| 110 | 426.4465 | 424.89 | 1.556477 | 129 | 397.4657 | 395.86 | 1.605746 |
| 111 | 424.0929 | 422.53 | 1.562918 | 130 | 395.8324 | 394.23 | 1.602386 |
| 112 | 423.5556 | 422 | 1.555622 | 131 | 394.1862 | 392.58 | 1.606248 |
| 113 | 422.0977 | 420.53 | 1.5677 | 132 | 392.527 | 390.93 | 1.596983 |
| 115 | 419.156 | 417.58 | 1.575988 | 133 | 390.8542 | 389.26 | 1.594226 |
| 116 | 417.6718 | 416.1 | 1.571828 | 134 | 389.1676 | 387.57 | 1.597595 |
| 117 | 416.1785 | 414.6 | 1.578542 | 135 | 387.4667 | 385.87 | 1.596691 |
| 118 | 414.6759 | 413.09 | 1.585929 | 136 | 385.7511 | 384.16 | 1.591095 |
| 119 | 413.1638 | 411.57 | 1.593778 | 137 | 384.0204 | 382.43 | 1.590369 |
| 120 | 411.6419 | 410.05 | 1.591874 | 138 | 382.2741 | 380.69 | 1.584053 |
| 121 | 410.11 | 408.51 | 1.599993 | 139 | 380.5117 | 378.93 | 1.581664 |
| 122 | 408.5679 | 406.97 | 1.5979 | 140 | 378.7327 | 377.15 | 1.582692 |
| 123 | 407.0154 | 405.41 | 1.605354 | 142 | 375.1228 | 373.54 | 1.582836 |
| 124 | 405.4521 | 403.85 | 1.602103 | 144 | 371.4399 | 369.85 | 1.589853 |
| 125 | 403.8779 | 402.27 | 1.607886 | 146 | 367.6786 | 366.08 | 1.59858 |
| 126 | 402.2924 | 400.69 | 1.602431 | 148 | 363.8332 | 362.22 | 1.613237 |
| 127 | 400.6955 | 399.09 | 1.605454 | 150 | 359.8973 | 358.26 | 1.637317 |
| 128 | 399.0867 | 397.48 | 1.606662 | 152 | 355.8635 | 354.19 | 1.673456 |

There are some deviations between the calculation density and physical property density, the average difference is 1.597479, and the revised formula 3 is:

$$\rho_{\mathrm{liquid}} = AB^{-(1-T/T_c)^n} - 1.597479. \quad (4)$$

Natural gas in gas space remains saturated state could be treated as an ideal gas, according to the gas state equation: PV=nRT n=m/M

$$\rho_{\mathrm{gas}} = \frac{m}{V} = \frac{PM}{RT}. \quad (5)$$

P: unit of state pressure is pa, T is for thermodynamic temperature K, R is for gas constant 8.314J·mol-1·K-1, M is for methane molar mass 16.414g/mol.

### *3.2. The Relationship of LNG Pressure and Temperature*

When LNG liquid stored in tanks, the gas space turns into a saturated vapor state. Saturated vapor pressure and temperature have a certain relationship described by Antoine equation[2].

Five-parameter Antoine equation:

$$\log P = A + \frac{B}{T} + C\log T + DT + ET^2. \tag{6}$$

A B C E are physical constants, T is for thermodynamic temperature K, unit of P is mmHg.

A=14.6667, B=-5.7097*102, C=-3.3373, D=2.1999*10-9, E=1.3096*10-5.

Simplified equation: $\lg(750P) = A - \frac{B}{T+C}$

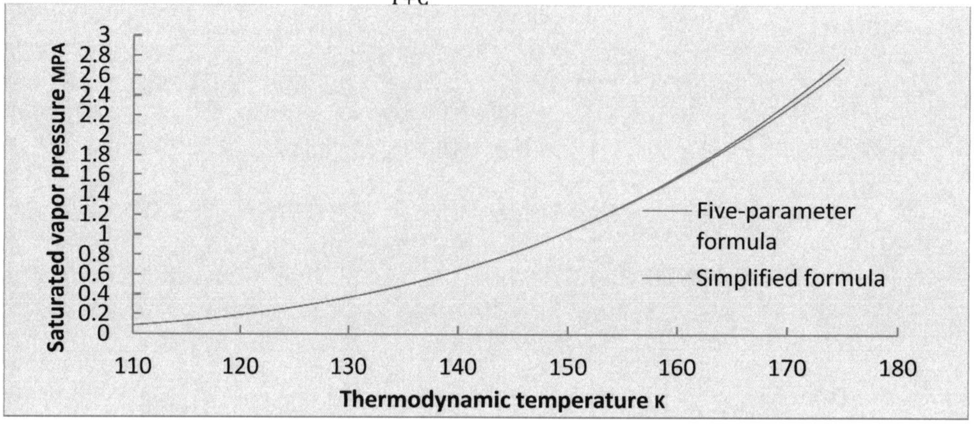


**Figure.6** Five-parameter Antoine equation and Simplified equation

From the change rules of temperature range of two equations in the coordinate system from 110K to 150K, there have no great difference. The relationship could be described by simplified equation which can be concluded that:

$$T = \frac{B}{A - \lg(750P)} - C\ . \tag{7}$$

A B C are physical constants, T is for Celsius temperature°C, P is the indication of measuring instrument whose unit is bar.

A=6.69561, B=405.420,C=267.777.

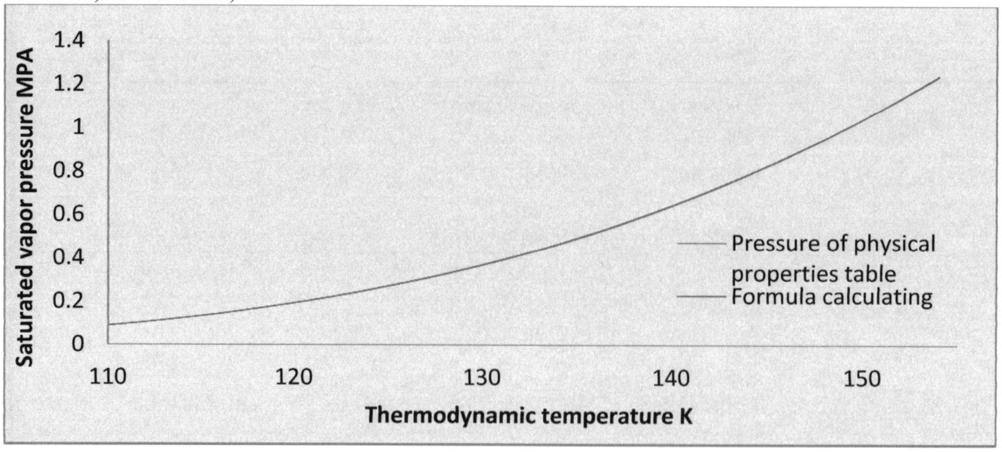


**Figure.7** Relations between saturation pressure and temperature calculated by physical properties actual data and simplified equation

From the results indicated in Figure 6 and the physical properties, the results almost the same. The calculation result from Formula 7 could be treated as an important parameter basis to calculate for different state of LNG.

IOP Conf. Series: Earth and Environmental Science **111** (2018) 012030 doi:10.1088/1755-1315/111/1/012030

### *3.3. The Relationship Between Liquid Level and Differential Pressure of LNG Tanks*

LNG tanks generally use differential pressure level meter to measure differential pressure, the volume or quality of LNG is confirmed through the liquid level comparison table provided by manufacturers(As shown in Table 2). Differential pressure level meter is an instrument which calculates the liquid level (pressure difference) by measuring the pressure difference from two different points in tanks. The comparison data is a calculation result of physical property parameters of water in a stationary state. However LNG constantly changes in tanks in the course of usage, the table is likely to produce a certain deviation in practice.

**Table.2.** Comparison table of tank liquid level in one manufacturer

3m³/1.1Mpa tank liquid level comparison table
Saturation liquid pressure 0bar

| Liquid indicator Water(mmWC) | Volume(L) | Liquid indicator Water (mmWC) | Volume(L) | Liquid indicator Water (mmWC) | Volume(L) |
|---|---|---|---|---|---|
| 0 | 0 | 120 | 471 | 300 | 1808.3 |
| 20 | 30.5 | 140 | 595 | 320 | 1916.2 |
| 40 | 87.6 | 160 | 727.2 | 340 | 2114.2 |
| 60 | 162.9 | 180 | 866 | 360 | 2213.5 |
| 80 | 253.3 | 200 | 1010.3 | 380 | 2406 |
| 100 | 356.6 | 220 | 1158.5 | 400 | 2493.2 |
| 120 | 471 | 240 | 1309.2 | 420 | 2672 |
| 140 | 595 | 260 | 1461.1 | 440 | 2744.2 |
| 160 | 727.2 | 280 | 1612.7 | 458 | 2844 |

Indicating value made by differential pressure level meter is the pressure difference produced by liquid level height H from two monitoring points. $P = \rho g H$

The unit of indicating value P is KPA, the liquid level H (unit m) is:

$$H = \frac{1000P}{\rho g} . \quad (8)$$

Combined with Formula 1 and Formula 4, tank pressure P and density ρ confirmed by Formula 4 after dynamic changes, a relatively accurate level height could be real-time calculated.

$$H = \frac{102P}{AB^{-(1-T/T_c)^n} - 1.597479} . \quad (9)$$

## 4. The Practical Application of Tanks for LNG-fueled Vessels

### *4.1. The Actual Storage Capacity of LNG Tanks*

Real-time sensors monitoring tank pressure and liquid level are generally equipped in LNG tanks. However, due to the structure features of tanks and temperature sensors, the actual storage temperature of LNG inner tanks could not be effectively monitored in time. While the storage temperature could be speculated according to tank pressure using the thermodynamic formulas as stated earlier (assuming the tank in vapor-liquid equilibrium saturation condition), and the density of natural gas from liquid to gas in different pressure condition could be confirmed. Parameters in different pressure conditions could be further calculated.

When length is L, tank pressure of which radius is R is P, the differential pressure is p, all parameters as follows:

Quality of natural gas in tanks: $M = M_{liquid} + M_{gas} = \rho_{liquid} V_{liquid} + \rho_{gas} V_{gas}$

The volume of natural gas in standard condition: $V_{standard} = \frac{M}{\rho_{standard}}$ $\rho_{standard}$=0.7167kg/m³

Calculating the actual storage capacity of LNG tanks and converting into gas volume in standard condition provide a data basis for calculating the energy consumption, instead of relying on flow meter to calculate gas consumption.

### *4.2. Calculation of Parameters in Practice*

Loading Limit (LL) and Filling Limit (FL) are introduced in the IGF Code. LL is the ratio of the maximum permissible liquid volume to the fuel tank loadable capacity. FL is the ratio of the maximum liquid volume in the fuel tank to the total fuel tank volume when the liquid fuel reaches the reference temperature. The two concepts raise strict requirements on the limits of practical use of storage tanks. Due to the variation conditions of LNG and the difference of filling gas sources, parameters as mentioned before will change to some extent, which requires a clear formula to provide a reference for monitoring system and energy management system of tanks.

The reference temperature as mentioned above is the corresponding temperature of fuel steam pressure under pressure relief valve setting pressure. The pressure relief setting of Type-C tank should not exceed 1.2MPA for inland LNG-fueled vessels. The corresponding temperature in pressure saturation condition according to physical properties and formula mentioned above is -120°C, the density of LNG is 353.8kg/m³. IGF Code and other domestic rules ask for 98% and 95% on Filling Limit respectively. The article adopts 98% FL under the consideration of the practical usage of Type-C tanks. Therefore the actual LL is confirmed as follows:

$$LL = FL\frac{\rho_R}{\rho_L} = \frac{347}{\rho_L} \ . \tag{10}$$

$\rho_R$ is the relative density of fuel under the reference temperature, that is 353.8 kg / m³. $\rho_L$ is the relative density of fuel under loading temperature, that is the density of LNG filing tanks which could be required from the filling party or confirmed according to the Formula 4 and 7 mentioned above.

## 5. Conclusion

LNG-fueled vessels have been widely employed in China and the technical level of hardware equipment is sufficient to meet the requirements of existing domestic vessels. However, unlike other traditional vessels using diesel as the shipping fuel, LNG-fueled vessels make stricter requirements on security and monitoring aspects due to its characteristics. The calculation of parameters mentioned in the article provides a reference for real-time monitoring of LNG tanks and the entire bunkering and energy management system. However, LNG is a compound with various components, parameters and formula involved in the article on the basis of methane physical properties. A deviation may occur in practice which requires to adjust promptly.

## 6. Reference

[1] Yanling Wang, Ming Li. Establishment of Corresponding Relationship between Level and Volume of Horizontal Tank with Elliptical Head. Journal of qiqihar university, March 2002;
[2] Fan Yang. Measurement method of liquid level in LNG tank[J]. Tianjin chemical. 2014.4.27(4):54-55.
[3] Weiming Zhou, Rongshun Wang, Yumei Shi. Advances and Prospects of Transportable Pressure Vessel Standards and Codes[J]. Pressure Vessel Technology2004.20(1):1-7.
[4] Huasheng Zou, Chunlai Huang. Analysis of rolling phenomena in the storage and transportation of liquefied natural gas[J]. oil and gas storage and transportation. 2006,25(11):13-15.
[5] Rui Yang. Risk analysis of liquefied natural gas and its safety protection scheme[J]. chemical management, 2018.8:57.
[6] Ruozhu Liu. Research on risk assessment and control for LNG storage system[D]. East of China university of science and technology.2011.
[7] Guobang Chen, Rui Bao, Yonghua Huang. Cryogenic engineering technology (data volume). Chemical industry press, 2005.12

# Study on Influence factors of Deposition Efficiency of Particles in Airflow Channel of Wall Absorbing Haze Automatically

**Chunzhen Qiao[1, a], Xiang Yuan[1, b], Zhen Zhang[1, c], Zhiyong Li[1, d]**

[1]School of Civil Engineering, North China University Beijing, China
E-mail: [a]joeczl221@sina.com, [b]yuanxiang_2007@163.com, [c]13661172615@163.com, [d]lizhiyongnbe@163.com

**Abstract.** The effect of velocity, channel's width, particle size and concentration on particles' deposition efficiency were simulated and analyzed by Fluent. The results show those particles' deposition efficiency increases with the increase of air velocity. When the air velocity is greater than zero point five meters per second, the deposition efficiency begins to decrease slightly. The deposition efficiency of the particles increases with the increase of the airflow channel width when the airflow channel size is reasonable. The particles' deposition efficiency increases with the increase of the concentration. When the particle's concentration reaches 250μg/m³, the deposition efficiency no longer changes.

## 1. Introduction

In recent years, most parts of our country have suffered from severe fog and haze. At the same time, the indoor air's quality is not optimistic [1]. A large number of inhalable particles enter the room and directly endanger the respiratory system of the human body. Can indoor particles be effectively and timely removed has become an important problem affecting people's health. Due to the complexity and long-term nature of China's economic restructuring, the haze weather in China is difficult to cure from the source in the short term. So making the building posse the function absorbing haze has become more and more urgent [2].

At present, the purification of indoor air is mostly using air conditioning air supply system or placing haze removal equipment in the building indoor. It is rare to use building envelope to absorb haze. Different traditional airtight wall, there are airflow channels in wall absorbing haze automatically on the premise of ensuring the basic load-bearing function. And the adsorption material plates are arranged on both sides of airflow channels. In this paper, a new type of wall based on industrialized building is presented. The function absorbing haze is integrated into the wall, and it utilizes the way of absorbing to remove inhalable particles with the help of the characteristic that the contact area between the wall and indoor air is large. The wall does not occupy the effective usable area of the building, and its adsorption area is large, so the absorbing haze efficiency can be greatly improved. Its diagrammatic sketch is shown in the Fig.1-1 [3].

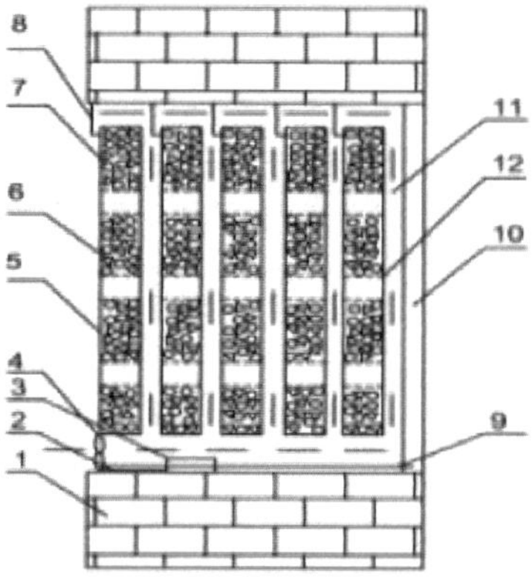


**Figure.1-1.** Diagrammatic sketch of structure of wall absorbing haze automatically
1—Wall, 2—Wire, 3—Battery, 4—Fan, 5—Metal mesh, 6—Adsorption material, 7—Phase Change material, 8—Bracket, 9—Wire, 10—Solar panel, 11—Airflow channel, 12—Hardware cloth.

Part of the particles [4] mixed in the airflow is absorbed by the absorption material in the metal mesh, and the part deposited in the channel after the airflow entered the wall. Therefore, the indoor air is purified. The deposition amount of particles in the airflow channel is mainly determined by gravity and inertial impact [5]. Gravity is derived from the mass size of particles, so the gravity of large particles is larger than that of small particles. However, most of the particles in the indoor air are PM2.5, and the effect of particle size on the deposition efficiency is negligible.

In general, the greater the air velocity is, the easier the particles collide and attach to the internal face of the airflow channel. At the same time, the collisions between particles and particles are also more frequent [6], and then the amount of particles deposited in the airflow channel is increasing. In addition, the airflow channel width and particles concentration also have a great influence on the deposition efficiency. In this paper, the influence factors of the deposition efficiency of particles in the channel are discussed based on the airflow structure and air purification performance of the wall absorbing haze automatically.

## 2. Relevant Theories and Wall Model

There are two kinds of collisions when particles move in the airflow channels, collisions between particles and collisions between particles and inner wall of the channels. Usually when the density of particles is sparse, there are almost no collisions between particles and particles. By comparing the magnitude of residual kinetic energy of particles and the attractive force of the internal wall surface, we can determine whether the particles will deposit in the channel after the particles collide with the surface of the channel. If the residual kinetic energy of the particles is greater than the attractive force given by the wall, the particles fly away from the wall. If not, the particles will deposit in the channel.

If the particles size is small and the continuum assumption is no longer applicable to the air flow near the particles, it is necessary to modify the drag force based on the continuous medium condition, as indicated in equation (1).

$$\Phi o=\Phi/X. \qquad (1)$$

In the above equation, Φois the modified drag coefficient, and X is slip correction factor.

According to RANS theory, the velocity of air flow can be expressed as a sum of time average velocity and a pulsating component. The eddy viscosity model of RNG k- epsilon turbulence is expressed as turbulent flow energy k and turbulent flow energy dissipation rate e, and they are solved with momentum equations. We assume that the turbulence of the airflow is in the same.

The deposition efficiency of particles in the airflow channel represents the ratio of the number of particles deposited in the channel and the total number of particles passed through the channels; it can be calculated by the following equation (2).

$$E=N/No. \qquad (2)$$

N represents the number of particles deposited in the airflow channel, and No represents the total number of particles passed through the channel.
A three-dimensional wall model [7] is established by using Fluent, and the calculation area is meshed in the form of Hex-Map. The wall model is shown in Fig. 2-1.

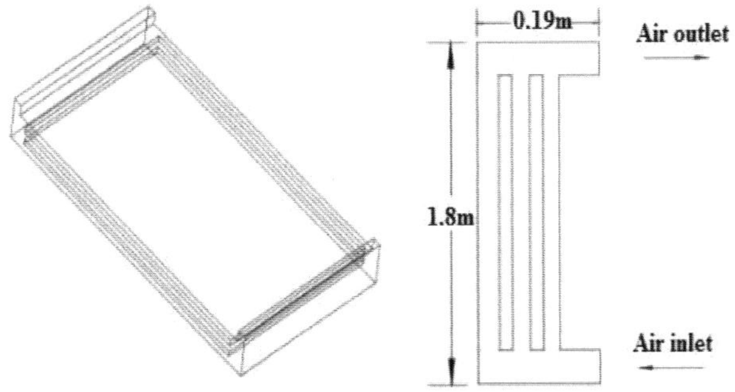


**Figure.2-1.** Diagrammatic sketch of the model of wall absorbing haze automatically
A particle trajectory analysis model based on Lagrange method is established, we comprehensively consider the effect of Drag force, and Gravity, Brown diffusion force, Thermophoretic force and Saffman lift on particles [8]. Hard sphere model is applied to collision process between particles and wall, and RNG k-ε turbulence eddy viscosity model is adopted for air flow. Momentum and turbulence equations are dispersed by the two order upwind scheme, and pressure and velocity are solved by SIMPLE coupling algorithm [9].
Firstly, the physical quantities involved in the model are iterated, and the number of iterations is 100. There are three types of boundary conditions in the wall model, velocity inlet, outflow and wall. These three kinds of boundary conditions involve velocity, pressure and density in XYZ directions. After the iteration, the residual curve is relatively stable. It shows that the numerical convergence result is reasonable and can carry out following simulation [10].

## 3. Influence factors of Deposition Efficiency

### *3.1. Air velocity's effect*

By analyzing the velocity change diagram of air in wall absorbing haze automatically, we can found that the deposition amount of particles in the airflow channel has a direct impact on the change of air velocity. This effect is mutual, so the air velocity also affects the deposition efficiency of particles [11].

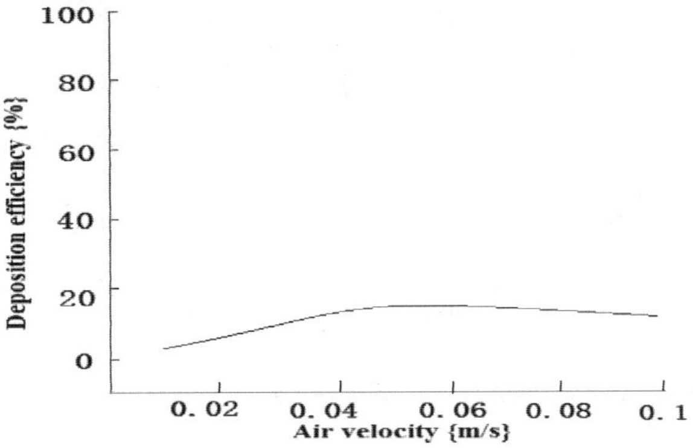


**Figure.3-1.** Variation of deposition efficiency of PM2.5 with the air velocity
It can be seen from Fig.3-1 that the deposition efficiency of particles (PM2.5) increases with the increase of air velocity. When the air velocity is greater than 0.5m/s, the deposition efficiency begins to decrease slightly. The reason for this change is that the larger the particle velocity is, the easier it is to

collide with the internal face of the channel. When the particle velocity is less than 0.05m/s, the initial kinetic energy is not large. The residual kinetic energy of the particles is less than the attractive force given by the wall after the particles collide with the surface of the channel, so the particles deposit in the channels and the amount of deposition is mainly affected by inertial impact. When the air velocity is greater than 0.05m/s, the probability of collision between the particles and the internal face of the channel increases, but the initial kinetic energy of the particles also increases. The residual kinetic energy of the particles is greater than the attractive force given by the wall, so the particles fly away from the surface of the channel, and the amount of deposition is mainly affected by gravity.

### *3.2. The effects of the airflow channel width and particle size*

The velocity of airflow at the entrance is set as 0.05m/s, and the particle density is set as 200μg/m³ [12] and the range of the particle size is 0.1μm~15μm. To simulate the deposition efficiency of the particles in the channel with width 20mm, 30mm, 40mm, 50mm and 60mm.

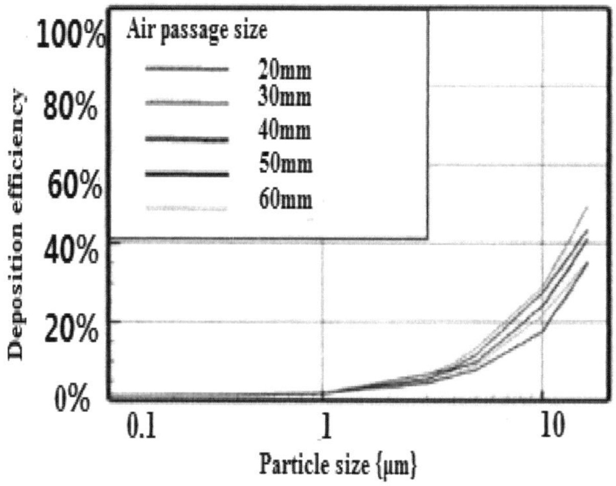


**Figure.3-2.** Deposition efficiency of particles with different size in airflow channel with different width

It can be seen from Fig.3-2 that the deposition efficiency of the particles that the size is less than 1μm is almost zero. This is because the mass of the particles is too small, and the effects of gravity and inertial impact are extremely limited. These particles almost do not deposit in the airflow channel. The deposition efficiency of the particles that the size ranges from 1μm to 5μm in the airflow channel with different widths is hardly different. The particles in indoor air are mainly composed of PM2.5, so the effect of particle size on deposition efficiency can be neglected. In this paper, the size of the wall model is based on the size of the wall of ordinary residential buildings, and its airflow channels' layout must ensure the basic load-bearing function of the wall. The six set of channel size parameters set in the simulation are in reasonable range. To analysis Fig. 4, we can found that the deposition efficiency of the particles increases with the increase of the airflow channel width when the airflow channel size is reasonable. This change is because the particles' Drag force and Brown diffusion force are relatively large in the wider airflow channel. To a certain extent, it effectively hinders the continuous movement of particles, so particles are easy to deposit.

### *3.3. The effect of the particles concentration*

Setting air velocity as 0.05m/s, then we simulate the deposition efficiency of PM2.5 that its concentration ranges from 100μg/m³ to 300μg/m³ in the airflow channel of 60mm width.

It can be seen from Fig.3-3 that the deposition efficiency of particles (PM2.5) increases with the increase of the concentration. When the particles concentration reaches 250μg/m³, the change of particles' deposition efficiency tends to be stable, and it is no longer affected by concentration. When the particle concentration is large, collisions are easy to occur between particles. Collision causes an increase in particles' internal energy, and then Thermophoretic force gets larger. Therefore, the effect of gravity on particle deposition is greatly weakened by Thermophoretic force. In this state, the particles are mainly attached to the airflow channel surface with the help of inertial impact.

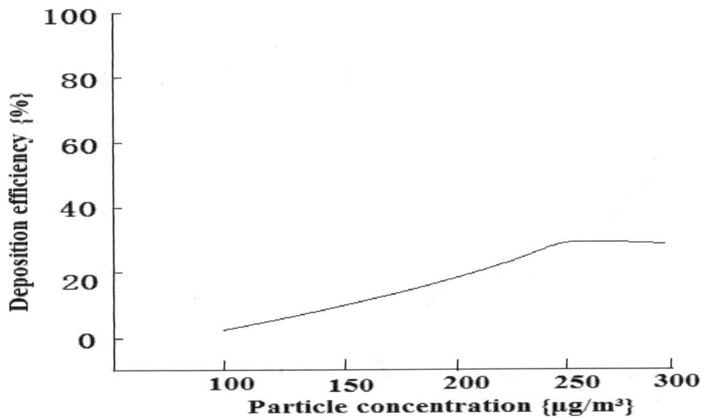


**Figure.3-3.** The change of PM2.5 deposition efficiency with its concentration

## 4. Conclusions

Different from traditional method removing haze, using building envelope to remove haze has the characteristics of energy saving, environmental protection and good economy. As a new type of absorbing haze, wall absorbing haze automatically is of great significance for improving the quality of building indoor air. Particles deposit in the airflow channels of wall absorbing haze automatically, which is part of the air purification process. Therefore, several factors affecting the particles' deposition efficiency were simulated and analyzed by Fluent. The results are as follows:

(1) The deposition efficiency of particles increases with the increase of air velocity. When the air velocity is greater than 0.5m/s, the deposition efficiency begins to decrease slightly.

(2) The deposition efficiency of the particles that the size is less than 1μm is almost zero. For indoor air, the effect of particle size on the deposition efficiency can be neglected.

(3) The deposition efficiency of the particles increases with the increase of the airflow channel width when the airflow channel size is reasonable.

(4) The deposition efficiency of particles increases with the increase of the concentration. When the particles concentration reaches 250μg/m$^3$, the change of particles' deposition efficiency tends to be stable, and it is no longer affected by concentration.

## 5. Acknowledgment

This work was financially supported by the Great Wall scholars reserve project, North China University of Technology (XN018018).

## 6. References

[1] ZHANG Yinping, DENG Qihong and QIAN Hua, "Progress report on indoor environment and health in China 2012," China Building Industry Press, 2012.
[2] YAO Li and XU Kuan, "The Pollution Situation and Countermeasures of Building Indoor Environment Under Haze-weather Condition," Journal of Green Science and Technology. vol. 8, 2016, pp49-51.
[3] ZHANG Zhen, "Airflow organization's simulation analysis of wall absorbing haze automatically," unpublished.
[4] LIU Zhirong, "Classification and nomenclature of atmospheric particulates," China Terminology. vol. 2, 2013, pp31-34.

[5] ZHANG Jie, "Characteristics study of aerosol particles settling in an air conditioning system," unpublished.
[6] FANG Tianyu, "Modeling of Airborne Particle Filtration in a Breathing Wal," unpublished.
[7] WANG Chengyue and FANG Tianyu, "Study on Trombe's air filtration performance," Science and Technology Innovation Herald. vol. 7, 2015, pp1-1.
[8] QIAO Chunzhen, ZHANG Zhen, YUAN Xiang and LI Zhiyong," Airflow organization' simulation study of wall absorbing haze automatically based on industrial Architecture," Symposiu of Chinese society of environmental science's annual conference on science and technology in 2017,Xiamen, in press.
[9] Gao R and Li A, "Modeling deposition of particles in vertical square ventilation duct flows," Building and Environment. vol. 1, 2012, pp245-252.
[10] Hinds W, "Aerosol technology: properties, behavior and measurement of airborne particles," John Wiley & Sons, 2013.
[11] SUN Lei, "Study on Seismic Performance of the Self-insulation Structural Wall with Inner Frame," unpublished.
[12] WU Qian, "Study on the relationship between indoor air flow rate and human comfort and physiological stress," unpublished.

IOP Conf. Series: Earth and Environmental Science **111** (2018) 012032 doi:10.1088/1755-1315/111/1/012032

# 2-DE Compared with iTRAQ-based Proteomic Analysis of the Functional Regulation of Proteins in Rhodococcus sp. BAP-1 Response to Fluoranthene

**Jie Xu[1,a], Hongqi Wang[1,b*], Dekang Kong[1]**

[1]Beijing Normal University, College of water sciences, Beijing, China
E-mail: [a]742615244@qq.com, [b*]whongqi@126.com

**Abstract.** Although the degradation pathways of Polycyclic aromatic hydrocarbons (PAHs) have been extensively studied in many bacteria, the variations in the expression levels of the key functional regulation of proteins during catabolism are still not quantitatively understood. In this study, we compared two proteomic methods, that one is two-dimensional gel electrophoresis (2-DE), a traditional widely used way and the other is isobaric tags for relative and absolute quantization (iTRAQ), an innovative approach, in order to analyze the functional regulation at the protein level in high effective fluoranthene-degrading bacteria named *Rhodococcus* sp. BAP-1. The number of differentially expressed proteins identified using iTRAQ is much larger than employing 2-DE. Response to fluoranthene, the key over expressed proteins in BAP-1 were NADPH-dependent FMN reductase, 30S ribosomal protein S2, S-ribosylhomocysteinase, etc.; the significant down-regulated proteins were cytochrome ubiquinol oxidase subunit, NAD(P) transhydrogenase subunit alpha, 5-methyltetrahydropteroyltriglutamate-homocysteine methyltransferase, et al.

## 1. Introduction

In the past few years, with the rapid development of chemical industry and auto industry, oil, one of the most important non-renewable energy resources, has been widely used as a kind of important fossil energy. During oil exploitation and petroleum transportation, oil spill occurs inevitably and makes serious damage for soil, water, air and other natural resources. Polycyclic aromatic hydrocarbons (PAHs), which could lead to toxicity, mutagenicity and carcinogenicity to human being, are primary pollution components among oil. The principle processes for their successful removal is generally considered microbial remediation. In our previous study, Rhodococcus sp. BAP-1 had a good degradation capacity in the presence of 20 mg/L of fluoranthene. Also, in Ahmed's research [1], Rhodococcus sp. CMGCZ was found to be capable to degrade high concentration of fluoranthene. Although the degradation pathways of PAHs have been extensively studied in many bacteria, the variations in the expression levels of the key functional regulation of proteins during catabolism are still not quantitatively understood.

Comparative proteomics, which detects the relative abundance of differentially expressed proteins under different conditions, focuses on up- or down-regulated proteins to analyze the characteristics of the expressed proteins among different samples under different conditions. Two-dimensional gel electrophoresis (2-DE), a traditional proteomic technology, separates thousands of proteins according to their isoelectric point by isoelectric focusing (IEF) in the first dimension and according to their molecular weight (MW) by sodium dodecyl sulphate polyacrylamide gel electrophoresis (SDS-PAGE) in the second dimension. Differential proteomics, which uses Isobaric tags for relative and absolute quantization (iTRAQ) to detect the abundance of differential expressed proteins, pay more attention to

IOP Conf. Series: Earth and Environmental Science **111** (2018) 012032 doi:10.1088/1755-1315/111/1/012032

up or down regulated proteins in order to analyze characteristics of expressed protein among different samples under different scenes [2].

## 2. Materials and methods

Rhodococcus sp. BAP-1 (GenBank entry: JX683682, CGMCC ID: 7.68) in duplicates were cultured and harvested at specific time points (1d and 3d). Protein extraction from two biological replicates was done consecutive day of sampling. Proteins were extracted from BAP-1 treated with fluoranthene using the Total Protein Extraction Kit (BestBio Company, China) following the manufacturer's protocol. After undergoing digestion, proteins were separated by 2-DE or iTRAQ, separately. Then, peptides experimented fractionation and liquid chromatography electrospray ionisation tandem mass spectrometry (LC-ESI-MS/MS) proteomic analysis.

## 3. Results

### *3.1. Differential protein plots by 2-DE*

Proteins were extracted from Rhodococcus sp.BAP-1 induced by fluoranthene (3mg/L) using the Total Protein Extraction Kit (BestBio Company, China). The differential proteins' expression of *Rhodococcus* sp. BAP-1 after the first and the third day was separated by 2-DE (Fig.3-1), and then selected 11 significantly differential proteins pots were analyzed by matrix-assisted laser desorption/ionization time-of-flight mass spectrometry (MALDI-TOF/MS) (Fig.3-2)

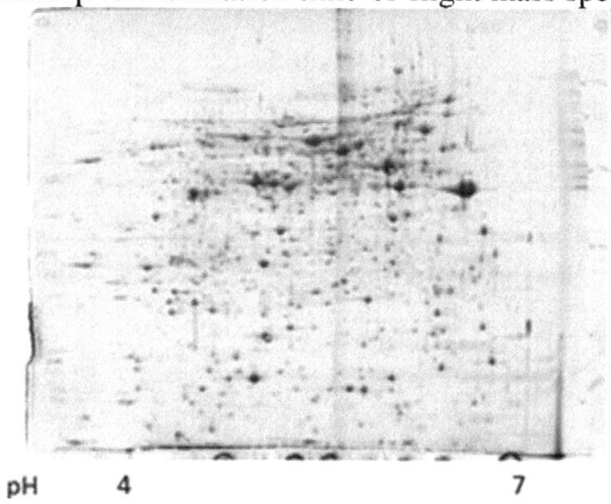


**Figure.3-1.** Two-dimensional electrophoretic map of *Rhodococcus* sp. BAP-1 induced by fluoranthene on the first day (a)

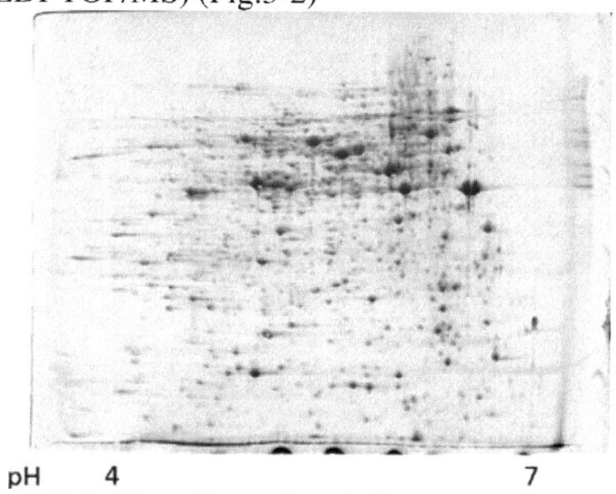


**Figure.3-2.** Two-dimensional electrophoretic map of Rhodococcus sp. BAP-1 induced by fluoranthene on the third day(b)

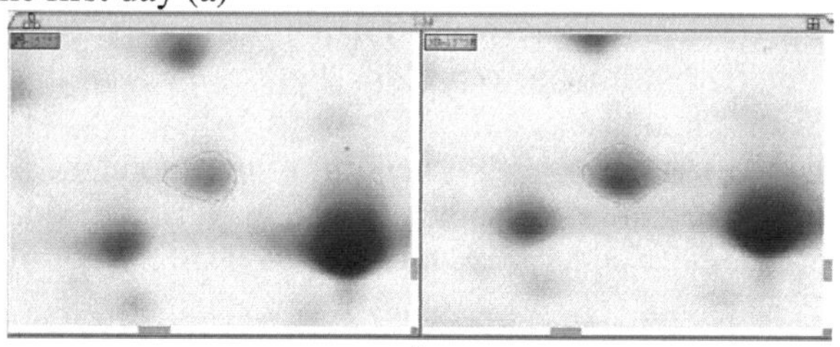

8

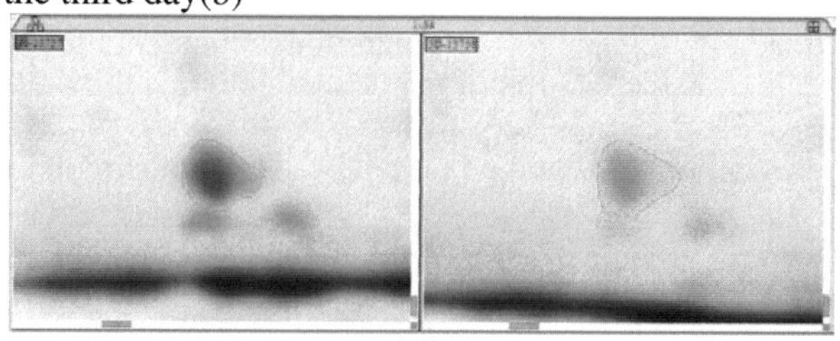

117

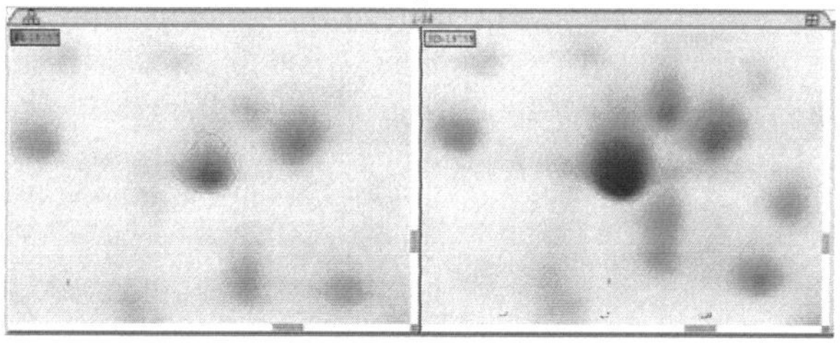

189

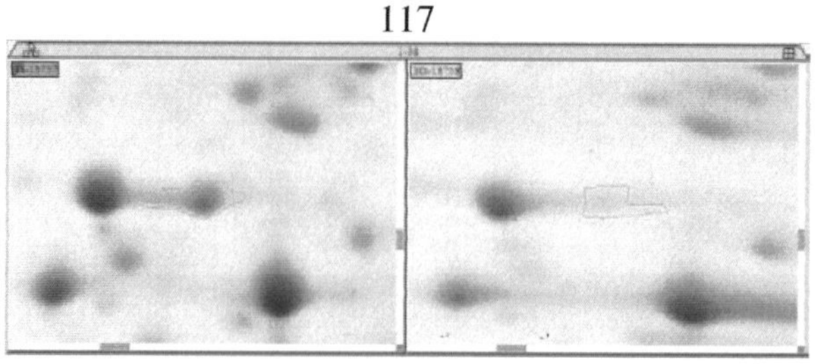

306

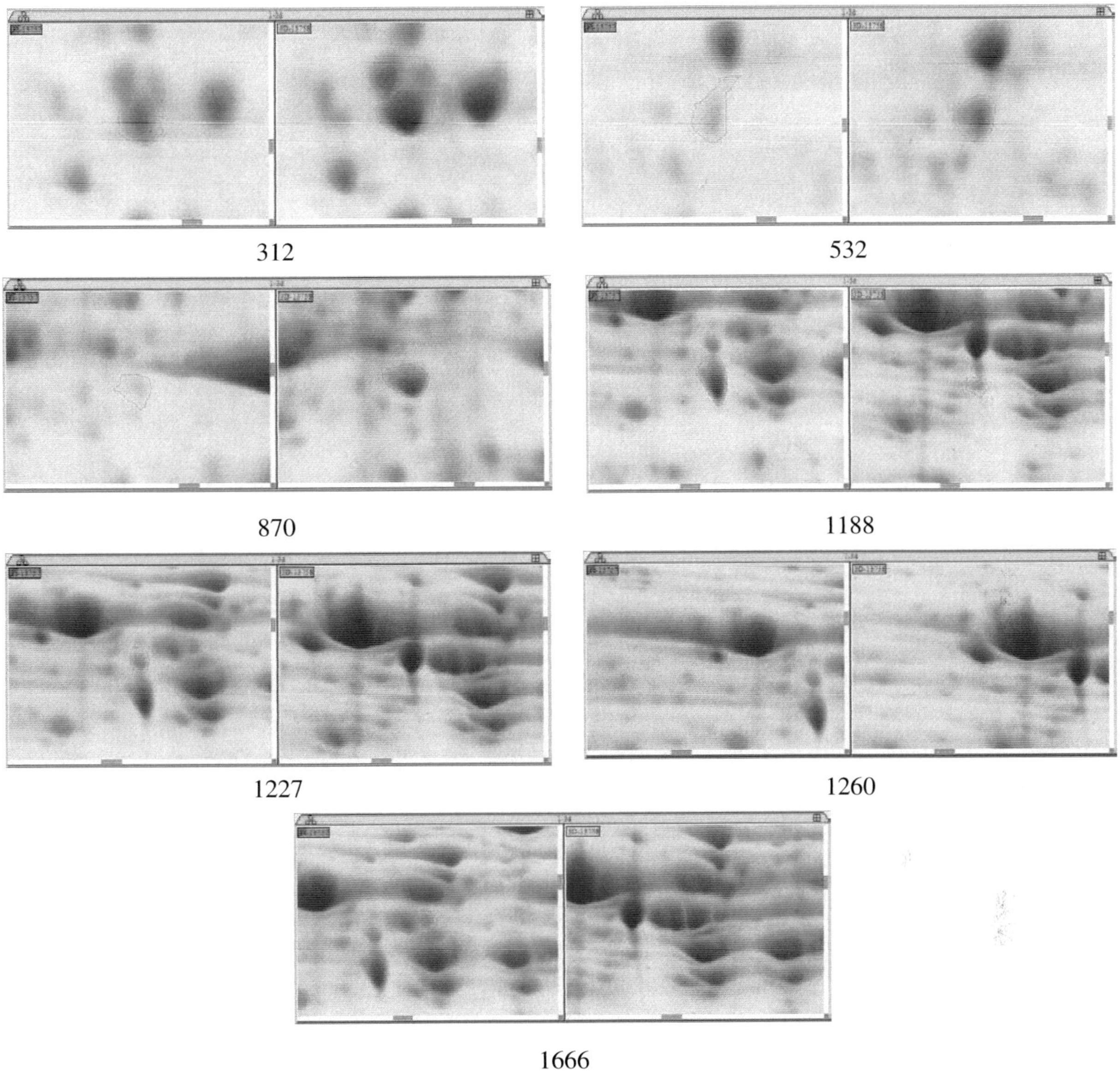


**Figure.3.3.** significantly differential protein pots in the 2-DE gel map

*3.2. Differential protein plots by iTRAQ*

iTRAQ labels 113,114,115 and 116 have been used to label fluoranthene-exposed time of 1d', 1d'', 3d' and 3d'', respectively. We set 115/113 (3d'/1d') and 116/114 (3d''/1d'') as comparison clusters and then took the average as 3d/1d. We filtered the results by applying the statistical criteria of only protein a fold-change higher than 50% (ratio of either <0.66 or >1.5, p<=0.05) with at least two unique peptides in all two replicate experiments as the significantly altered relative levels. Then, we were able to detect 591 proteins, including 400 up-regulated and 191 down-regulated proteins. The number of proteins with increased levels were much more than proteins with decreased levels, demonstrating that Rhodococcus sp. BAP-1 exposure to fluoranthene preferred to induce more new or low-expressed proteins compared with before in order to adapt to new situation, rather than restrain or weaken original proteins which play the role of sustaining life activities.

**Table.3-1.** The Top10 up-regulated Differentially Proteins in BAP-1 Induced by Fluoranthene

| **Protein ID** | **Description** | **Ratio (3d/1d)** |
|---|---|---|
| Unigene50806_All | NADPH-dependent FMN reductase | 10.00 |
| Unigene14485_All | 30S ribosomal protein S2 | 10.00 |
| CL2776.Contig3_All | S-ribosylhomocysteinase | 10.00 |
| Unigene15123_All | iron ABC transporter ATP-binding protein | 10.00 |
| Unigene44773_All | Nucleoside-diphosphate kinase | 10.00 |
| Unigene20021_All | ATP synthase F1 subunit alpha | 10.00 |
| CL302.Contig6_All | xylose isomerase domain-containing protein | 10.00 |
| Unigene3703_All | DinB family protein | 9.87 |
| CL983.Contig2_All | 348 3212 phosphopyruvate hydratase | 9.82 |
| CL1764.Contig4_All | Glycine hydroxymethyltransferase | 9.66 |
| CL199.Contig4_All | isocitrate dehydrogenase NADP-dependent | 9.58 |

**Table.3-2.**The Top10 Down-regulated Differentially Proteins in BAP-1 Induced by Fluoranthene

| **Protein ID** | **Description** | **Ratio (3d/1d)** |
|---|---|---|
| CL344.Contig11_All | cytochrome ubiquinol oxidase subunit | 0.14 |
| Unigene15264_All | NAD(P) transhydrogenase subunit alpha | 0.19 |
| CL2066.Contig2_All | 5-methyltetrahydropteroyltriglutamate-homocysteine methyltransferase | 0.19 |
| Unigene9285_All | formate acetyltransferase | 0.19 |
| Unigene9332_All | quinoprotein glucose dehydrogenase | 0.20 |
| Unigene21188_All | aldehyde dehydrogenase | 0.20 |
| Unigene3186_All | ribonucleotide-diphosphate reductase subunit alpha | 0.21 |
| Unigene18249_All | acyl-CoA dehydrogenase | 0.21 |
| Unigene9514_All | glycerol-3-phosphate dehydrogenase | 0.22 |
| Unigene24119_All | pyruvate decarboxylas | 0.24 |

We separately selected the top10 changed (up- or down-) expressed proteins for further analysis (Table.3-1 and Table .3-2).

Dehydrogenase plays an important role in the oxidation of n-alkane and ensured an adequate metabolic flux into the downstream catabolic pathway [3, 4]. It is possible that the bacteria increased the expression of isocitrate dehydrogenase and decreased quinoprotein glucose dehydrogenase, aldehyde dehydrogenase, acyl-CoA dehydrogenase and glycerol-3-phosphate dehydrogenase to provide energy in order to enhance its survival in a restricted environment. Substrate-binding protein (SBP)-dependent transporters are extensively present in bacteria and are involved in diverse processes, such as nutrient uptake, quorum sensing and multidrug resistance, which were considered as the primary choices for high-affinity uptake of nutrients[5, 6]. Iron ABC transporter ATP-binding proteins were probably the members responsible for specific transportation systems for fluoranthene or other PAHs. Ribosomal proteins (RPs) are significant in development and gene expression [7, 8]. The expression of RP genes are largely regulated by the environment and cell's development. RPs have been found to be involved in DNA repair, cell development, protein biosynthesis, regulation, and cell differentiation [9]. The up-regulation of 30S ribosomal protcin S2 hclps thc bacteria adapting to restricted PAHs conditions, reduce toxicity, and maintain the normal metabolism of cells.

### *3.3. Comparison between 2-DE and iTRAQ.*
In this study, the number of differentially expressed proteins identified using iTRAQ is much larger than employing 2-DE. Owing to abundant data, we could launch functional analysis based on Clusters of Orthologous Groups (COG) categories, the gene ontology (GO) classification system and the Kyoto Encyclopedia of Genes and Genomes (KEGG) database [10].

## 4. Conclusions
As always, 2-DE, which can provide visual glue map, is a conventional proteomic method. iTRAQ has been increasingly applied in proteomic studies due to the advantages of this methodology over 2D gel electrophoresis. There are several reasons accountable for the superiority of iTRAQ compared with 2-DE. First, iTRAQ allows the analysis of multiple samples (up to eight) simultaneously, whereas 2-DE allows only one sample at a time. iTRAQ greatly improves the processing efficiency as proteins can be extracted from multiple biological replicates and analyzed in a single experiment. The ease of use and its multiplexing capabilities make it an effective approach in many scientific fields [11]. In addition, with iTRAQ labelling, proteins can be easily identified by tandem mass spectrometry automatically, unlike 2-DE, wherein protein spots of interest must be excised from two or three replicated gels depending on the image analysis software being used. Additionally, iTRAQ can identify more proteins, even proteins with low copy number, membrane proteins, or other proteins that are difficult to detect by 2-DE. In the future, combining proteomics with genomics, transcriptome and metabolomics is main directions for our research.

## 5. Acknowledgment
This paper was sponsored by the National Natural Science Foundation of China (No. 41372232, No.41772234).

## 6. References
[1] R.Z. Ahmed, N. Ahmed. "Isolation of *Rhodococcu*s sp. CMGCZ capable to degrade high voncentration of fluoranthene" , Water Air Soil Pollut, 227 (2016).
[2] J. Han, P. Gao, S. Zhao, X. Bie, Z. Lu, C. Zhang, F. Lv, "iTRAQ-based proteomic analysis of LI-F type peptides produced by *Paenibacillus polymyxa* JSa-9 mode of action against *Bacillus cereus*", J Proteomics, 150 (2017), pp. 130-140.
[3] J.B. van Beilen, S. Panke, S. Lucchini, A.G. Franchini, M. Rothlisberger, B. Witholt, "Analysis of *Pseudomonas putida* alkane-degradation gene clusters and flanking insertion sequences: evolution and regulation of the alk genes", Microbiology, 147 (2001), pp.1621-1630.
[4] H. Liu, W.B. Sun, R.B. Liang, L. Huang, J.L. Hou, J.H. Liu, "iTRAQ-based quantitative proteomic analysis of *Pseudomonas aeruginosa* SJTD-1: A global response to n-octadecane induced stress", J Proteomics, 123 (2015), pp.14-28.
[5] A.H.F. Hosie, D. Allaway, C.S. Galloway, H.A. Dunsby, P.S. Poole, "*Rhizobium leguminosarum* Has a Second General Amino Acid Permease with Unusually Broad Substrate Specificity and High Similarity to Branched-Chain Amino Acid Transporters (Bra/LIV) of the ABC Family", Journal of Bacteriology, 184 (2002), pp.4071-4080.
[6] G.H. Thomas, "Homes for the orphans: utilization of multiple substrate-binding proteins by ABC transporters", Mol Microbiol, 75 (2010), pp.6-9.
[7] M.E. Byrne, "A role for the ribosome in development", Trends Plant Sci, 14 (2009), pp.512-519.

[8] N. Kondrashov, A. Pusic, C.R. Stumpf, K. Shimizu, A.C. Hsieh, S. Xue, J. Ishijima, T. Shiroishi, M. Barna, "Ribosome-mediated specificity in Hox mRNA translation and vertebrate tissue patterning", Cell, 145 (2011), pp.383-397.
[9] S.C. Chantha, B.S. Emerald, D.P. Matton, "Characterization of the plant Notchless homolog, a WD repeat protein involved in seed development", Plant Mol Biol, 62 (2006), pp.897-912.
[10] R.N Carvalho, T.Lettier. "Proteomic analysis of the marine diatom. *Thalassiosira pseudonana* upon exposure to benzo(a)pyrene", Carvalho and Lettieri BMC Genomics,12 (2012).
[11] J. Noirel, C. Evans, M. Salim, J. Mukherjee, S.Y. Ow, J. Pandhal, T.K. Pham, C.A. Biggs, P.C. Wright, "Methods in Quantitative Proteomics: Setting iTRAQ on the Right Track", Curr Proteomics, 8 (2011) pp.17-30.

# Analysis on the Application and Characteristics of Chinese Patent Medicines Containing Dried Rehmanniae Radix

**Hui Guo [1, a], Yanyan Miao [1, b], mingsan Miao [1]**

[1]Henan University of traditional Chinese Medicine, Zhengzhou, 450046
E-mail: [a]guohuiygzh@163.com, [b]miaomingsan @163.com

**Abstract**. Dried Rehmanniae Radix has sweet taste, and its drug property is cold.It acts on heart, liver and kidney.It has the effect of clearing heat and cooling blood, nourishing yin and promoting fluid production.The active constituents of dried Rehmanniae Radix are mainly iridoid glycosides, polysaccharides, oligosaccharides and so on.This article sorted and analyzed the application forms, efficacy, applicable symptomsh and the frequency of the use of single traditional Chinese medicine in the Chinese patent medicines containing dried Rehmanniae Radix in the Chinese Pharmacopoeia of 2015.This method provides a train of thought for the further study of the pharmacological constituents of dried Rehmanniae Radix, and supplements the pharmacodynamics of Chinese herbal medicine of dried Rehmanniae Radix. It also provides ideas for the improvement of dried Rehmanniae Radix prescription and the new usage of its old prescription.

## 1. Introduction

Rehmannia is fresh or dry root of Scrophulariaceae plant. It's mainly produced in Henan, and its quality is better in Henan Province of Wenxian County, Qinyang, Wuzhi, and Boai County. In the fall to dig rehmannia and to remove the reed head, fibrous roots mud for fresh use; or to slowly bake to about 80% dry rehmannia root. The former was said the "fresh rehmannia", and the latter was called dried Rehmanniae Radix. Dried Rehmanniae Radix was first recorded in "The Shen Nong Ben Cao Jing" .It has sweet taste and its drug property is cold. It acts on heart, liver and kidney. It has the effect of qingreliangxue and yangyinshengjin[1].The current research on pharmacological action on dried Rehmanniae Radix showed that it has neuroprotective effect, reinforcement learning and memory ability, immune enhancement, inhibiting pulmonary fibrosis, anti-inflammatory and antipyretic and haemostatic, removing blood stasis and promoting blood circulation, blood glucose, anti-tumor, anti-oxidation, anti-aging, protecting the gastric mucosa, bacteriostasis, estrogen-like effect[2-5].In the application of dried Rehmanniae Radix, it is often mixed with other Chinese herbs to form a prescription for the treatment of some diseases. The traditional Chinese medicine is under the guidance of TCM theory to Chinese herbal medicine as raw materials, production in accordance with the provisions of the prescription, production technology and quality standard, the utility model has the advantages of convenient carrying, convenient use, and good safety. Different from the single target therapy of Western medicine, Chinese patent medicine has the characteristics of multiple components, multiple levels and multiple targets. Combined with the above advantages, Chinese patent medicines are more and more popular among the majority of people. This article sorted and analyzed the application forms, efficacy, applicable symptoms and the frequency of the use of single traditional Chinese medicine in the Chinese patent medicines containing dried Rehmanniae Radix in the Chinese Pharmacopoeia of 2015.In this way, we hope to promote the innovative research and clinical application of Chinese patent medicine containing dried Rehmanniae Radix in the field of traditional Chinese medicine.

## 2. The records of Chinese patent medicines containing dried Rehmanniae Radix in Chinese Pharmacopoeia

There are 152 Chinese patent medicines containing dried Rehmanniae Radix in Chinese Pharmacopoeia. According to the dosage form, these Chinese patent medicines are classified into pills, granules, tablets, mixture, capsules, ointments, liniment and aerosol, a total of 8 kinds, of which pills were the most, a total of 47 species. Details of the 8 formulations are shown in Table.2-1. Only 7 of the Chinese patent medicines containing dried Rehmanniae Radix can be used for external use, and the rest are all taken orally, indicating that dried Rehmanniae Radix is mostly used as part of a prescription for oral administration.

**Table.2-1.** The classification of the dosage forms of Chinese patent medicines containing dried Rehmanniae Radix

| the dosage forms | pills | granules | tablets | capsules | ointments | mixture | liniment | aerosol |
|---|---|---|---|---|---|---|---|---|
| quantity | 47 | 29 | 28 | 18 | 8 | 19 | 2 | 1 |
| Rate (%) | 30.9 | 19.1 | 18.4 | 11.8 | 5.3 | 12.5 | 1.3 | 0.7 |

## 3. Characteristic analysis of forms of dried Rehmanniae Radix producing Chinese patent medicines

In 152 Chinese patent medicines containing dried Rehmanniae Radix which is recorded in the Chinese Pharmacopoeia of 2015, xinyuan capsule and jingwanhong ointment belong to the secret prescription. In addition to them, dried is made of the other Chinese patent medicines by four forms of decoction, fine powder of crude medicine , alcohol extract and oil-fried thing. The statistical results are shown in Table.3-1. It is found that dried Rehmanniae Radix mainly participates in making Chinese patent medicine by the form of decoction. The present study shows that the main bioactive substances in dried Rehmanniae Radix are a water-soluble component, which is consistent with above rules. However there are three other forms of dried Rehmanniae Radix used as intermediate. Taking alcohol extract as an example, we can speculate that there may be some heat unstable components in dried Rehmanniae Radix which can exert some pharmacological activity. That would be the basis and direction for further study of pharmacological active ingredients in dried Rehmanniae Radix.

**Table.3-1.** The four forms by which dried Rehmanniae Radix is made of the other Chinese patent medicines

| | decoction | fine powder of crude medicine | alcohol extract | oil-fried thing |
|---|---|---|---|---|
| quantity | 92 | 10 | 45 | 3 |
| Rate (%) | 61.3 | 6.7 | 30 | 2 |

## 4. Characteristic analysis of functional classification of the Chinese patent medicines containing dried Rehmanniae Radix

According to the records of efficacy of qingreliangxue and yangyinshengjin about dried Rehmanniae Radix in the Chinese Pharmacopoeia of 2015,the 152 Chinese patent medicines containing dried Rehmanniae Radix are broadly classified. There are 43 Chinese patent medicines with the effect of qingre, accounting for 28.3%,and 14 Chinese patent medicines with the effect of liangxue, accounting for 9.2%,and 29 Chinese patent medicines with the effect of yangyin, accounting for 19.1%,and 9 Chinese patent medicines with the effect of shengjin, accounting for 5.9%.The Chinese patent medicine is characterized by multiple level and multiple target, and its clinical use of drugs not only play a single effect, but at the same time play two or more effects.Therefore,there are 12 Chinese patent medicines exerting qingreliangxue,11 Chinese patent medicines exerting qingreyangyin,4 Chinese patent medicines exerting qingreshengjin,3 Chinese patent medicines exerting liangxueyangyin,and 8 Chinese patent medicines exerting yangyinshengjin in the 152 Chinese patent medicines containing dried Rehmanniae Radix.In addition, related reports reported that dried Rehmanniae Radix has the modern pharmacological action of hematopoietic, suggesting that it also has the effect of yangxue, and there are 18 Chinese patent medicines with this effect, accounting for 11.8%.According to some books of traditional Chinese medicine recorded the effect of runchang of dried Rehmanniae Radix, there are 2 Chinese patent medicines with this effect, accounting for 1.3%.Be similar to prepared rehmannia root which can nourish the liver and kidney, dried may have the effect of nourishing the liver, kidney and heart for acting on the heart, liver and kidney. This provides the basis for further study on the modern pharmacological action of dried Rehmanniae Radix.Dried Rehmanniae Radix has the modern pharmacological action of homeostasis, which corresponds Chinese patent medicines containing dried Rehmanniae Radix, including zhihongchangpi pills, eye pills, zanglian pills and gongning granules. Some modern pharmacological actions of dried Rehmanniae Radix have not been applied in a prescription, which provides certain scientific basis for the new development of some old prescriptions. The above mentioned effects of Chinese patent medicines containing dried Rehmanniae Radix were summarized. The statistical results are shown in Table.4-1.

According to table.4-1. the Chinese patent medicines containing dried Rehmanniae Radix mainly have the function of qingre, yangyin and yingxue, which reveals the application rule of dried Rehmanniae Radix in prescription. In addition, dried Rehmanniae Radix may have the effect of nourishing heart, liver or kidney, which provides a direction for further research on the modern pharmacological action of it.

**Table.4-1.**Pharmacodynamic classification of Chinese patent medicines containing dried Rehmanniae Radix

| Efficacy | qingre | liangxue | yangyin | shengjin | yangxue | runchang | zhixue | nourishing the liver, kidney or heart |
|---|---|---|---|---|---|---|---|---|
| quantity | 43 | 14 | 29 | 9 | 18 | 2 | 4 | 32 |
| Rate (%) | 28.3 | 9.2 | 19.1 | 5.9 | 11.8 | 1.3 | 2.6 | 21.1 |

## 5. Characteristic analysis of using frequency of single herb in the Chinese patent medicines containing dried Rehmanniae Radix

The using frequency of single herb in the Chinese patent medicines is counted. The single herbs are used more than 30 times as the standard, the results are as follows. There are 64 Chinese patent medicines containing Chinese angelica,64 Chinese patent medicines containing Radix liquiritiae,54 Chinese patent medicines containing Radix Ophiopogonis,47 Chinese patent medicines containing Scutellaria baicalensis,38 Chinese patent medicines containing Ligusticum wallichii,48 Chinese patent medicines containing Radix Scrophulariae, 34 Chinese patent medicines containing Radix paeoniae Alba,34 Chinese patent medicines containing Paeoniae Radix and 32 Chinese patent medicines containing Radix platycodonis. The statistical results are shown in Table.5-1.

It can be seen from the Table.5-1 that these 8 kinds of Chinese herbs are used in 152 kinds of Chinese patent medicines containing dried Rehmanniae Radix. The use of frequency from high to low is Chinese angelica, Radix liquidities, Radix Ophiopogonis, Radix Scrophulariae, Ligusticum wallichii, Radix Paeoniae Alba, Paeoniae Radix, Radix platycodonis. It is suggested that dried Rehmanniae Radix is often combined with traditional Chinese medicine of buyi, qingre or huoxue. And it provides the basis for the improvement of the prescription of dried Rehmanniae Radix with buyi, qingre and huoxue effect. The compatibility rule of dried Rehmanniae Radix has also been reported in other literatures [6].

**Table.5-1.**The using frequency of single herb in the Chinese patent medicines containing dried Rehmanniae Radix

| single herb | Radix liquiritiae | Chinese angelica | Radix Ophiopogonis | Ligusticum wallichii | Rradix Scrophulariae | Radix paeoniae Alba | Paeoniae Radix | Radix platycodo-nis |
|---|---|---|---|---|---|---|---|---|
| The using frequency | 64 | 64 | 54 | 38 | 48 | 34 | 34 | 32 |

## 6. Conclusion

Dried Rehmanniae Radix is widely distributed in China, and it is one of the four Huai medicines in Henan. It is a commonly used Chinese herbal medicine. Some achievements have been made in the study of pharmacological action and active constituents of dried Rehmanniae Radix. At present, research shows that dried Rehmanniae Radix has neuroprotective effect, reinforcement learning and memory ability, immune enhancement, pharmacological action of inhibiting pulmonary fibrosis, anti-inflammatory and antipyretic and hemostatic, removing blood stasis and promoting blood circulation, blood glucose, anti-tumor, anti-oxidation, anti-aging, protect the gastric mucosa, antibacterial, estrogenic effects. The bioactive constituents of dried Rehmanniae Radix are iridoid glycosides, polysaccharides, oligosaccharides, etc. It has been preliminarily found that the active components of dried Rehmanniae Radix can be used as hemostatic agents, which are related to carbohydrates and iridoid glycosides [7]. In lowering blood sugar, the active site P-BP-F has been found to have significant hypoglycemic effect [4]. Catalpol, one of the effective components of dried Rehmanniae Radix, has the functions of lowering blood sugar, protecting nerve, protecting heart and resisting tumor [8]. Study on polysaccharides of dried Rehmanniae Radix oligosaccharides and the discovery of dried Rehmanniae Radix polysaccharides can strengthen immunity, anti fatigue, anti anxiety, treatment of obesity; diabetes induced bone marrow mesenchymal stem cells to differentiate into neuron like cells. Dried Rehmanniae Radix oligosaccharides can improve glucose metabolism disorder, protect liver, and improve the survival ability of adipose derived mesenchymal stem cells [9].

In this paper, statistics on the Chinese patent medicines containing dried Rehmanniae Radix in the 2015 edition of Chinese Pharmacopoeia are carried out. It is found that the administration mode of Chinese medicine containing dried Rehmanniae Radix is less than that of topical use, and summarized the characteristics of the 152 kinds of Chinese patent medicines containing dried Rehmanniae Radix in the application form and efficacy, and analyzed the characteristics of these two aspects. The results show that the main application form of dried Rehmanniae Radix in the preparation of Chinese patent medicine is water decoction, and the traditional Chinese medicine containing dried Rehmanniae Radix mainly played the role of qingre, yangyin and yangxue.Compared with the single function of traditional Chinese medicine rehmannia root function and the active ingredients and modern pharmacological effects. It is found that dried Rehmanniae Radix may have the effect of nourishing the liver and kidney deficiency. However, it does not reflect the modern pharmacological effects of dried Rehmanniae Radix. The former provides the direction for the further study of the modern pharmacological action of dried Rehmanniae Radix, and the latter provides the direction and thought for the new use of dried Rehmanniae Radix. This paper also sums up the 152 kinds of Chinese medicine containing Rehmannia single herb in the frequency of use, and analyzes the characteristics. The results show that dried Rehmanniae Radix is combined with Chinese medicines of buyi, qingre and huoxue, which provide the basis for giving full play to its efficacy and improving the related prescription.

## 7. Acknowledgement

Foundation project: national international cooperation base (2016-151), Henan industry university research project (142107000039)

## 8. References

[1] National pharmacopoeia commission. Pharmacopoeia of the People's Republic of China [M].

Beijing: China Medical Science and Technology Press, 2015.
[2] Zhao Runsheng, Zhang Yuan, Miao Xue Xue, Li Biao. Effects of Radix Rehmanniae on Hemorrheology in blood stasis model rats [J]. Chinese materia medica pharmacology and clinic, 2006, 22 (Z1): 123.
[3] Xu Hong, Song Changchun, Zhao Jingshan, Zhang Yixin, Miao Hui, Wei Cuiping, Zhou Hua. Influence on plasma endothelin in blood stasis model rats serum levels of Rehmannia born [J]. Hebei TCM journal, 2008, 23 (02): 5-6.
[4] Guo Lin, Miao Ming three. Chemistry, pharmacology and application characteristics of raw Rehmannia glutinosa [J]. Chinese Journal of traditional Chinese medicine, 2014, 29 (03): 375-377.
[5] Zheng Xiaoke, Liu Zhaoyan, Jiang Yun, Zhang Na, Feng Zhiyi, Wang Xiaolan, Feng Sucheng Fu, health. With the experimental study of [J]. Dihuang estrogenic activity screening of China Pharmaceutical Journal, 2013, 48 (21): 1831-1836.
[6] Li Weimin, Li Weihong. Study on compatibility rules of Rehmannia glutinosa in formula [J]. Chinese Journal of traditional Chinese medicine, 2011, 26 (03): 483-485.
[7] Wang Mei, Zhang Lijuan, Guo Dongyan. A preliminary study on [J]. Lishizhen medicine and Chinese medicine of Rehmannia glutinosa Libosch pharmacodynamic material basis, 2011, 22 (08): 1938-1939.
[8] Huang Chuanjun, Zhao Fangzheng, Zhang Caiqing. Shengdihuang effective component research progress in mechanisms of Catalpol pharmacological [J]. Shanghai Journal of traditional Chinese medicine, 2017, 51 (02): 93-97.
[9] Wang Junming, Guo Xiaona, Zhang Yueyue, Feng health, Cui Ying. Rehmannia polysaccharide and oligosaccharide [J]. China progress in pharmacological study of Journal of Gerontology, 2015, 35 (22): 6603-6605.

IOP Conf. Series: Earth and Environmental Science **111** (2018) 012034 doi:10.1088/1755-1315/111/1/012034

# Simultaneous Removal of Thallium and EDTA by Fenton Process

**Ruibing Xu[1,a], Xuexia Huang[2,b], Huosheng Li[3,c], Minhua Su[1,d*], Diyun Chen[1,e*]**

[1]Guangdong Provincial Key Laboratory of Radionuclides Pollution Control and Resources, Guangzhou University, Guangzhou, China
[2]School of Environmental Science and Engineering, Guangzhou University, Guangzhou, China
[3]Key Laboratory for Water Quality Security and Conservation of the Pearl River Delta, Guangzhou University, Guangzhou, China
E-mail: [a]rebeccahsu@foxmail.com, [b]huangxuexia66@163.com, [c]hilihuo@163.com, [d]mhsu@gzhu.edu.cn, [e]cdy@gzhu.edu.cn

***Abstract.*** The wastewater containing heavy metals and organic pollutants is widely discharged from industries. Because of the coexistence of heavy metals and organic pollutants, the treatment of such wastewater is very difficult. Fenton process is considered to be one of the most effective approaches for the degradation of organic pollutants in aqueous solution due to the strong oxidative ability of hydroxyl radical which generated from the Fenton process. Apart from this, heavy metals are able to be removed during Fenton process owning to the synergic effect of coagulation and precipitation. In this work, pollutants of thallium and EDTA were successfully removed via the Fenton process. A series of single-factor experiments were designed and performed to achieve an optimal reaction conditions for the removal of both thallium and EDTA. Results showed that the removal efficiencies of thallium and TOC could be as high as 96.54% and 70.42%, respectively. The outcomes from our study demonstrate that Fenton process is a promising method for the purification of wastewater containing thallium and EDTA.

## 1. Introduction

Thallium (Tl) is an extremely toxic heavy metal [1, 2]. Its toxicity is known to be greatly higher than that of lead (Pb), cadmium (Cd) and mercury (Hg) [3]. Even if at low concentration, the toxic effect of thallium on the organisms is still significant, posing substantial risks to ecosystem and human health [4, 5]. A standard from US Environmental Protection Agency (USEPA) requires that the maximum contaminant level of Tl in drinking water is 2 $\mu$g/L, while a more stringent limit of 0.1 $\mu$g/L is adopted in China [6]. The release of Tl from industries (e.g., electrician industry, metallurgical industry and paper industry, etc.) often accompanied by organic pollutants, such as ethylenediamine tetraacetic acid (EDTA). In a wastewater stream, the complex of EDTA-chelated thallium exhibits much higher toxicity to environment and human because of its high mobility and recalcitrant feature [7, 8].

To remove metal-organic complex from wastewater stream, a number of techniques, such as electrolysis, photocatalysis, ion exchange and reverse osmosis [9-11], have been developed. Though these techniques have some positive outcomes, a number of significant drawbacks inevitably displayed, including poor treatment capability for organic wastewater with high concentration, high processing costs, and potential secondary pollution. Recently, Fenton's reagent method has been attracted an increasing attention in the degradation of organic contaminants in aqueous medium. The reaction mechanism of Fenton's reagent system generally involves a series of complex reactions (1)-(7)

forming hydroxyl radical (·OH) which can effectively degrade organics and finally convert these organics into carbon dioxide ($CO_2$) and water ($H_2O$) [12].

$$Fe^{2+} + H_2O_2 \rightarrow Fe^{3+} + \cdot OH + OH^-, \quad (1)$$

$$Fe^{2+} + \cdot OH \rightarrow Fe^{3+} + OH-, \quad (2)$$

$$Fe^{3+} + H_2O_2 \rightarrow Fe^{2+} + H^+ + HO_2\cdot, \quad (3)$$

$$O_2\cdot + H_2O_2 \rightarrow O_2 + H_2O + \cdot OH, \quad (4)$$

$$H + \cdot OH \rightarrow R\cdot + H_2O, \quad (5)$$

$$R\cdot + Fe^{3+} \rightarrow R^+ + Fe^{2+}, \quad (6)$$

$$R\cdot + O_2 \rightarrow ROO + \ldots\ldots \rightarrow CO_2 + H_2O. \quad (7)$$

Fenton's reagent method is considered to be a reliable wastewater treatment method with many outstanding properties, such as fast reaction speed, easy to operate and environmentally friendly [13]. Fenton's reagent method has been successfully applied in treating different types of industrial wastewater, for instance, landfill leachate, dyeing wastewater, alkyl benzene and phenol wastewater [10, 14, and 15]. Owning to the strong oxidation capability of hydroxyl radical, it can be expected that Tl (I) is able to be oxidized into Tl (III). Compared with Tl (I), Tl (III) is much easily precipitated and thus finally removed from aqueous solution. Many efforts on the application of Fenton's reagent method to treat organic matters and generic heavy metal-organic complex have been conducted. However, to our best knowledge, the study on the simultaneous removal of both Tl and EDTA is very rare.

In the present work, we successfully employed Fenton's reagent method to eliminate both Tl and EDTA from aqueous solution. The removal efficiency of Tl and EDTA was studied under different treatment conditions such as various $H_2O_2$ concentration and reaction pH.

## 2. Materials and experimental methods

### *2.1. Materials*

All the reagents used in this study were analytical grade and used without further purification. Thallium nitrate ($TlNO_3$) was purchased from National iron & steel research institute, China. Others were purchased from Guangzhou chemical reagent factory, China. The stock solution of (1000 mg/L) Tl (I) was obtained by dissolving $TlNO_3$ with deionized water. As to 1000 mg/L EDTA stock solution for the Fenton reaction, certain amount of $C_{10}H_{14}N_2O_8Na_2\cdot 2H_2O$ was dissolved into deionized water. The Tl (I) working solution (50 $\mu$M) and EDTA working solution (500 $\mu$M) were prepared daily by serial dilution in 40 ml solution.

### *2.2. Fenton reaction*

All the Fenton reactions were performed in 50 ml beakers while each beaker's experimental solution was 40 ml in total. The Fenton reaction system was kept stirring on temperature-controlled stirrers and the reaction temperature was maintained at 298 K. The detailed experimental conditions of influencing factors of $H_2O_2$ concentration and reaction pH were described in the following sections.

#### *2.2.1. Effect of $H_2O_2$ concentration*

The tests were carried out in 50 ml beakers that contained 0.409 ml Tl(I) stock solution and 7.445 ml EDTA stock solution (final concentration of 50 $\mu$M and 500 $\mu$M after dilution in 40 ml solution, respectively), and then deionized water was added to 40 ml to form the synthetic wastewater. 0.24 g

$FeSO_4 \cdot 7H_2O$ (21.58 mM) was dissolved into the synthetic wastewater under continuous stirring. For the variation of molar ratio of $H_2O_2/Fe^{2+}$ (0:1-10:1), certain amount of $H_2O_2$ was added. Afterward, using $HNO_3$ and NaOH solutions (in various concentrations) to adjust the reaction pH to be 3.5. After 2 h reaction, the final pH was adjusted to be 11 by adding Ca $(OH)_2$ powder. 10 mL final reaction solution was withdrawn by syringes and filtrated with 0.45 $\mu$m PDVF filter. Finally, 0.2 mL $HNO_3$ solution (50%, v/v) was added into the filtrate for preservation and further analysis.

*2.2.2. Effect of reaction pH*

As to the tests to study the effect of reaction pH, most of the experimental procedures were identical with those for the effect of $H_2O_2$ concentration. 0.24 g $FeSO_4 \cdot 7H_2O$ (21.58 mM) and 0.220ml $H_2O_2$ (53.96 mM) were dissolved into the synthetic wastewater. For the variation of pH, we used $HNO_3$ and NaOH solutions (in various concentrations) to adjust reaction pH values in the range of 2.0-7.0.

*2.3. Evaluation of removal efficiency*

Tl concentration was determined by an atomic absorption spectrometer (Thermo Scientific, USA). The content of EDTA is expressed by measuring the concentration of TOC. The TOC was determined by a liqui TOCII spectrometer (Elementar, Germany). The removal rate $\eta$ (%) of Tl and TOC was calculated by the following equation:

$$\eta = \frac{(C_0 - C_e)}{C_0} \times 100\% \tag{8}$$

Where $C_0$ and $C_e$ (mg/L) are the initial and equilibrium Tl (or TOC) concentrations of the working solutions, respectively.

## 3. Results and discussion

*3.1. Effect of $H_2O_2$ concentration on Tl and EDTA removal*

Fig.3-1 showed the effect of $H_2O_2$ concentration on the removal rate of Tl and EDTA via Fenton process. It can be seen that $H_2O_2$ concentration was of little effect on the Tl removal rate, as only 4% removal rate of Tl increased (from 95.5% to 99.5%) when the dosage of $H_2O_2$ concentration increased from 0 to 215.8 mM. The removal of Tl is most likely due to the processes of precipitation, adsorption and coagulation. With a certain amount of $Fe^{2+}$ in the solution, Ca $(OH)_2$ was employed to adjust the flocculation pH so that Tl was effectively removed via adsorption and coagulation. On the other hand, as for the removal rate of TOC, by increasing $H_2O_2$ concentration, the removal rate of TOC tended to increase rapidly but slow down under the conditions with high $H_2O_2$ concentrations (Fig. 3-1). With a high amount of $H_2O_2$ in the solution, more $\cdot OH$ will be generated because of the combination of $H_2O_2$ and $Fe^{2+}$. Thus, the organic matter can be oxidized and degraded effectively, achieving a higher removal rate of TOC. The amount of $H_2O_2$ significantly affects the oxidation effectiveness of Fenton process. In the case of with very high concentration of $H_2O_2$, the excess $H_2O_2$ in the system would react with $\cdot OH$ to produce $HO_2\cdot$, shown as the following equations (9)-(10). These reactions would result in invalid decomposition of $H_2O_2$ [16] thus inhibiting the generation of $\cdot OH$ and reducing the oxidation efficiency.

$$\cdot OH + H_2O_2 \rightarrow HO_2\cdot + H_2O, \tag{9}$$

$$\cdot OH + HO_2\cdot \rightarrow H_2O + O_2. \tag{10}$$

*3.2. Effect of reaction pH on Tl and EDTA removal*

Fig.3-2. showed the effect of reaction pH on the removal rate of Tl and EDTA via Fenton process. Reaction pH affected directly the mechanism of oxidation and was remarkably influenced Tl and EDTA removal. When there was a change in the pH of the solution, the rate of production of hydroxyl radical would be influenced. At low pH, the generation of ·OH is favored and thus the degradation of EDTA and the removal of Tl are effective. When pH=2.5, TOC got a maximum removal rate of 70.42% and Tl removal rate of 96.54%. However, with reaction pH increased, the generation of ·OH is tended to be inhibited due to the interaction between $Fe^{2+}$ and $H_2O_2$. When pH is larger than 3.5, a portion of $Fe^{2+}$ would convert into $Fe^{3+}$ and then reacted with $OH^-$ to form Fe $(OH)_3$. This process would invalidate the catalytic actives of Fenton system [17]. While pH is above 5.0, $H_2O_2$ would decompose into $H_2O$ and $O_2$, which further impeded the Fenton reaction [18]. The reaction rates of Fenton process were much faster in acidic medium than that in alkaline medium.

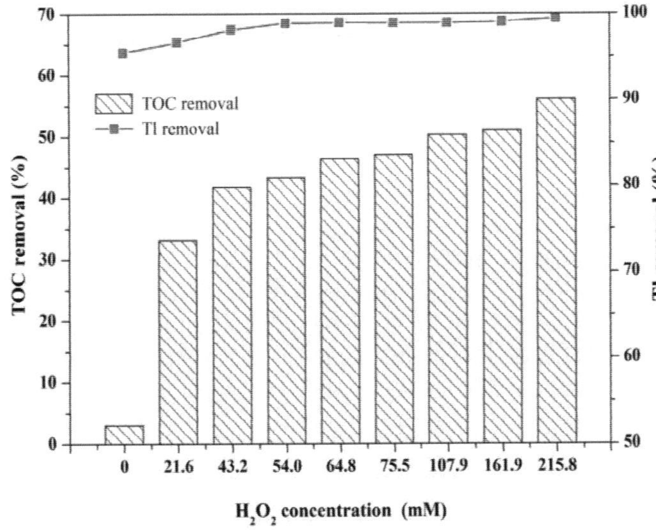


**Figure.3-1.** Effect of $H_2O_2$ concentration on the removal rate of Tl and EDTA by Fenton process (Reaction conditions: initial Tl concentration: 50 μM, initial EDTA concentration: 500 μM, $Fe^{2+}$ dosage: 21.58 mM, reaction pH: 3.5, reaction time: 2 h, flocculation pH: 11.0).

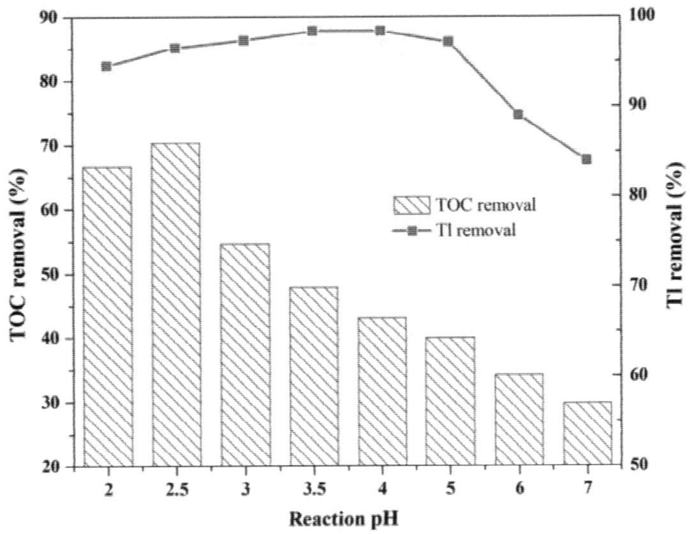


**Figure.3-2.** Effect of reaction pH on the removal rate of Tl and EDTA by Fenton process (Reaction conditions: initial Tl concentration: 50 $\mu$M, initial EDTA concentration: 500 $\mu$M, $Fe^{2+}$ dosage: 21.58 mM, $H_2O_2$ dosage: 53.96 mM, reaction time: 2 h, flocculation pH: 11.0)

## 4. Conclusions

This study successfully demonstrated that both Tl and EDTA were effectively removed by a Fenton process. The optimum processing conditions for Tl and EDTA removal was 53.96 mM $H_2O_2$ and reaction pH of 2.5. The removal efficiencies of Tl and TOC were 96.54% and 70.42%, respectively. The removal mechanism of Tl is mainly due to the oxidation, adsorption, coagulation and precipitation process and the removal mechanism of EDTA is mainly ascribed to the strong oxidation capability of newly generated hydroxyl radical in the Fenton process.

## 5. Acknowledgment

This work was supported by National Natural Science Foundation of China (51708143, 41372364, U1501231, U1612442), the Project of Radioactive Pollution Control and Resource Reuse Key Laboratory (2012A061400023), and Guangzhou University's Training Program for Excellent New-recruited Doctors (YB201710), High Level University Construction Project (Regional Water Environment Safety and Water Ecological Protection).

## 6. References

[1] Voegelin A, Pfenninger N, Petrikis J, Majzlan J, Plötze M, Senn AC, Mangold S, Steininger R, Göttlicher J. Thallium speciation and extractability in a thallium- and arsenic-rich soil developed from mineralized carbonate rock. Environmental Science & Technology, 2015, 49(9):5390.

[2] Xiao T, Yang F, Li S, Zheng B, Ning Z. Thallium pollution in China: A geo-environmental perspective. Science of the Total Environment, 2012, 421-422(1):51.

[3] Wan S, Ma M, Lv L, Qian L, Xu S, Xue Y, Ma Z. Selective capture of thallium(I) ion from aqueous solutions by amorphous hydrous manganese dioxide. Chemical Engineering Journal, 2014, 239:200-206.

[4] Xiao T, Guha J, Liu C Q, Zheng B, Wilson G, Ning Z, He L. Potential health risk in areas of high natural concentrations of thallium and importance of urine screening. Applied Geochemistry, 2007, 22(5):919-929.

[5] Zhang G, Fan F, Li X, Chen Y. Superior adsorption of thallium(I) on titanium peroxide: performance and mechanism. Chemical Engineering Journal, 2017.

[6] Agency E P. 2006 Edition of the drinking water standards and health advisories. Environmental Protection Agency, 2006.

[7] Ghosh P, Samanta A N, Ray S. Reduction of COD and removal of $Zn^{2+}$, from rayon industry wastewater by combined electro-Fenton treatment and chemical precipitation. Desalination, 2011, 266(1–3):213-217.

[8] Voglar D, Lestan D. Electrochemical treatment of spent solution after EDTA-based soil washing. Water Research, 2012, 46(6):1999-2008.

[9] Ju F, Hu Y. Removal of EDTA-chelated copper from aqueous solution by interior microelectrolysis. Separation & Purification Technology, 2011, 78(1):33-41.

[10] Madden T H, Datye A K, Fulton M, Prairie M R, Majumdar S A, Stange B M. Oxidation of metal–EDTA complexes by $TiO_2$ photocatalysis. Environmental Science & Technology, 1997, 31(12):3475-3481.

[11] Fernández Y, Marañón E, Castrillón L, Vázquez I. Removal of Cd and Zn from inorganic industrial waste leachate by ion exchange. Journal of Hazardous Materials, 2005, 126(1-3):169.

[12] Sychev A Y, Isak V G. Iron compounds and the mechanisms of the homogeneous catalysis of the activation of $O_2$ and $H_2O_2$ and of the oxidation of organic substrates. Russ Chem Rev. Russian Chemical Reviews, 1995, 64(12):1105-1129.

[13] Shyh F K, The H W, Yen H L. Decolorization and degradation of 2,4-dinitrophenol by fenton's reagent. Journal of Environmental Science & Health Part A, 1999, 34(4):935-950.

[14] Lopez A, Pagano M, Volpe A, Di PA. Fenton's pre-treatment of mature landfill leachate. Chemosphere, 2004, 54(7):1005-1010.

[15] Gümüş D, Akbal F. Comparison of fenton and electro-Fenton processes for oxidation of phenol. Process Safety & Environmental Protection, 2016, 103:252-258.

[16] Kavitha V, Palanivelu K. The role of ferrous ion in Fenton and photo-Fenton processes for the degradation of phenol. Chemosphere, 2004, 55(9):1235-43.
[17] Gallard H, Laat J D, Legube B. Spectrophotometric study of the formation of iron(III)-hydroperoxy complexes in homogeneous aqueous solutions. Water Research, 1999, 33(13):2929-2936.
[18] Chen R, Pignatello J. Role of quinone intermediates as electron shuttles in Fenton and photoassisted Fenton oxidations of aromatic compounds. Environmental Science & Technology, 1997, 31(8):2399-2406.

IOP Conf. Series: Earth and Environmental Science **111** (2018) 012035 doi:10.1088/1755-1315/111/1/012035

# Analysis on Zero Energy Consumption Strategy for Office Buildings Lighting in Lianyungang Area

**Dongmei Wu**

Huaihai Institute technology
Email: 1043618481@qq.com

**Abstract.** In recent years, the energy-saving environmental protection has aroused the people's high concern, and set off a new application practice in China. By analyzing the advantages of the illumination condition in Lianyungang area and combining the content and form of office space, the author puts forward a series of ways and means of energy saving in office building lighting, in order to provide a way for reference to the goal of building Zero energy consumption in the office space environment under the background of green architecture.

## 1. Lianyungang Area Lighting Conditions Advantage

Lianyungang natural illumination resources are abundant, according to gb/t5ZeroZero33-2ZeroZero1 "Architectural Lighting Design Standards", it is known that the area belongs to the Ⅳ area of light climate zone, and the average illuminance of other districts is 22klx≤eq<24klx, which can be illuminated by the annual average illumination of 23klx.The area's natural lighting can replace daytime electric lighting, averaging at least 1Zero hours of daylight (6:ZeroZero-18:ZeroZero) in winter (8:ZeroZero-16:ZeroZero), natural lighting does not require electricity, and all parts of the lighting system can be recycled.

## 2. Method Strategy of Zero Energy Consumption in Space day Lighting Illumination of Office Building under Green Building Background in Lianyungang Area

### *2.1. Reasonable Layout of Office Space--Natural Lighting*

Design first consider the layout of the building space, arrange office space with high frequency use in the south and east of construction, while arrange office space that not often used, such as conference rooms, toilets, staircases in the north of the building, which can effectively improve the space utilization of natural resources and reduce the power consumption of artificial lighting.

### *2.2. Coordinated Configuration of Natural Lighting and Artificial Illumination-Complementary Dynamic Illumination*

For example, most of the office space lighting is by the side window light, the average lighting coefficient required is usually 2%-3%, when the window area occupies the external wall area about 1/3, and the area near window that is the range about 1.5 times higher than the depth of the window can be used in natural light lighting. However, the natural illumination below the prescribed standard of the area must be artificial illumination auxiliary, in the light coefficient of less than Zero.5% of the area, it mainly relies on artificial illumination, this will form three different lighting areas: natural lighting area, mixed lighting area and artificial lighting area.

The overall lighting of the work area is the rational disposition of the natural light and indoor artificial lights injected into the window, in order to create a comfortable and harmonious office environment, the whole illumination of the space is coordinated by dynamic adjustment. The coordination method is divided into two kinds: illuminance balance and brightness balance, illumination balance refers to daytime indoor natural light exposure window, for the deep space of office space, in order to make the depth of illumination and near window illumination balance, the artificial illumination near the window can be reduced to make the whole space illumination uniformity; Brightness balance refers to the daytime interior window brightness is very high, for the people in room, the window near ceiling and walls appear dark, which will make people feel the entire space dark, so it must make indoor artificial lighting and natural light window brightness balance. When the brightness of the window decreases, the indoor artificial illumination is correspondingly reduced to meet the sensory needs of the office scene, and the illumination design adopts the dynamic control of the photosensitive device to reduce the light energy consumption and save energy.

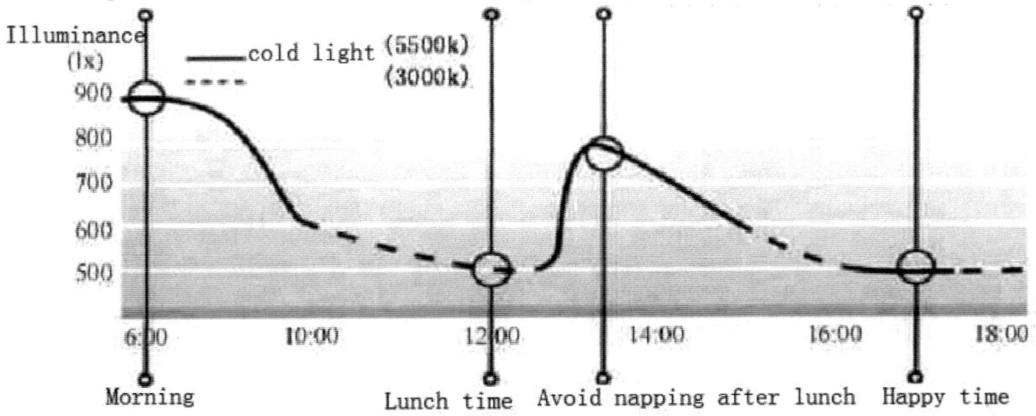


**Figure.1** One day of different moments of dynamic lighting program model

**Figure.2** Solar windmill lamps daytime power generation and night lighting

*2.3. Rational Configuration of Dimming Control System-Adjusting Type Dynamic Illumination*

Some office space parts are equipped with photoelectric sensor to control dimming system, such as: manager's office, conference room and so on, the privacy of the building interior semi-private office space can be based on natural light changes in a timely manner to continuously adjust the light and shade, so that the office area of natural lighting and artificial lighting can maintain the total illuminance, to save lighting power consumption reduction.

In addition, the study found that long time in the same artificial lighting environment of the work will lead to a decline in sensory sensitivity, especially the modern computer automation office and traditional office make it more easy to produce visual fatigue and affect the quality and efficiency of the work, so the overall artificial light environment of office space can simulate the dynamic regulation of the movement of natural daylight, according to the need of human visual senses, adjust the intensity of illumination and the change of color temperature dynamically, use intelligent control to carry on dynamic illumination. As shown in picture, this is a dynamic illumination scheme model at different times in one day.

### *2.4. Refraction Lighting-Using Prism-Type Crystal Glass Windows to Improve the Uniform Distribution of Indoor Illumination*

As shown in picture, the building uses a crystal prism window, one side is the crystal glass plane and the other side is the serrated prism surface, the saw-tooth mirror angle is 9ZeroZero, as shown in picture, the middle gap between two sides of saw-tooth interlock is small, and glass materials can be derived from crystal, because of the crystal hardness, high melting point, good temperature resistance, light transmittance, chemical stability, and Lianyungang has a wealth of crystal, quartz and other resources, which is named as China's crystal capital, the crystal is used for exterior lighting decoration of buildings can highlight the regional characteristics. As shown in picture, mosaic crystal wall, crystal partition indicates that the domestic crystal has been gradually developed by handicraft jewelry for architectural decoration, all the physical properties of crystal are stronger than glass, especially the piezoelectric effect, so it is more suitable for solar modules, in the Rainbow Church, designed by Japanese designer Tokujin Yoshioka, there exists a glass window with 5ZeroZero crystal prisms, There is also a crystal prism inverted cone in the building of the Reichstag of Berlin. Foreign designers have widely used crystal prism glass windows extensively for architectural decoration.

**Figure.3** Crystal Prism Inverted Cone of the Berlin Congress building

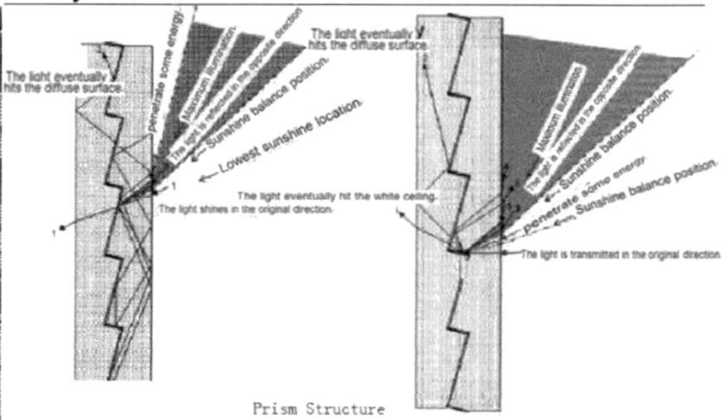


**Figure.4** Crystal Prism Glass Side view

**Figure.5** Rainbow Church

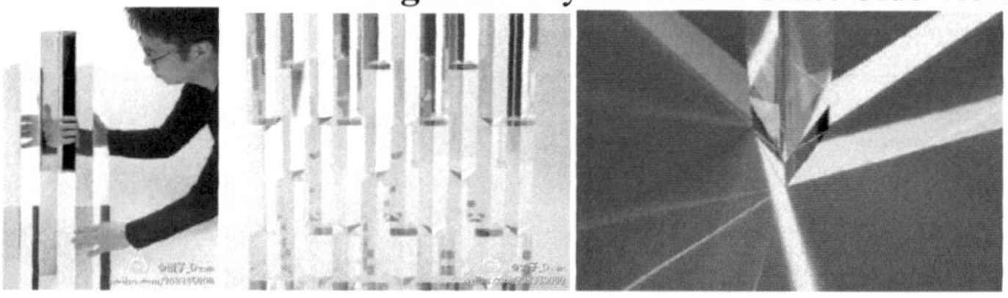

**Figure.6** Crystal Prism window

**Figure.7** Crystal partition

**Figure.8** Crystal Mosaic Wall

**Figure.9** Crystal Glass Column

The use of Crystal prism window refraction can change the direction of incident light exposure, so that sunlight can travel into the depths of space. Through the prism window refraction, even if the building spacing is very small, more light can be obtained. Because of the constant variation of the sun's altitude angle, the angle direction of the crystal prism is flexibly set by the photoelectric sensor, keeping track of the changes of the sun's orbit, and in different seasons and time, the prism angle direction is also different. Crystal prism window is generally installed in the upper part of the window, in thc position abovc thc people's view, which can avoid the view of the scene to produce blurred landscape deformation.

### *2.5. Reflective Lighting-Using Reflective Material to Guide the Light into the Interior*

In order to give full play to the effect of natural lighting illumination, the design must study the device which will lead to the deep space of natural light, and avoid glare when the device is used. By brushing or installing reflective materials on the walls of the building's office area, or setting up a reflector in the window inside and outside the line of light, the brightness distribution of indoor lighting can be improved at the same time, it also can prevent the vicinity of the area by direct sunlight to produce glare interference work, as well as to avoid the anti-light plate shielding people's view. As shown in the picture, several design schemes using reflective materials for day lighting.

### *2.6. Direct Scattering and Transmission Lighting-High Efficiency and Energy-Saving Environmental Protection of Solar Cell Module Wall*

The Zero energy-consuming building of the outer wall of the whole solar cell module is itself a large solar radiation receiver, which is made up of a series of monocrystalline silicon solar cells, with high current density and smaller size. The solar panel is formed by the aluminum alloy frame, the front plate and the bottom plate lamination.The metal frame of the assembly has a mounting hole that is easy to use and has a solid structure. The 3mm thick tempered glass front plate can provide efficient protection against environmental and mechanical effects, with high transmittance to direct and scattering light, and can make indoor light with diffuse effect and direct effect, in addition, seal floor can effectively prevent moisture intrusion and seawater infiltration corrosion.

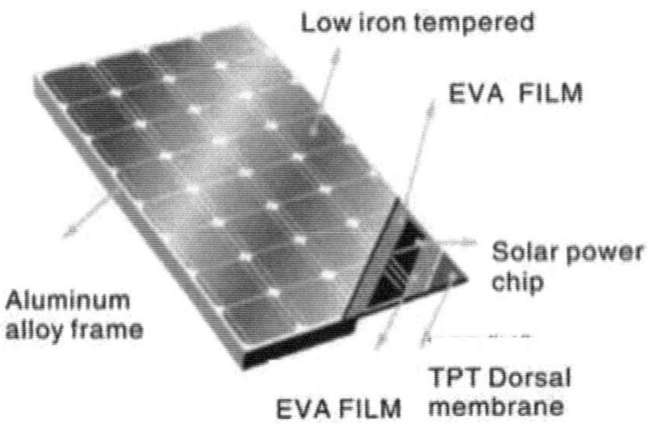

**Figure.10** Solar cell components Appearance effect of artificial freshwater island

## 3. Analysis of Zero Energy Consumption Strategy of Day Lighting Illumination in Office Space under Green Building Background

In terms of the lighting of office buildings in Lianyungang area, it is required to distribute the natural light evenly, to prevent direct sunlight, and to avoid the glare of specular reflection from glass wall. Therefore, the overall illumination of the building is mainly skylight refractive window lighting, partial illumination in some areas is supplemented by one side lighting, when the depth of the office space is 6m-8m, and the window area is about 1/4 to 1/5 of the floor, the minimum day lighting coefficient of the working area is 1.5% to meet the requirement of the light environment of the employees ' office work. The space illumination uniformity of the one side day lighting is poorer, the illumination of area near the window is more than 1Zero times higher than that of the inner wall, therefore, the design should set up windows on the inner wall to supplement the natural illumination. For the south-facing space, the windows should be arranged in the middle of the side window, and set up a parallel window visor on the high side to prevent direct sunlight, the daily illumination of the office area in the window can also be lowered appropriately, as a part of the natural lights being reflected to indoors, the minimum value of day lighting coefficient in the office area can be slightly improved, thus improving the overall space office lighting level. When the high brightness of the side window in the space wall forms specular reflection and produce glare, at this time, at the distance of 1.5M to the wall, the windows in the local work area within the scope of the external wall should not be opened, the area can be supplemented by appropriate artificial lighting to assist in order to improve the brightness of the wall, also, by making the wall decorated with matte diffuse reflective material, the glare reflected by the light and shadow can also be weaken .

Simulated daylight illumination test on Zero energy-efficient building office space, this paper analyzes and discusses the organic combination of solar lighting and sunshade system deflection system: In summer, the sun's height is larger than the winter, the sunrise and sunset rays of office space of eastern-western wall are incident angle, relative to the southern wall, it is more likely to produce light heat. While for the western window, the height angle of the solar incident in summer is lower than that of the south, and the incident sunlight is easier to reach the working area, resulting in glare effect work. Therefore, for the west wall, in addition to the internal shading, it is mainly in the block plate shading, if the use of horizontal sun visor panels, it needs to be parallel with windows, and low incidence angle of solar irradiation can be controlled by combining with active sun shading.

## 4. Concluding Remarks

As shown in the picture, the application research of lighting design can take a dynamic form, and imagine the architecture as a transitional space filled with various scales in the city--the transitional space in which light and shadow converge. This paper studies the model experiment of Zero energy-consuming buildings based on the green building background, interprets the changes of the content and form of the new office space and the standard requirements of the new office lighting. The paper also studies and explores more natural forms of space and natural light resources. In the study of Zero energy-efficient building office space lighting program, in the exploration and practice of lighting design, the paper has always been committed to the local style perfect integration into the design. Nowadays, many local factors influence people's deep thinking on design, such as architectural form, material selection, geographical environment and climatic conditions, especially the corresponding lighting and ventilation. With the development of building technology, no matter whether daylight illumination or artificial lighting design, they all will have a broader stage.

## 5. References

[1] Xie Hao. The use of sunlight and the construction of architectural space [J]. Shanghai Building Materials, 2Zero11 (3): 33-35.
[2] Ren Yuanhui. Office building characteristics of the times and lighting applications [J]. Electrical Engineering Applications, 2Zero11 (2): 45-48.
[3] Zhang Jingqiu. Office building and urban development and the relationship between the human environment [J]. Decorative, 2Zero12 (11): 36-39.
[4] Wu Dongmei.Study on the night lighting of harbor city at home and abroad [J].
[5] Lai Zili. Building planning and planning. [J]. Science and Technology Information, 2Zero12 (26): 21-24.
[6] Yuanjin. Office lighting status and future [J]. Intelligent Building and Urban Information, 2ZeroZero7 (6): 55-59.
[7] Sun Daming, Feng Dabin. China's latest green building one hundred cases [M]. China Construction Industry Press, 2Zero11.
[8] Wu Dongmei.Design of zero energy consumption lighting scheme in office space based on green building concept [J]. Peony, 2Zero15 (2Zero).

# Fast Reliability Assessing Method for Distribution Network with Distributed Renewable Energy Generation

**Fan Chen[1], Shaoxiong Huang[2], Jinjin Ding[1], Bo Gao[1], Yuguang Xie[1] and Xiaoming Wang[1]**

[1]Anhui Electrical Power Research Institute of State Grid, Hefei, China
[2]Dispatching Control Center of Anhui Electric Power Company of State Grid, Hefei, China
Email: chan_nanjing@163.com

**Abstract.** This paper proposes a fast reliability assessing method for distribution grid with distributed renewable energy generation. First, the Weibull distribution and the Beta distribution are used to describe the probability distribution characteristics of wind speed and solar irradiance respectively, and the models of wind farm, solar park and local load are built for reliability assessment. Then based on power system production cost simulation probability discretization and linearization power flow, a optimal power flow objected with minimum cost of conventional power generation is to be resolved. Thus a reliability assessment for distribution grid is implemented fast and accurately. The Loss Of Load Probability (LOLP) and Expected Energy Not Supplied (EENS) are selected as the reliability index, a simulation for IEEE RBTS BUS6 system in MATLAB indicates that the fast reliability assessing method calculates the reliability index much faster with the accuracy ensured when compared with Monte Carlo method.

## 1. Introduction

As distributed renewable energy generation is widely connected to the distribution network, the one way radiation network has turned into a network with consumers interconnected and power sources scattering[1,2]. Considering the intermittency and randomness of new energy generation-such as wind power and solar power, the security and stability of power system is significantly affected. Therefore, to study power supply reliability of distribution grid with distributed new-energy generation appears to be particularly important.

Previous work has made large number of achievements. Ref[3] adopts the clustering algorithm and establishes the reliability analysis model based on analytical method. Ref[4] addresses the reliability evaluation issue by using Monte Carlo method based on Latin hypercube sampling. Ref[5] utilizes the Copula theory to build joint probability distribution model of output power of wind farm and solar park, and set up the model of wind-solar hybrid power system on that basis. Ref[6] makes use of the mapping relationship between estimation points of distributed generation and power load and power supply reliability, and implements points estimation for reliability of distributed network. Ref[7] proposes the reliability probability distribution model of distribution grid with DG through random function for reliability indexes of distributed network in system level and node layer.

The Monte Carlo method is frequently used to assess the power supply reliability of new energy generation[8,9], due to the advantage of processing random problems and high accuracy. However this method is very time-consuming on the other side. During the planning stages of power system with new energy generation, power supply reliability is evaluated by comparing among several alternative plans.

Generally this type of evaluation has no use for high accuracy and requires a small time cost, as the system network topology can be adjusted. Hence, the traditional Monte Carlo method appears inefficient under that circumstances.

This paper proposes a fast reliability assessing method for distribution network with distributed renewable energy generation. First, the Weibull distribution and the Beta distribution are used to describe the probability distribution characteristics of wind speed and solar irradiance respectively, and the models of wind farm, solar park and local load are built for reliability assessment. Then based on power system production cost simulation probability discretization and linearization power flow, a optimal power flow objected with minimum cost of conventional power generation is to be resolved. Thus a reliability assessment for distribution grid is implemented fast and accurately. The Loss Of Load Probability (LOLP) and Expected Energy Not Supplied (EENS) are selected as the reliability index, a simulation for IEEE RBTS BUS6 system in MATLAB indicates that the fast reliability assessing method calculates the reliability index much faster with the accuracy ensured when compared with Monte Carlo method.

## 2. Models of Wind Farm, Solar Park and Powr Load

### *2.1. Probability Models of Wind Speed and Solar Irradiance*

Using Weibull distribution with double parameters to describe the probability distribution characteristics of wind speed[8,10], the probability density function is given as follows:

$$f(v)=\frac{k}{c}\left(\frac{v}{c}\right)^{k-1}\exp\left(-\left(\frac{v}{c}\right)^{k}\right) \tag{1}$$

where $v$ is wind speed, $k,c$ is shape parameter and scale parameter of Weibull distribution respectively.

Using Beta distribution to describe the probability distribution characteristics of solar irradiance in a short period[9], the probability density function can be given as

$$f(r)=\frac{\Gamma(\alpha+\beta)}{\Gamma(\alpha)\Gamma(\beta)}\cdot\left(\frac{r}{r_{\max}}\right)^{\alpha-1}\cdot\left(1-\frac{r}{r_{\max}}\right)^{\beta-1}. \tag{2}$$

where $\Gamma(\cdot)$ is the Gamma function, $r$ is the actual irradiation intensity, $r_{\max}$ is the max irradiation intensity, $\alpha$ and $\beta$ are shape parameters of Beta distribution.

$\alpha$ and $\beta$ can be calculated as follows:

$$\alpha=\mu\left[\frac{\mu(1-\mu)}{\sigma^{2}}-1\right]. \tag{3}$$

$$\beta=(1-\mu)\left[\frac{\mu(1-\mu)}{\sigma^{2}}-1\right]. \tag{4}$$

where $\mu$ and $\sigma2$ are the mean value and variance of irradiation intensity in a certain period. $\alpha$ and $\beta$ have different values in different seasons and climates.

### *2.2. Model of Wind Farm*

The active power output of wind turbine generation is closely related to wind speed, which can be presented in Figure 1.

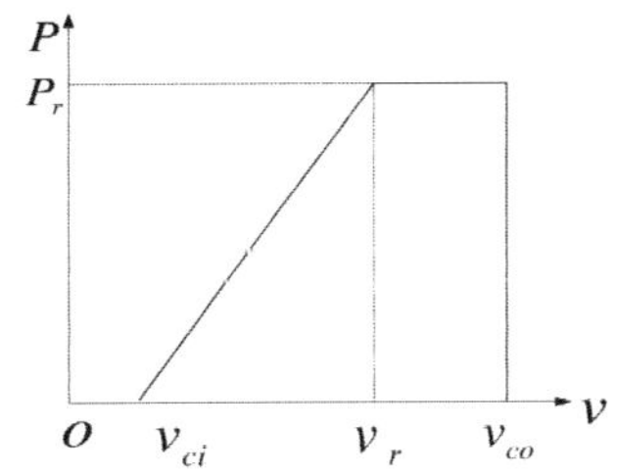


**Figure.1** Output curve of wind turbine generator

The piecewise function between output power of wind turbine generation and wind speed can be given as

$$P=\begin{cases} 0 & (V \le V_{ci} \ or \ V \ge V_{co}) \\ P_r \dfrac{V-V_{ci}}{V_r-V_{ci}} & (V_{ci} \le V \le V_r) \\ P_r & (V_r \le V < V_{co}) \end{cases} \tag{5}$$

where $V$ is the wind speed at wind turbine hub's height, $V_{ci}$ is the cut-in wind speed, $V_{co}$ is the cut-out wind speed, $V_r$ is the rated wind speed, and $P_r$ is the rated output power of wind turbine generation.

A steady-state simplified model is adopted in this paper(namely the PZ model)[8]. Taking wind turbine generators as PQ buses in power flow calculation, and considering active power of wind turbine generators equals to mechanical power of wind turbine and only relates to wind speed, we can get the whole wind farm's output active power as

$$P_f = n \bullet P_w \tag{6}$$

where $P_f$ is the output active power of wind farm, $P_w$ is the output active power of a single wind turbine generator, and $n$ is the amount of wind turbine generators in operation.

The probability density function of wind farm output power can be described as follows combined with the probability density function of wind speed[10]:

$$f(P_f)=\frac{k}{nk_1c}\left(\frac{P_f-nk_2}{nk_1c}\right)^{k-1}\exp\left(-\left(\frac{P_f-nk_2}{nk_1c}\right)^k\right). \tag{7}$$

Noticing that equation (7) is Weibull distribution with triple parameters, where $k_2=-k_1V_{ci}$ , $k_1=P_r/(V_r-V_{ci})$

## 2.3. Model of Solar Park

Assuming that photovoltaic modules works at maximum power point, the solar irradiance of installation site is $r$ at the certain time point $t$, the conversion efficiency of photovoltaic module equals to $\eta$ , the area of a single photovoltaic cell panel is $A$, and the amount of photovoltaic cell panels of whole solar park is $n$, we can calculate the output power $P_m$ of solar park at the certain time point $t$ as follows[9]:

$$P_m = n \bullet \frac{rA\eta}{860.4}(kW). \tag{8}$$

The solar park can be addressed in this manner in power flow calculation[9]: solar park can be calculated as an ordinary power plant , and solar park buses are regarded as PV buses during the daytime while solar park buses are regarded as PQ buses when the output equals to 0(considering that active power is 0).

According to the probability density function of solar irradiance, we can see that the probability density function of solar park output active power also appears Beta distribution, which can be described as

$$f(P_m)=\frac{\Gamma(\alpha+\beta)}{\Gamma(\alpha)\Gamma(\beta)}\cdot(\frac{P_m}{P_{\max}})^{\alpha-1}\cdot(1-\frac{P_m}{P_{\max}})^{\beta-1}. \tag{9}$$

where $P_m = n \bullet \frac{rA\eta}{860.4}(kW)$., namely the maximum output power of solar park.

## 2.4. *Model of Power Load*

Considering that power load forecasting is somewhat uncertain during the power system planning stage, this paper adopts the normal distribution to signify load forecasting results(only consider active power), which can well reflect the uncertainty of load forecasting[11].

$$f(P)=\frac{k}{\sqrt{2\pi\sigma_P}}\exp\left(-\frac{(P-\mu_P)^2}{2\sigma_P^2}\right). \tag{10}$$

where $\mu$ is the expectation, $\sigma^2$ is the variance, and $P$ is the active power.

# 3. Fast Reliability Assessing Method

## 3.1. *Power System Production Cost Simulation Probability Discretization*

The power supply elements in power system can be called resources, such as conventional power generator, wind farm and solar park. While the power-consuming elements can be called demands, such as power load. As Ref[12,13] indicate, we can deduce the distribution law by taking available capacity of resources and demands elements as random variables.

Assuming that the discretization step of common capacity of all elements is $\Delta C$, the capacity of element $m$ is $C_m$,we can discretize element $m$ as follows.

Let the equation be written as

$$S_m=[C_m/\Delta C]. \tag{11}$$

where $[x]$ denotes the largest integer less than $x$.

Therefore, element $m$ has ( $S_m+1$ ) states, where the available capacity of the $i$ th state can be given as

$$C_m^i=i\Delta C,\ \ 0\le i\le S_m. \tag{12}$$

Every state of element has some probability. Taking the $i$ th state of element $m$ in the initial circumstances as $PR_m^{(0)}(i)$, hence $PR_m^{(0)}(i)$ ( $i=0,1,2...S_m$ ) is to be a sequence of probability discretization. Assuming that wind farm, solar park and power load are mutually independent, we can calculate every state's probability of every single element by using equations (7), (9) and (10).

## 3.2. *Linearization Power Flow of Distribution Grid*

The DistFlow method is commonly used to calculate power flow of distribution network[14]. We can get the calculation formula of distribution network power flow by linearizing the DistFlow equation[15], as is shown below:

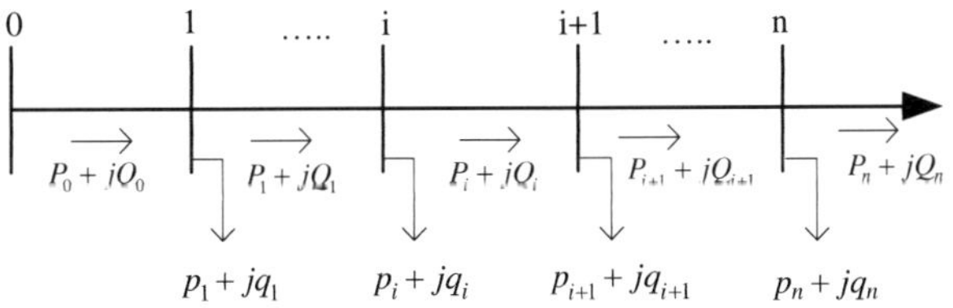


**Figure.2** Diagram of radial distribution network

$$\begin{cases} P_{i+1} = P_i - p_{i+1} \\ Q_{i+1} = Q_i - q_{i+1} \\ V_{i+1} = V_i - (r_i P_i + x_i Q_i)/V_0^2 \\ p_i = p_i^D - p_i^G,\ q_i = q_i^D - q_i^G \end{cases} \tag{13}$$

where $P_i$ and $Q_i$ are active power and reactive power from bus $i$ to bus $i+1$, $p_i^D$ and $q_i^D$ are active load and reactive load of bus $i$, $p_i^G$ and $q_i^G$ are active supply and reactive supply of bus $i$, $r_i$ and $x_i$ are resistance and reactance between bus $i$ and bus $i+1$.

### 3.3. Principle and Flow of Fast Reliability Assessing Method

Assume that $\Omega_{REW}$ is set of all wind and solar generations, $\Omega_G$ is set of all conventional generations, $\Omega_L$ is set of all power loads, $\Omega_F$ is set of all transmission lines, $\Omega_N$ is set of all buses in power system. For element $\xi \in \Omega_{REW} \cup \Omega_L$, the probability of operating in state $i_\xi$ is noted as $\Pr(i_\xi)$, $1 \le i_\xi \le S_\xi + 1$.

Considering constrained boundary conditions of power system, the optimal power flow objected with minimum cost of conventional power generation can be expressed as follows

$$\begin{cases} \min \sum_{\xi \in \Omega_G} c_\xi(i_\xi) + c_\omega(\omega) \\ s.t.\ \sum_{\xi \in \Omega_G} i_\xi + \sum_{\xi \in \Omega_{REW}} i_\xi - \sum_{\xi \in \Omega_L} i_\xi + \omega = 0 \\ f_l^{\min} \le f_l \le f_l^{\max}, l \in \Omega_F \\ 0.95 \le V_l \le 1.05, l \in \Omega_N \\ \omega \ge 0 \end{cases} \tag{14}$$

where $c_\xi(\cdot)$ is cost function of element $\xi(\xi \in \Omega_G)$, $f_l$ is power flow of line $l$, $f_l^{\min}$ and $f_l^{\max}$ are the maximum and minimum power flow of line $l$, $V_l$ is per-unit voltage of bus $l$, $\omega$ is slack variable($\omega \ge 0$) and $c_\omega(\cdot)$ is cost function of slack variable. Generally $c_\omega(\cdot)$ is far more than $c_\xi(\cdot)$.

To improve computation speed, linearization power flow method in section 2.2 is adopted to calculate $f_l$. Considering that power flow of most transmission lines would not cross the boundaries in practical circumstances, we can accelerate the optimal power flow problem (14) in the following way:

i0) Let LOLP=0, EENS=0;

i1) Resolve the optimal power flow problem neglecting constrained conditions of power system, and get the output power of all conventional power generations $i_\xi^{(0)}(\xi \in \Omega_G)$;

$$\begin{cases} \min \sum_{\xi \in \Omega_G} c_\xi(i_\xi) + c_\omega(\omega) \\ s.t. \sum_{\xi \in \Omega_G} i_\xi + \sum_{\xi \in \Omega_{REW}} i_\xi - \sum_{\xi \in \Omega_L} i_\xi + \omega = 0 \\ \omega \geq 0 \end{cases} . \quad (15)$$

i2) If $\omega > 0$, the system will lose power load and then flip to i5); if $\omega = 0$, calculate linearization power flow in terms of $i_\xi^{(0)}(\xi \in \Omega_G)$ and $i_\xi\ (\xi \in \Omega_{REW} \cup \Omega_L)$;

i3) If line power flow is not out of limit, then the system will lose no power load; otherwise, if power flow of line $h$ is out of limit, add the power flow constraints of line $h$ to equation (15) as follows (if voltage of is out of limit, address as the similar way):

$$\begin{cases} \min \sum_{\xi \in \Omega_G} c_\xi(i_\xi) + c_\omega(\omega) \\ s.t. \sum_{\xi \in \Omega_G} i_\xi + \sum_{\xi \in \Omega_{REW}} i_\xi - \sum_{\xi \in \Omega_L} i_\xi + \omega = 0 \\ f_h^{\min} \leq f_h \leq f_h^{\max}, h \in \Omega_F \\ \omega \geq 0 \end{cases} . \quad (16)$$

i4) Replace equation (15) with (16), repeat flow i1) to i3), and get $i_\xi^{(1)}(\xi \in \Omega_G)$. If all lines cross no boundaries and supply and consumption get balanced, then the system will lose no power load in certain states combination of wind power, solar power and load; otherwise flip to i5);

i5) The system will lose power load under this state combination of wind power, solar power and load, and the probability of this state combination can be calculated as $\prod_{\xi \in \Omega_{REW} \cup \Omega_L} \Pr(i_\xi)$, and the losing power load is to be $\omega \cdot \Delta C$;

i6) Calculate the LOLP and EENS as the following equation

$$\begin{cases} LOLP^1 = LOLP^0 + \prod_{\xi \in \Omega_{REW} \cup \Omega_L} \Pr(i_\xi) \\ EENS^1 = EENS^0 + \omega \end{cases} . \quad (17)$$

where $LOLP^0$ and $LOLP^1$ are the earlier value and updating value of LOLP; $EENS^0$ and $EENS^1$ are the earlier value and updating value of EENS.

Based on the above, the calculation flow of fast reliability assessing method in this paper can be designed in following steps:

j1) Set the discretization step capacity of all elements' capacity $\Delta C$ (the greatest common divisor of all elements' capacity can be selected as $\Delta C$);

j2) Get every element's capacity probability–discretized, and calculate the certain probability of every single state;

j3) Implement step i0) to i6) for all the possible combinations of wind power, solar power and power load;

j4) Exit calculation, and get LOLP and EENS.

## 4. Simulation Analysis

A IEEE RBTS BUS6 test system is adopted to verify the validity of fast reliability assessing method proposed in this paper. Select the radial network topology of A IEEE RBTS BUS6 system, and only consider 30-buses network to study the case[9]. The topology of simulation system is shown in Figure 3.

The conventional power generation is connected to the 33kV bus, wind farm is connected to bus 18, solar park is connected to bus 6. The concerning parameters are adopted with the practical parameters of certain wind farm and solar park in AnHui province.

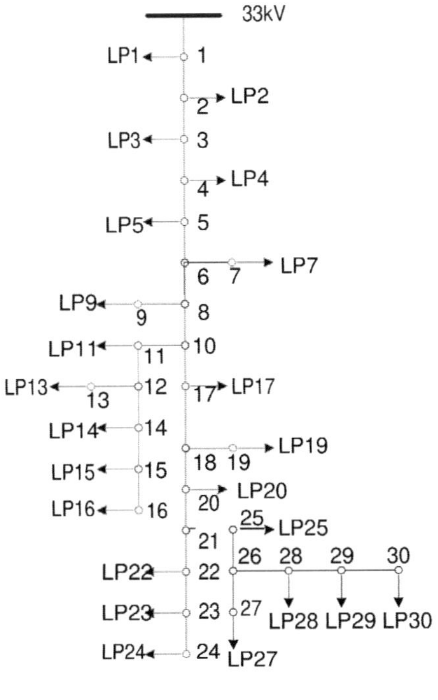


**Figure.3** Topology of IEEE RBTS BUS6 system

The parameters of wind farm are given as follows: the total capacity is 5×600kW, the cut-in wind speed, cut-out wind speed and rated wind speed are 3m/s, 25m/s and 16m/s respectively, the impedance of transmission line between wind farm and bus 18 is (12.6+j24.96) Ω, the air density of wind farm is 1.2245kg/m3, and the sweeping area of a wind turbine blade is 1840m2.

The parameters of solar park are given as follows: the total capacity of solar park is 600 kW, the latitude of photovoltaic module installation site is 33.38°, the power capacity of a photovoltaic module is 180 wP, the geometry size of a photovoltaic cell panel is 1580mm×793mm×50mm, the cell panels are installed in the south direction and inclined angle is , the revised atmospheric transparency is 0.7, and rated conversion efficiency of photovoltaic module is 13%.

The parameters of power load: according to Ref[8], calculate load forecasting uncertainty by using normal distribution.

Simulation is conducted in MATLAB with a PC of 2.5GHz CPU clock speed and 2GB RAM. To make a good comparison, the Monte Carlo method is also used to calculate the power supply reliability for the case. Table 1 shows the reliability indexes and computation time by using the two methods.

**Table.1.** Value of reliability index and computation time

| method | LOLP | EENS/MWh | time/s |
|---|---|---|---|
| Monte Carlo method | 0.010141 | 60.7939 | 3321 |
| proposed method ($\triangle C$=0.5MW) | 0.011137 | 64.7509 | 361 |
| proposed method ($\triangle C$=1.0MW) | 0.012671 | 75.3455 | 194 |

The computation efficiency of proposed method and Monte Carlo method is compared in Figure 4 and Figure 5.

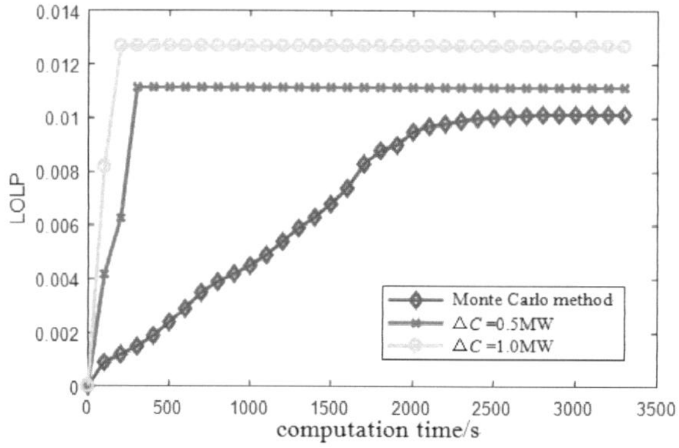


**Figure.4** Computation efficiency of LOLP

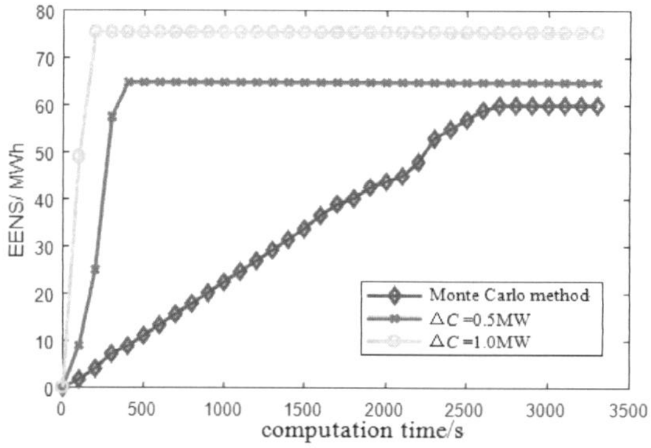


**Figure.5** Computation efficiency of EENS

We can conclude from the above simulation results as follows:

1) The convergence speed of Monte Carlo method is rather slow, and the calculation consumes 3321s (about 55.35min). In comparison, the proposed method in this paper calculate significantly fast while maintain a rather high accuracy.

2) When discretization step $\Delta C = 0.5\text{MW}$, the computation time is 361s, which account for 10.87% of Monte Carlo method's consuming-time, and the calculation errors are 9.82% and 6.51% respectively compared with the one by using Monte Carlo method.

3) When discretization step $\Delta C = 1.0\text{MW}$, the computation time is even faster, reducing to 194s and accounting for 5.84% of the one by Monte Carlo method, and the calculation errors get larger to some degree, increasing to 24.95% and 23.94%.

4) A appropriate discretization step is critical to calculation accuracy and computation time. On one side, if $\Delta C$ is selected rather big, the calculation accuracy would greatly decrease although the computation speed would be very fast. On the other side, if $\Delta C$ is selected rather small, the computation time would last much longer although the calculation accuracy would be greatly improved. Hence, a suitable discretization step can get a good balance between the two concerning factors, and would significantly improve the efficiency of power system planning.

## 5. Conclusion

A fast reliability assessing method for distribution network with distributed renewable energy generation is proposed in this paper in terms of the low efficiency of Monte Carlo method. The method is based on power system production cost simulation probability discretization and linearization power flow, and resolve a optimal power flow objected with minimum cost of conventional power generation. A appropriate discretization step can get high calculation accuracy and fast computation speed, making this method an efficient and practical way to evaluate the power supply reliability of power system.

## 6. Acknowledgment
The work is supported by the National Key Research and Development Plan(2016YFB0900400) and Science Project of Anhui Electric Power Company(521205160021).

## 7. References
[1] Wang Jian, Li Xingyuan, Qiu Xiaoyan. Power system research on distributed generation penetration[J]. Automation of Electric Power Systems, 2005, 29(24): 90-96.
[2] Liu Chuanquan, Zhang Yan. Distribution network reliability considering distribution generation[J]. Automation of Electric Power Systems, 2007, 31(22): 46-49.
[3] Jiang Wen, Yan Zheng, Yang Jianlin. Wind farm raliability model based on analytical method[J]. Electric Power Automation Equipment, 2010, 30(10): 79-83.
[4] Jiang Cheng, Wang Shuo, Wang Baoqing. Probabilistic reliability assessment of power system containing wind power based on latin hypercube sampling[J]. Transactions of China Electrotechnical Society, 2016, 31(10): 193-206.
[5] Zhao Jichao, Yuan Yue, Fu Zhixin, et al. Reliability assessment of wind-PV hybrid generation system based on Copula theory[J]. Electric Power Automation Equipment, 2013, 33(1): 124-129.
[6] Lu Jingjing, Zhao Yuan, Zhao Yongshuai, et al. A point estimation method for reliability evaluation of distribution network with distributed generation[J]. Power System Technology, 2013, 37(8) : 2250-2257.
[7] Tian Hongxun, Yuan Rong, Zhao Yuan. Computation of probability distribution for reliability indices of distribution system containing distribution generation[J]. Power System Technology, 2013, 37(6) : 1562-1569.
[8] Wu Yichun. Research on reliability and planning of power system with wind farm[D].Hefei University of Technology, 2006.
[9] Chen Minjiang.The sequential Monte-Carlo simulation and reliability in photovoltaic system[D]. Hefei University of Technology, 2004.
[10] Wang Chengshan, Zheng Haifeng, Xie Yinghua, et al. Probabilistic power flow containing distributed generation in distribution system[J]. Automation of Electric Power Systems, 2005, 29(24) : 39-44.
[11] G. Tina, S. Gagliano, S. Raiti. Hybrid solar/wind power system probabilistic modeling for long-term performance assessment [J]. Solar Energy, 2006,80(5): 578–588.
[12] Kang Chongqing, Xia Qing, Xiang Niande, et al. Sequence-based analysis of probabilistic production cost simulation[J]. Proceeding of the CSEE, 2002, 22(4) : 8-12.
[13] Kang Chongqing, Xia Qing, Xu Wei, et al.Uncertainty analysis of power systems[M].Science Press, 2011.
[14] M. E. Baran, F. F. Wu. Network reconfiguration in distribution systems for loss reduction and load balancing[J]. IEEE Transactions on Power Delievery, 1989, 4(2):1401–1407.
[15] Wang Z, Chen B, Wang J, et al. Coordinated energy management of networked microgrids in distribution systems[J]. IEEE Transactions on Smart Grid, 2015, 6(1): 45-53.

IOP Conf. Series: Earth and Environmental Science **111** (2018) 012037 doi:10.1088/1755-1315/111/1/012037

# The Evaluation Method of the Lightning Strike on Transmission Lines Aiming at Power Grid Reliability

**Jianfeng Wen [1], Jianwei Wu [1], LiandongHuang[1], Yinan Geng [2] and zhanqing Yu [2]**

[1]Jiangmen Power Supply Company of Guangdong Power Grid Co., Ltd, Guangdong 529000, China
[2]Department of Electrical Engineering, Tsinghua University. Beijing 100084, China
Email: 79581819@qq.com , gengyinan@tsinghua.edu.cn

**Abstract.** Lightning protection of power system focuses on reducing the flashover rate, only distinguishing by the voltage level, without considering the functional differences between the transmission lines, and being lack of analysis the effect on the reliability of power grid. This will lead lightning protection design of general transmission lines is surplus but insufficient for key lines. In order to solve this problem, the analysis method of lightning striking on transmission lines for power grid reliability is given. Full wave process theory is used to analyze the lightning back striking; the leader propagation model is used to describe the process of shielding failure of transmission lines. The index of power grid reliability is introduced and the effect of transmission line fault on the reliability of power system is discussed in detail.

## 1. Introduction

Lightning stroke is the main cause of transmission line trip-out [1-2]. More than 50% power system faults are caused by lightning stroke in China according to statistics. Transmission line lightning stroke failure causes damage to transmission equipment on the one hand, line successive failures caused by lightning stroke also threaten safe and reliable operation of power grid, which affects the development of social economy seriously. Lighting protection of transmission lines is emphasized in power system, and relevant scholars conducted a lot of research. However, there are still following disadvantages in power transmission and transformation system lighting protection. Firstly, reduction of lighting trip-out rate is regarded as the objective, the electric network reliability problem caused by lightning stroke trip-out is rarely designed and studied directly. Secondly, the index of electric trip-out rate is distinguished according to voltage grade. The function difference of the lines is not considered. The same lightning trip-out rate index is adopted aiming at lines with the same voltage level and different capacity and importance. The protection level of key lines is relatively weak due to the above disadvantages, and general lines have relatively excessive protection design. Therefore, it is necessary to study the power transmission and transformation system lightning protection method for guaranteeing the electric network reliability, and study the influence of lightning stroke failure on the security reliability of power grid.

Jiangmen is located in southern China, and it is one of regions with the strongest and most frequent lightning activities in China [3]. It is important to develop lightning protection in Jiangmen power grid, which can be used for reference in power grid construction and operation in other multi-thunder regions in China.

In the paper, Jiangmen power grid is adopted as the research object for studying the transmission line lightning stroke analysis method for electric network reliability. A transmission line lightning stroke trip-out rate evaluation method is proposed based on full-wave process theory lightning back-striking calculation method and leader progression theory lightning shielding failure calculation method, wherein tower models, dielectric strength, ground resistance, terrain and lightning activity along the line, etc. are considered comprehensively. Electric network reliability index is introduced for fully discussing the influence of transmission line lightning stroke failure on electric network reliability. The lightning stroke failure rate of 500kV, 220kV and 110kV typical transmission lines are evaluated, and differentiated lightning stroke trip-out rate indexes are proposed.

## 2. Transmission Line Lightning Stroke Analysis Method

### *2.1. Analysis Method of Transmission Line Back-Striking*

Wherever Times is specified, Times Roman or Times New Roman may be used. If neither is available on your word processor, please use the font closest in appearance to Times. Avoid using bit-mapped fonts if possible. True-Type 1 or Open Type fonts are preferred. Please embed symbol fonts, as well, for math, etc. The lightning stroke process is transiently analyzed based on the theory of full wave process. Meanwhile, the actual wave process of ground power transmission and transformation system as well as underground grounding device and grounding system is considered. The line back-striking performance when lighting direct stroke on the tower top. Matrix eigenvalue and characteristic vector principles are used. Matrix similarity transformation is adopted for simplifying wave process on multi-conductor lines with mutual electromagnetic contact into n mutually independent modulus, which is equivalent to single conductor wave process. A multi-wave impedance model is adopted for the tower in simulation [4-5]. The tower is divided into a main support and branch supports. It is assumed that each part is evenly distributed. The wave impedance is calculated through their own dimension and geometry function [6]; current source model is adopted for lightning current model, and the impedance size is 300 Ω, and bi-exponent pulse is adopted for lightning current [7]:

$$i(t) = \frac{I_0}{\lambda}\left(e^{-t/\tau_2} - e^{-t/\tau_1}\right) = \frac{I_0}{\lambda}\left(e^{-\alpha t} - e^{-\beta t}\right) \quad (1)$$

$$\lambda = \exp\left[-\frac{\alpha}{\beta}\left(\frac{n\beta}{\alpha}\right)^{1/(n+1)}\right] \quad (2)$$

In the formula: I0 refers to lightning current maximum value, λ refers to current correction factor τ1 and τ2 are time constants determining current rise time and delay time, τ1<τ2, and α=1/τ1, β=1/τ2. It is concluded through calculation that λ=0.966159, α=14623, β=1883107 aiming at 2.6/50 μs lightning current, and n is 3.

The leader progression method is regarded as insulator flashover criterion [8-9], air discharge physical mechanism is utilized for calculating the leader discharge in air gaps, when the leader length in the gap reaches the gap length, insulation flashover occurs. Gap breakdown time $T_B$ can be regarded as the sum of injection starting time $t_p$, injection development time $T_s$, ionization front propagation time $T_i$, leader progression time $T_L$ and gas heating time $T_g$, namely breakdown time can be written into follows:

$$T_B = t_p + T_S + T_i + T_L + T_g \quad (3)$$

In the formula, ionization front propagation time $T_i$ is very short compared with other time duration. It is generally included into leader progression time $T_L$; gas heating time $T_g$ experimental value should be shorter than 0.1 μs, therefore $T_g$ also can be ignored in calculation. In addition, injection starting time $t_p$ also can be ignored generally in actual calculation, or can be directly included into injection development time $T_s$. The leader progression method is regarded as insulation flashover criterion for judging whether the insulation is flashed over or not under the effect of non-standard shock wave, and its flashover voltage and flashover time should be determined.

The calculation formula of the line back-striking trip-out rate is shown as follows:

$$n = N_L \eta P g \quad (4)$$

In the formula: $N_L$ refers to the line ground flash frequency per100km year, $\eta$ refers to arc over rate, $P$ refers to line statistic lightning withstand level probability, and $g$ refers to lightning striking rate against poles.

*2.2. Analysis Method of Transmission Line Shielding Failure*

Basic thought of leader progression model: the field intensity of the ground and ground objects are gradually increased during downward development process of lighting. When the field intensity is increased to a certain value, upward head-on leaders are produced on the ground objects. When the upward and downward leaders meet the breakdown condition during relative development, lightning strike occurs. When a lightning leader model is used for analyzing the line shielding failure characteristics, the model parameters and criterion with higher influence on the calculation result can be divided into the following four aspects: lightning leader model, head-on leader starting criterion, upward leader model and final breakdown criterion, etc. [10].

In the paper, it is believed in the model that the electric charge is evenly distributed in the lightning leader, total electric charge in the lightning channel $Q$ is related with lightning current amplitude $I$, which can be calculated according to formula (5) [11-12], In the formula, $Q$ unit is C, and $I$ unit is kA. In addition, one-point electric charge is configured at the head of the lighting leader.

$$Q = 76 \times 10^{-3} \cdot I^{0.68} \quad (5)$$

The development direction of the lightning channel is random to certain extent in downward development process. However, the statistical rules show that lightning channel is always developed along the direction with the maximum electric field. In lightning fractal simulation, the electric field is regarded as the main factor for judging lightning development direction. In the paper, it is believed that the lightning leader is always developed along the direction with the maximum electric field strength.

In the paper, the starting field strength of critical radius object surface corona is regarded as the starting criterion of the upward leader. When the surface field strength is stronger than the corona starting field strength under corresponding radius [13]. Because the field intensity is the most direct cause of air ionization, thereby producing an upward leader, the criterion is common to different gap types, and the influence of the conductor working voltage can be considered.

The electrical characteristics of the upstream leader channel are represented by the channel voltage drop, and the voltage drop is calculated by the formula given by Rizk [14]. The upward leader is always developed to the head of the lightning leader at the speed, which is equivalent to 0.25 times of the downward leader.

The distance is gradually decreased, and the gap field intensity is increased gradually with the relative development of the upward and downward leaders. When the distance meets certain condition, breakdown occurs. The average field strength between the lighting leader head and the upward leader or the ground object is regarded as the final breakdown criterion [15]. The critical breakdown field strength between the lightning leader and the head-on leader or the ground object (conductor ground wire and tower) is 500 kV/m. Since the capacity of triggered lightning of the earth is poor, the critical breakdown field strength between the lightning leader and earth is 750 kV/m.

The above-mentioned lightning leader model is utilized for calculating the ground lightning position of the lightning leader which is located in one place with certain lightning leader (hit conductor, ground wire or earth). The method of calculating line shielding failure trip out rate refers that the area along the line is divided into square areas then the lightning current amplitude scope of possible shielding failure of the lighting leader in the area can be calculated. The shielding failure frequency in the area within one year can be expressed as follows:

$$a = 0.1 N_g \int_{y_1}^{y_2} \int_{I_1}^{I_2} p(I) \mathrm{d}I \mathrm{d}y \quad (6)$$

IOP Conf. Series: Earth and Environmental Science **111** (2018) 012037 doi:10.1088/1755-1315/111/1/012037

In the formula $a$ refers to shielding failure rate, time/ (100km year), $N_g$ refers to ground flash density, times/ (km$^2$ year), $p$ ($I$) refers to lightning current amplitude probability density function, $y_1$ and $y_2$ represent upper and lower bounds of the lateral range from ground lightning position to the transmission line, $I_1$ and $I_2$ represent the minimum and maximum currents during shielding failure.

### *2.3. Calculation Method of Lightning Stroke Trip-out Rate*

Line lightning stroke trip-out rate is related to line tower models, dielectric strength, ground resistance as well as terrain and lightning activity along the line etc., they respectively affect line back-striking trip-out rate and shielding failure trip-out rate, thereby determining line lightning protection performance. Tower model, dimension, dielectric strength, terrain and grounding conditions of each base tower are different in the actual line, and the lightning protection performance is also different. However, since there are many towers in the actual line, each base tower respectively undergoes trip-out rate analysis one by one with huge workload. Therefore, it is necessary to analyze overall lightning protection performance of lines according to the following feasible method: the lines are divided into several categories according to the statistic rules of all parameters affecting line lightning protection performance. Line lightning stroke trip-out rate in reach category is calculated. On the basis, the overall lightning protection performance of the line is obtained through the weighted sum method.

## 3. Power Gird Reliability

Power grid reliability refers that power system can continuously meet system operation restraints and measurement of power user load demands for short term under real-time operation mode and external working environment. Power generation and transmission system reliability is measured through quantitative reliability index. The indexes generally include PLC (Probability of Load Curtailments), EFLC (Expected Frequency of Load Curtailments), EDLC (Expected Duration of Load Curtailments), ADLC ADLC (Average Duration of Load Curtailments), ELC (Expected Load Curtailments), EENS (Expected Energy Not Supplied), BPII (Bulk Power Interruption Index), BPECI (Bulk Power Energy Curtailment Index) and SI (Severity Index), etc.

## 4. Evaluation of Typical Transmission Line

### *4.1. Typical Line Basic Parameters*

In the paper, respectively one 500 kV, 220 kV and 110 kV line with more frequent lightning stroke trip-out accidents in Jiangmen is selected to analyzing and studying the lightning protection performance of the three typical lines. Three typical phase conductor line and ground wire parameters are shown in table I. The sag should be selected according to the typical value of standard DL/T 620-1997 'Over-voltage protection and insulation coordination of AC electrical apparatus'. The insulator parameters are selected according to the typical value of the line, and $U_{50\%}$ is converted according to standard DL/T 620-1997. Tower models in each line with the highest proportion or the most representative tower models in the aspect of tower characteristics are calculated. Single-circuit erection is adopted for 500 kV line, the adopted towers mainly include straight line cup type towers and angled towers, as shown in Fig.1. Two typical tower structures and dimensions selected in the paper are shown in Fig 2. 110 kV lines include double-circuit, three-circuit and four -circuit erection modes. The adopted towers include double-circuit straight line tower, double-circuit angled tower, three-circuit straight line tower (1 base), four-circuit straight line tower (1 base), etc. The structure and model of two typical towers selected in the paper is shown in Fig.3.

### *4.2. Typical Line Calculation Result*

Lightning stroke trip-out rate calculation results of of the 500 kV typical line is 1.03time/ (100km year), the back-striking trip-out rate is 0.053time/ (100km year), and total trip-out rate is 1.08time/ (100km year); for 220 kV typical line, shielding failure trip-out rate is 0.83 times/ (100km year), back-striking trip-out rate is 1.45 times/ (100km year), and total trip-out rate is 2.28 times/ (100km year); for 110 kV typical line, shielding failure trip-out rate is 1.40time/ (100km year), back-striking trip-out rate is 2.36time/ (100km year), and total trip-out rate is 3.76 times/100 km year.

**Table.1.** Parameters of phase conductor and ground wire

| Line | | diameter (mm) | bundle No | Insulator length (m) | $U_{50\%}$ (kV) |
|---|---|---|---|---|---|
| 500 kV | Conductor | 23.94 | 4 | 4.3 | 2385 |
| | Ground wire | 16 | 1 | - | - |
| 220 kV | Conductor | 33.6 | 2 | 2.2 | 1292 |
| | Ground wire | 16 | 1 | - | - |
| 110 kV | Conductor | 23.94 | 1 | 1.2 | 700 |
| | ground wire | 9.6 | 1 | - | - |

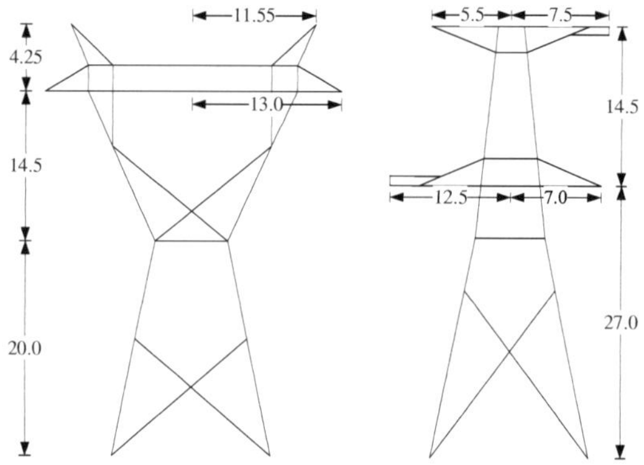


(a) ZB1 (b) JT

**Figure.1** Typical tower parameters of 500 kV lines

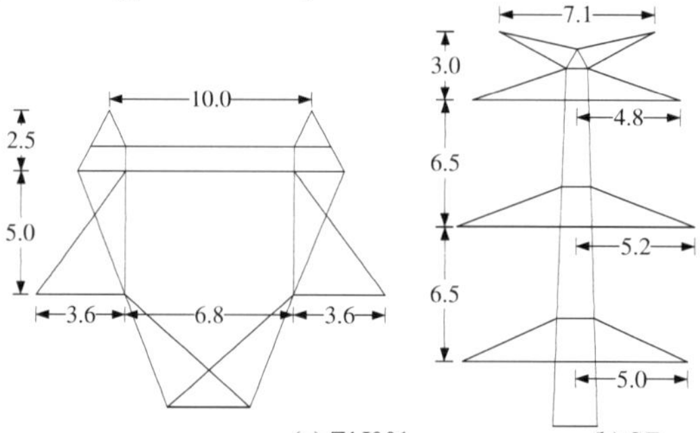


(a) Z1J001 (b) SZ

**Figure.2** Typical tower parameters of 220 kV lines

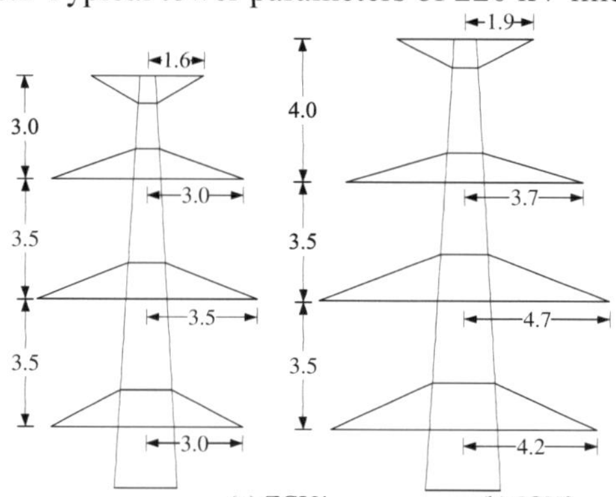


(a) ZGU1 (b) JGU3

**Figure.3** Typical tower parameters of 110 kV lines

### *4.3. Reliability Evaluation with Consideration of Lightning Stroke Risks*

Lightning occurs in summer mostly, and electric network reliability is evaluated under lightning stroke failure rate for Jiangmen power grid during summer in the project. Time sequential simulation method is obtained to obtain the electric network reliability evaluation results of Jiangmen power grid power generation and transmission system under lightning stroke failure rate as shown in the table. Calculation results in the table show that the reliability level in summer mode under lightning operation mode is reduced, wherein the PLC of Guangzhou power grid under the operation mode is 0.002404, PLC is reduced by 0.000207 under lighting operation mode; ELC and EENS are respectively 89.86MW/year and 4820.05MWh/year, and the system SI is 17.55 system score /year; The SI of ELC, EENS and system is respectively increased by 23MW/year, 459MWh/year and 1.67 system scores/year.

Calculation results in the table show that the reliability level in summer mode under lightning operation mode is reduced, wherein the PLC of Guangzhou power grid under the operation mode is 0.002404, PLC is reduced by 0.000207 under lighting operation mode; ELC and EENS are respectively 89.86MW/year and 4820.05MWh/year, and the system SI is 17.55 system score /year; The SI of ELC, EENS and system is respectively increased by 23MW/year, 459MWh/year and 1.67 system scores/year.

**Table.2.** Electric network reliability evaluation result under lightning stroke failure rate

| **Index** | **Regular summer mode** | **Summer mode under lightning stroke** |
|---|---|---|
| PLC PLC | 0.002197 | 0.002404 |
| EFLC (time/year) | 10.91 | 11.75 |
| EDLC (hour/year) | 19.25 | 21.06 |
| ADLC (hour/time) | 1.77 | 1.79 |
| ELC (MW/year) | 66.70 | 89.86 |
| EENS (MWh/year) | 4360.62 | 4820.05 |
| BPII (MW/MW.yr) | 0.00 | 0.01 |
| BPECI (MWh/MW.year) | 0.26 | 0.29 |
| SI (system branch /year) | 15.88 | 17.55 |

## 5. Conclusion

Lightning stroke characteristics of transmission lines with frequent lightning strokes in Jiangmen are analyzed. The line corridor ground lightning density is higher than 10time/ (km2• year). The evaluation result shows that the 500 kV line shielding failure trip-out rate is far higher than the back-striking trip-out rate, which is consistent with the operation result; 220 kV transmission line shielding failure trip-out rate is similar to back-striking trip-out rate. The relation thereof is related to tower ship, terrain and other factors. 110 kV transmission line back-striking trip-out rate is high relatively, and the back-striking trip-out rate in some terrains is prominently higher than shielding failure.

Power grid reliability analysis is introduced. The influence of transmission line lightning stroke failure on electric network reliability is considered. Quantitative reliability index is used for evaluating the changes of electric network reliability index before and after lightning stroke failure. The result shows that lightning stroke risks lead to higher influence on electric network reliability, transmission line lightning risk evaluation should be developed from pure lightning stroke trip-out rate to consideration of electric network reliability.

## 6. References

[1] HU Yi, LIU Kai, WU Tian, et al. Analysis of Influential Factors on Operation Safety of Transmission Line and Countermeasures. High Voltage Engineering, 2014, 40(11): 3491-3499.
[2] SIMA Wenxia, ZHENG Haoyuan, YANG Qing, et al. Three-dimensional Method of Calculating Shielding Failure Flashover Times of Transmission Line in Complex Terrain Area. High Voltage Engineering, 2014, 40(3): 662-668.
[3] LU Guojun, XIONG Jun, CHEN Jiahong, et al. Analysis of Ground Flash Density and Lightning Parameters of Guangzhou in 1999 ~ 2008. High Voltage Engineering, 2008, 35(12): 2930-2936.
[4] ZHANG Yongji, SIMA Wenxia, ZHANG Zhijin. Summary of the Study of Tower Models for Lightning Protection Analysis. High Voltage Engineering, 2006, 32(7): 93-98.
[5] Hara T, Yamamoto O. Modeling of a transmission tower for lightning surge analysis. Institution of Engineering and Technology, Generation, Transmission and Distribution, 1996, 143(3): 283-289.
[6] YUAN Zhaoxiang, LI Hu, XIANG Ling. Influence of Different Tower Models on the Lightning Back –Strike Intruding Wave Overvoltage for UHV Substation. High Voltage Engineering, 2008, 34(5): 867-871.
[7] F.Heidler, J.M.Cvetic, B.V.Stanic. Calculation of lightning current parameters. IEEE Transactions on Power Delivery, 1999, 14(2): 399-404.
[8] HUANG Weigang, ZHOU Dongpin. Study and discussion on insulation flashover criterion for lightning performance calculation. Electric Power, 1999, 32(11): 59-63.
[9] ZENG Rong, GENG Yinan. Lightning shielding failure model of transmission line based on leader progress model. High Voltage Engineering, 2008, 34(10): 2041-2046.
[10] WANG Zhiyong, YU Zhanqing, LI Yu, et al. Lightning Shielding Failure Characteristics of UHVDC Transmission Line Using Leader Progress Method. High Voltage Engineering, 2011, 37(9): 2178-2184.
[11] Dellera L, Garbagnati E. Lightning stroke simulation by means of the leader progression model. I. Description of the model and evaluation of exposure of free-standing structures. IEEE Transactions on Power Delivery, 1990, 5(4): 2009-2022.
[12] Dellera L, Garbagnati E. Lightning stroke simulation by means of the leader progression model. II. Exposure and shielding failure evaluation of overhead lines with assessment of application graphs. IEEE Transactions on Power Delivery, 1990, 5(4): 2023-2029
[13] ZENG Rong, GENG Yinan, LI Yu, et al. Lightning Shielding Failure Model of Transmission Line Based on Leader Progress Model. High Voltage Engineering, 2008, 34(10): 2041-2046.
[14] Rizk F A M. Modeling of transmission line exposure to direct lightning strokes. IEEE Transactions on Power Delivery, 1990, 5(4): 1983-1997.
[15] He J, Tu Y, Zeng R, et al. Numeral analysis model for shielding failure of transmission line under lightning stroke. IEEE Trans. on Power Delivery, 2005, 20(2):815-822.

IOP Conf. Series: Earth and Environmental Science **111** (2018) 012038 doi:10.1088/1755-1315/111/1/012038

# The Forecast of Baoding's Industrial Structure in Different Scenarios

**Wei Li[1], Hongzhi Zhang[1, a]*, Shu Zhang[1].**

[1] School of Economics and Management, North China Electric Power University, No. 619 Yonghua Street, Baoding, Hebei 071003, China.
E-mail:[a*] ncepuzhz2016@163.com.

**Abstract.** In November 12th, 2014, Chinese government officially announced reaching the peak of CO2 emissions around 2030. Baoding, as one of the first low-carbon pilot cities in China, bore an important mission and set an example to other cities to reach the peak of carbon emissions before 2030. This paper employed a computable general equilibrium model to simulate the industrial structure and social welfare of Baoding in different scenarios. And the calculation results have shown that on the principle of the optimal industrial structure and the minimum social welfare loss and a consideration of industrial economic development, the best time of CO2 emissions peak for Baoding was 2025 when its total emissions would reach 2.84 million tons. Finally, some suggestions were given to guide this city to reduce carbon emissions in industrial development.

## 1. Introduction

In November 12, 2014, according to "the Joint Declaration on Climate Change" by the United States and China at Beijing, China announced for the first time that the peak of $CO_2$ emissions would be occurred around 2030[1]. It is necessary for all low-carbon cities to play an exemplary role in the process of emission reduction, leading the other cities nationwide to achieve this commitment together. Baoding, a city of Hebei province, was one of the first low-carbon pilot cities and should undertake mission of reducing emissions firstly. However, there`s still no research about the peak time and quantities of carbon emission in Baoding and no report on how the industrial structure should be formed to reach the scheduled peak of $CO_2$.

This paper focused on Baoding`s industrial structure in different scenarios, the results were conducted with computable general equilibrium (CGE) model. Various industrial structures were simulated, which found out proportions of various industries in Baoding under different scenarios. Besides, on the foundation of the principle of maximum social welfare, an optimal time that occurred the peak of $CO_2$ emissions in Baoding was concluded.

Computable general equilibrium model (CGE) has been applied in many policy issues over the past 20 years. Researches conducted by CGE model can be classified into several aspects: fiscal and taxation policies, literatures like Dong and Whalley (2009) [2], Bao et al. (2011) [3]; international trade, researches like Li and Zhang (2012) [4]; relationship among economy, environment and energy, such as Jiang et al. (2009) [5], Bi et al. (2013) [6].

And there`s some researches on economic development and industrial reform, especially under the situation of carbon emissions reduction. Cheng et al. (2003) [7] built a CGE model to study the development strategy of high-tech industries in Hubei province. Li (2008) [8] and Liu (2013) [9] used dynamic CGE model to measure the impact of energy and environment policies on macroeconomic variables and industrial development in China. Liu et al. (2015) [10] analysed impacts of raising

China`s emission standard for thermal power plants by an extended CGE, and they found that these changes can trigger increase in economic output for industrial sectors. The research based on CGE model by Jin (2017) [11] focused on the impact of energy and environment policies on regional economic development and industrial structure. And it was suggested that regional industrial upgrading can be promoted by energy and environmental regulation, but it also need different policies to alleviate negative impacts on regional economy.

In this paper, an energy-economy CGE model was utilized to simulate the situation when Baoding reaches its peak of carbon emissions, and explore economic development and industrial structural changes in different scenarios. In addition, according to the calculation results in different scenarios, the optimal situation of $CO_2$ peak in Baoding was concluded by comparing social welfare.

## 2. Methodology

The energy-economy CGE model of Baoding is consisted of 3 blocks: the production block, income-expenditure block, market block. And these macroeconomics systems of mathematical equations can be calculated by General Algebraic Modelling System (GAMS), which is a type of modelling system software to solve mathematical programming problems.

### *2.1 Production block*

The structure of production block was shown in Figure. 2-1. the first block represents the structure of the production functions, which is shown in Figure.2-2. We assume that, each producer (represented by an activity) aims to maximize profits, which are maximized subject to a production technology and each activity produces one or more commodities according to fixed yield coefficients. A commodity may be produced by more than one activity. A nested constant elasticity of substitution (CES) function has been applied for production activities with multiple nested CES functions.

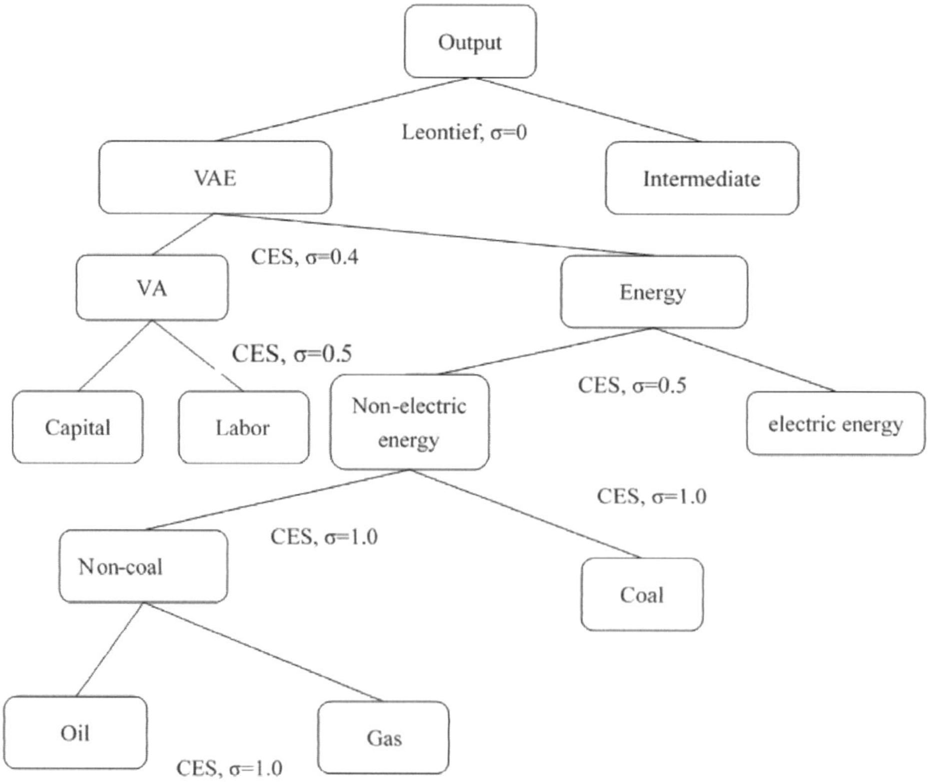


**Figure.2-1.** The Production Block of CGE Model.

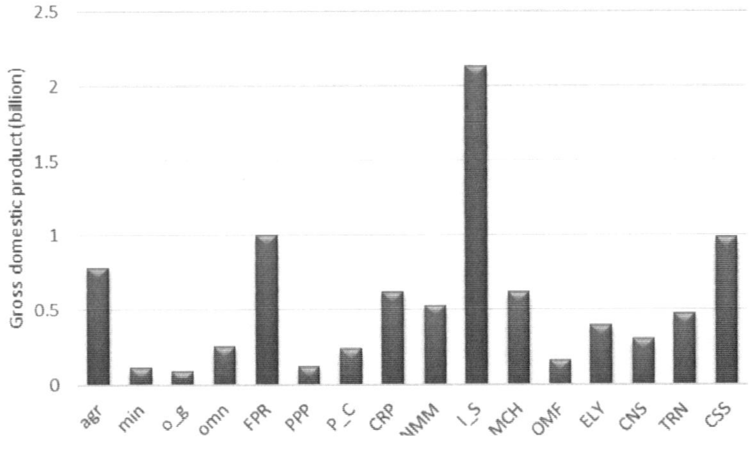


**Figure.2-2.**GDP of 14 Industries in 2030 of BAU Scenario.

*2.2 Income-expenditure block*

The second block is income-expenditure block, which is constituted households, enterprises, the government, and the rest of the country. The households receive incomes from the factors sold to the enterprises under the assumption that there are no self-consumption behaviors in households. The households use their income that after taxes, savings, and transfers to other institutions to support the Consumption of commodities, throughout the different combination of various commodities to achieve utility maximization. Cobb-Douglas (CD) functions have been applied to interpret the household's behaviors in this CGE model. Domestic enterprises determine how much of each factor the economy devotes to the production of each commodity for the purpose of profit maximization. Enterprises incomes are allocated to direct taxes, savings, and transfers to other institutions. In this study, government behavior mainly contains taxation and government consumption. All the tax revenue the government receives from direct tax, indirect tax, tariff and transfer payments is totally used for the government consumption. The government consumes all the commodities in the market at a fixed consumption tendency.

*2.3 Market block*

As an open macroeconomic model, the difference between domestic commodities and imported and exported commodities should have been considered. Therefore, Armington Assumption is used to interpret the imperfect substitutability relationship between domestic commodities and imported commodities and exported commodities. CES functions have been applied to represent the Armington Composite Commodities (ACC), which is composed of imported commodities and relevant domestic commodities, shown as Eq. (3) and Eq. (4).

$$QD_i = [\frac{\gamma_i^{\eta_i}\delta d_i PQ_i}{PD_i}]^{\frac{1}{1-\eta_i}} QQ_i \tag{1}$$

$$QM_i = [\frac{\gamma_i^{\eta_i}\delta m_i PQ_i}{(1+\tau_i^m)PM_i}]^{\frac{1}{1-\eta_i}} QQ_i \tag{2}$$

Where $QD_i$ interprets demand for domestic sales of commodity i; $QM_i$ represents quantity of imported commodity i. $\gamma_i$ stands for the scale coefficient of ACC production function. $\delta m_i$ and $\delta d_i$ are imported and domestic scales share coefficients of ACC production function, respectively.

$PQ_i$ is the price of composite commodity i. $PD_i$ and $PM_i$ are the imported price of commodity i and the domestic sales price of commodity i, respectively. $\tau_i^m$ is customs duty rate of commodity i. $\eta_i$ is the substitution parameter depended on substitution elasticity $\sigma_i$ ($\eta_i = (\sigma_i - 1)/\sigma_i, \eta_i \leq 1$), $\sigma_i$ is the substitution elasticity of imported commodity and domestic sales commodity in ACC production function.

In consideration of supply problem of exported commodities and domestic sales commodities, Constant Elasticity of Transformation (CET) functions have been used to interpret this conversion process, shown as Eq. (5) and Eq. (6).

$$QE_i = [\frac{\theta_i^{\phi_i} \xi e_i (1 + \tau_i^z) PZ_i}{PE_i}]^{\frac{1}{1-\phi_i}} QZ_i \tag{3}$$

$$QD_i = [\frac{\theta_i^{\phi_i} \xi d_i (1 + \tau_i^z) PZ_i}{PD_i}]^{\frac{1}{1-\phi_i}} QZ_i \tag{4}$$

Where $QE_i$ the quantity of is exported commodity i. $\xi e_i$ and $\xi d_i$ are exported and domestic scales share coefficients of ACC production function, respectively. $QZ_i$ represents the quantity of domestically produced commodity i. $PE_i$ is the exported price of commodity i; $PZ_i$ is the domestic production price of commodity i. $\tau_i^z$ interprets the rate of added-value tax. $\theta_i$ is the scale coefficient of the i-th commodity conversion function. $\phi_i$ is the substitution parameter depended on substitution elasticity $\psi_i$ ($\phi_i = (\psi_i + 1)/\psi_i, \psi_i \geq 1$), $\psi_i$ is the substitution elasticity of exported commodity and domestic sales commodity in ACC production function.

## 3. Material

Material data were adopted from the 42 sector input-output table of Hebei Province in 2007 and the energy consumption balance table in China Energy Statistical Yearbook (2008). According to the data in Input-output Table of Hebei Province, Baoding was divided into 16 productive activity: agriculture (agr), coal mining industry (min), oil and gas mining industry (o_g), mineral mining and other quarrying (omn), paper, paper products and pulp industry (ppp), light industry includes food products and agricultural products manufactory (fpr), petroleum and coal refinery industry (p_c), Chemical, Plastic and Rubber products industry (crp), Nonferrous metal, metal and metal product (nmm), steel and metal manufacturing industry (i_s), machinery and machinery equipment (mch), other manufacturing industry (omf), production and supply of electricity, heat and water supply (ely), construction industry (cns), transportation industry (trn) as well as Community ,and Services industry (css). In addition, the statistical data of energy consumption in Baoding was from energy balance in statistical yearbook. And the Cross-entropy Method (RAS) was used to balance the SAM tables of Baoding.

## 4. Scenarios

Bbefore calculating CGE model, three scenarios was designed to figure out the exact time when $CO_2$ peak appears. And by simulating total $CO_2$ emissions of Baoding in three scenarios, the peak time and the quantity of $CO_2$ was found out, which is in Table 4-1. All of these scenarios were based on politic documents which were the 13th Five-year Plan, Joint Statement on Climate Change in China and the United States, Implementation Plan of Greenhouse Gas Emission Control in 13th Five-year Plan in Hebei Province, Annual Report on Urban Development of China No.8.

Through economy-energy CGE model, the carbon emissions of Baoding in 2007-2030 were simulated. Under the corresponding peak situation, Baoding`s industrial structure and production were put forward.

**Table.4-1.** Different types of scenario in this paper.

| | | Energy intensity | GDP | population | Urbanization | Third industry share | Coal proportion | Time of $CO_2$ peak | Quantity of $CO_2$ |
|---|---|---|---|---|---|---|---|---|---|
| | 2017-2020 | -4.00% | 9.00% | 0.61% | 2.65% | 2.50% | -6.59% | | |
| S1[a] | 2021-2025 | -4.50% | 8.00% | 0.14% | 0.90% | 2.00% | -0.50% | 2025 | 2837.71 |
| | 2026-2030 | -5.00% | 6.50% | 0.12 | 0.90% | 2.00% | 0.50% | | |
| | 2017-2020 | -4.10% | 9.00% | 0.61% | 2.65% | 2.50% | -6.59% | | |
| S2 | 2021-2025 | -18.50% | 7.10% | 0.14% | 0.90% | 2.00% | -0.50% | 2027 | 3555.00 |
| | 2026-2030 | -20.50% | 7.10% | 0.12 | 0.90% | 2.00% | 0.50% | | |
| | 2017-2020 | -3.66% | 9.00% | 0.61% | 2.75% | 2.50% | -4.59% | | |
| S3 | 2021-2025 | -18.00% | 7.10% | 0.14% | 1.00% | 2.00% | -0.50% | 2029 | 4544.64 |
| | 2026-2030 | -20.00% | 7.10% | 0.12 | 1.00% | 2.00% | -0.50% | | |

[a] S was short for Scenario.

## 5. Results

As shown in Fig.2-2, Baoding still maintains a highly energy-consuming structure with heavy industry as its core in 2030 under the situation of no policy to adjust energy structure and reduce carbon emission.

In BAU scenario, gross output of 16 major industries in Baoding reaches 884.38 billion RMB in 2030 and it has formatted five pillar industries: a light industrial manufacturing industry depended on food and textile production, steel and metal manufacturing industry, automobile manufacturing machinery manufacturing industry, building materials industry and transportation industry including transportation equipment manufacturing. In addition, output of the first industry (agriculture) and the third industry (modern service industry) are far lower than that of the second industry. Which means in BAU scenario, Baoding become a city extremely relied on heavy industry and naturally an emitters with high $CO_2$ emissions.

### *5.1 Industrial analysis*

Figure.5-1. showed GDP and percentage of 16 industries of Baoding when it reaches $CO_2$ peak in 2025, 2027 and 2029 respectively. Obviously, Baoding`s industrial structure in three scenarios has huge difference comparing with BAU scenario. In three scenarios, in 2030, Baoding formed a sustainable industrial structure of service industry as its core, with light industry, power industry, construction industry and transportation industry supplementary.

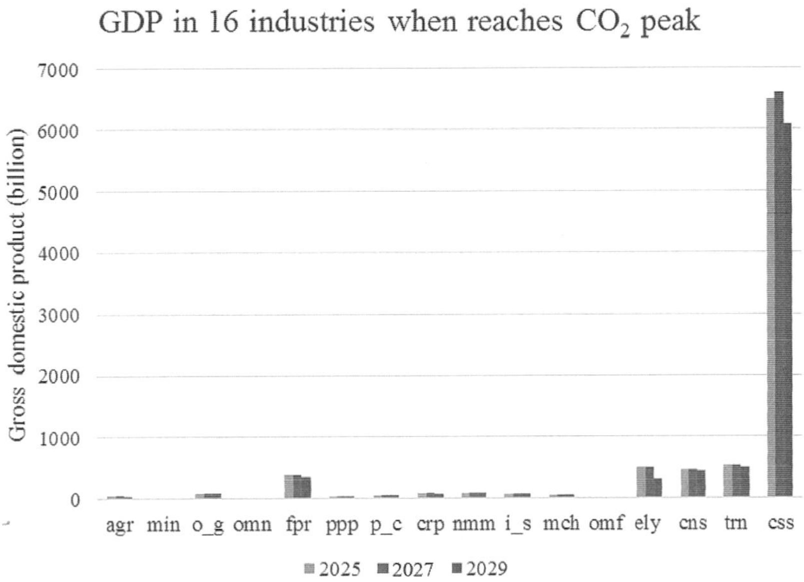

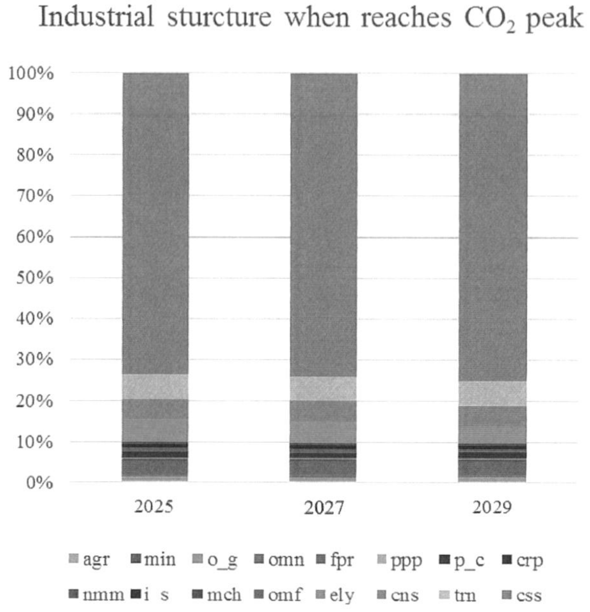


**Figure.5-1.** Baoding`s Industrial Structure in 2030 When Reaches $CO_2$.

When $CO_2$ emissions in Baoding reach its peak in 2025, 2027 and 2029, the proportion of 14 sectors was as follows:

(1) The third industry output becomes 75.07%, 74.09% and 73.60% respectively;

(2) Transportation industry (including automobiles and other transportation equipment), accounts for 5.90%, 6.07%, 5.97%;

(3) Power electricity industry is 5.40%, 5.83% and 5.53% respectively;

(4) Construction industry accounts for 5.05%, 5.44%, 5.11%;

(5) Light industry includes food products and agricultural products manufactory accounts for 4.28%, 4.31%, 4.24% respectively;

(6) Other sectors like agriculture, coal mining industry, mineral mining and other quarrying, paper, paper products and pulp industry, steel and metal manufacturing industry, and other manufacturing industry accounts for less than 1%.

In the scenario of reaching $CO_2$ peak in 2025, gross production of 16 industries will be 881.4 billion RMB, while in the scenario of 2027 and 2029 gross production will be 891.6 billion RMB and 811.0 billion RMB. That means the sooner Baoding`s industrial structure adjusted to sustainable development of taking the third industry as mainstays, the earlier this city will enter a green and high-speed way to economic growth.

*5.2 Social welfare analysis*

The utility index was used to compare improvements of social welfare among different equilibrium solutions in three scenarios. The Hicks value equations were used to quantify the utility base price, and thus the calculation results can be compared in quantitative. In this paper, we compared the 2030 welfare gap in each scenario to indicate the welfare deficiency caused by policies on carbon emission reduction.

**Table.5-1.** Comparing Social Welfare of Three Scenarios.

| | 2025 | 2027 | 2029 |
|---|---|---|---|
| EV | -2664340 | -3722890 | -8087852 |

As shown in Table.5-1, values of EV in scenarios of 2025, 2027 and 2029 were all negative, which indicated that policies on reducing carbon emissions surely cause certain welfare losses. It conforms to researches of domestic and foreign scholars on the carbon emission reduction policy influence. As a conclusion, taking changes of Baoding`s industrial structure in 2030 into consideration, whether focus on the production output or the value of social welfare, the scenario that Baoding reaches $CO_2$ peak in 2025 is the optimal situation.

## 6. Discussion

The manufacturing of light industry based on food processing and textile industry has always been one of the core industries in Baoding, whose output increase by 855% compared to 2016 in three scenarios. That means the manufacturing of light industry in Baoding need to maintain steady economic growing all through. However, industries like the power, construction and transportation need not only to keep the speed of growth but also to accelerate adjustment of their internal structure. Taking power generation as an example, proportion of electricity generated from non-fossil energy has always been emphasized in the process of sustainable development. As called "electricity valley" in China, Baoding takes the lead of renewable energy power generation interiorly for years. Based on results of economic-energy CGE model, we believe that on the one hand the power industry in Baoding should keep steady growth on total electricity generating capacity to meet social demands. On the other hand, the proportion of renewable power generation must be augmented to optimize the structure of energy generation and finally reduce carbon emissions in this city.

To achieve the goal of reaching CO2 peak with the least social welfare losses in 2025, some suggestions on industrial development planning was made in Baoding for next decades.

Firstly, the second industrial structure ought to be adjusted by promoting traditional industries upgrading and develop low-carbon industries vigorously, such as renewable energy generation and high-tech industries. And based on the results, by 2030, output of light industry includes food products and agricultural products manufactory (fpr) as well as production and supply of electricity, heat and water (ely) should account for 4.24% and 5.53% in Baoding`s gross production. While other industries in the second industry need to control their output under 1% of the total production.

Secondly, the proportion of the third industrial output should be improved intentionally. Emphasis should lay on services industry (css) and transportation industry (trn). On the basis of calculation, in 2030, output of these two industries should account for 73.60% and 5.97% in Baoding`s gross production. This means a massive economic stimulus package aimed at the modern services industry need to plan and implement soon.

All in all, the aim of industrial reform is to transform the current high-carbon industrial structure into the structure leading by the third industry which was characteristic of lower energy consumption and less carbon emissions.

## 7. References

[1] Joint statement on climate change between the leaders of China and the United States. http://finance.chinanews.com/gn/2015/09-26/7545884.shtml

[2] Dong Y and Whalley J 2009. How Large are the Impacts of Carbon Motivated Border Tax Adjustments. *NBER Working Paper* 15613.

[3] Bao Q, Tang L, Yang L and Qiao H 2011. How Will Energy-saving Technology Change the Impacts of Carbon Border Adjustments on China: An Estimation Based on Dynamic Recursive Computable General Equilibrium Model. *Journal of Systems Science and Mathematical Sciences* **2** 175-186

[4] Li J and Zhang Y 2015. A Quantitative Analysis on Economic Impact of Potential Green Barrier of International Trade for China: Case Study of Carbon Tariff with SIC-GE Model. *Journal of International Trade* **05** 105-118

[5] Jiang K, Hu X, Zhuang X and Liu Q 2009. China 2050 low carbon scenario and low carbon development approach. *Sino-Global Energy* **6** 1-7

[6] Bi Q, Fan Y, Cai S and Xia Y 2013. Analysis of China's primary energy demand scenarios based on the CDECGE model. *China population, resources and environment* **23** 41-48

[7] Cheng H, Zhang Z and Huang W 2003. A Study of CGE Model of Policy Analysis in High-tech Industry Based on Social Accounting Matrix. *Journal of Wuhan University of Technology* **27** 332-336

[8] Li P 2008. The general equilibrium analysis of China's energy and environmental policies. Xiamen: Xiamen University
[9] Liu Y 2013. The simulation research and policy choice of energy consumption,carbon emissions and economic growth:a dynamic CGE approach. Changsha: Hunan University
[10] Liu Y, Hu X and Feng K 2015. Economic and environmental implications of raising China's emission standard for thermal power plants: An environmentally extended CGE analysis. *Resources Conservation and Recycling* **121** 64-72
[11] Jin Y 2016. The research on the impact of energy-environment policy on regional development-based on CGE model. *International Conference on Energy Engineering and Environment Projection*

2017 International Conference on Advanced Environmental Engineering (ICAEE2017) IOP Publishing
IOP Conf. Series: Earth and Environmental Science **111** (2018) 012039 doi:10.1088/1755-1315/111/1/012039

# Analysis and Application of Antagonism Compound Prescription Compatibility

**Mengyan Li [1], Can Wang [1], Ming Bai [1], Mingsan Miao [1, a*]**

[1]Henan University of Chinese Medicine Zhengzhou 450046
E-mail: [a]miaomingsan @163.com.

**Abstract.** Deer horn glue is deer family animals deer or red deer horn made of solid plastic animal medicine, according to Chinese medicine "seven emotions together" theory, the antler and other Chinese herbal medicines compatibility can be better play its Medicinal value. In this paper, the chemical composition, pharmacological effects, compatibility analysis, clinical application and classic ancient prescriptions of antler are reviewed in recent years.

## 1. Introduction

Antlers for the first time appeared in the "Shennong Bencao Jing", is the deer Branch of the sika deer or red deer sawing section, washed, decoction and decoction made of plastic pieces, the product is flat square or Ding, Its color is translucent yellow-brown or red-brown, the upper part of the yellowish-white foam layer. [1] Crisp, crisp section, slightly sweet smell. Antlers sweet, salty, warm, go by the kidney. There are mainly warming liver and kidney, the effect of nourishing the blood, clinically used for liver and kidney deficiency caused by Yaoxisuanruan, impotence nocturnal emission, Consumptive thin, bleeding under the bleeding, hematochezia, hematochezia, swelling and pain. [2] As a traditional valuable animal medicine, has a long history and is widely used in modern clinical also has great medicinal value.

## 2. Antler chemical composition

Antlers mainly contain rich amino acids, such as: L-hydroxyproline, glycine, alanine and L-proline more than 10 kinds of amino acids, of which collagen is the main component hydroxyproline [3- 4]. Antlers also contain beta-3 subtype hemoglobin, antimicrobial peptide 1, peptidoglycan recognition protein[2] and other animal proteins containing Zn, Cu, Fe, Mn, calcium phosphate, calcium carbonate, magnesium phosphate, amino acids And nitrides and other trace elements, Zn, Cu, Fe, Mn these four kinds of trace elements and the antler effect of the drug there is a link [5]. Antlers also contain heavy metals, such as lead, cadmium, arsenic, mercury, etc., but the content of heavy metals in qualified antler products is in accordance with the standards of medicine. [6]

## 3. Antlers pharmacological effects

### *3.1. Anti-inflammatory, blood circulation*

Antlers on the xylene-induced ear swelling in mice, horn fork Ligand induced swelling of rat tissue has an inhibitory effect, antlers can also extend the hot plate caused by mouse pain caused by mouse licking time, Increased mouse writhing due to acetic acid, indicating that antlers have anti-inflammatory analgesic effect. [7] Antlers gum can significantly increase the cyclophosphamide-induced blood deficiency in whole blood red blood cells, platelets, leukocytes,

Tlymphocytes in the number of plasma prothrombin time and activated partial thromboplastin time values were significantly increased, Show that it has anti-inflammatory, blood function. [8]

### *3.2. Protect the breast*

Antlers can significantly reduce the mammary gland hyperplasia rat serum E2 and LH content, and the serum P content was significantly increased, FSH content also increased role, can significantly improve the mammary gland hyperplasia rat model of blood theology, decreased Nipple diameter and height, indicating antlers have a preventive effect on breast hyperplasia. [9]

### *3.3. Protect brain tissue*

Antlers can reduce cerebral infarct size, decrease the content of LDH and MDA, increase the activity of CAT, T-SOD and GSH-PX in rats with cerebral ischemia, and protect the morphology of rat brain [10].

### *3.4. Anti-osteoporosis*

Antlers can increase the bone mineral density in ovariectomized rats and increase the content of calcium and phosphorus in bone mineral, so that BGP in serum of ovariectomized rats increases significantly, ALP decreases significantly, NO and NOS increase significantly, organizational form. [11]

### *3.5. Impotence*

Antlers can enhance the reproductive capacity of male rats, can significantly shorten the potential limit of electrical stimulation-induced penile erection, can increase the weight of male rats seminal vesicles and prostate gland, males have to enhance the mating ability. [12]

In addition Antlers also protect gastric mucosa, blood and lipid-lowering effect of blood pressure can promote the expression of BMP-2 protein in osteoblasts, can improve phagocytic function of macrophages in patients with breast cancer.

## *4. Antlers plastic compound compatibility* [13]

**Table.4-1.** Antlers common compatibility and clinical efficacy

| Antlers plastic compatibility of traditional Chinese medicine | Compatibility efficacy and clinical application |
|---|---|
| Rehmanniae radix praeparata | Cape glue and Rehmannia match, yin and yang double fill, kidney yin yang rib. Can be used for the treatment of refractory immune thrombocytopenia recipe. |
| Angelicae sinensis radix | Antlers and Angelica combined with the treatment of virtual hypertension [14], the treatment of lumbar spinal stenosis blood deficiency anemia / hypoproteinemia [15]. |
| Astragali radix | And antlers gel with a yin and yang, the purpose of reaching yang and yin and yang, clinically applicable to the lack of qi and blood deficiency syndrome. |
| Atractylodis macrocephalae rhizoma | Bitter, sweet, bitter to spit, antler glue replenishing qi and blood, the combination of these two drugs commonly used in gynecological diseases. |
| Codonopsis radix | Codonopsis spleen benefits the lungs, nourishing fluid. Can regulate the body's immune function, antler glue Wenshen fill tonic, the two drug compatibility clinically available aplastic anemia. [16] |
| Ginseng radix et rhizome, Amomi fructose | Ginseng Dafu strength, Fumai solid off, strong heart Bushen, Antlers Jiao Yang help Yang, Amomum enteritis and kidney, kidney drugs into the spleen. Clinically available for blood deficiency hypertension. [17] |
| Testudinis carapacis et plastri colla, Asini corii colla | Gull deer cents glue, Yin fill, Jing yang. Clinical commonly used in the treatment of endocrine disorders caused by dysplasia, severe anemia, neurasthenia, osteoarthritis and sexual dysfunction and other diseases. Ejiao and then used, yin and yang, Bunao, priority, anti-epilepsy increased ability. |
| Drynariae rhizome, Eucommiae cortex | An antler glue, dried nausea for the king of medicine, dry nourishing the abdomen kidney impotence solid refinement of urine, the two king of medicine combined, there Jun Jun kidney yang, essence and marrow of the essence of fine, Jun minister phase Zuo. Warming liver and kidney, nourishing blood. Clinically used for Yaoxisuanru, consumptive weight loss, osteoporosis. [18] |
| Testudinis carapacis et plastri colla, Ginseng radix et rhizome, Lycii fructus | And antler glue all kinds of medicine, Buganyishenjian, Yang Tian Gui, yin and yang and make up, longevity, can promote type Ⅱ collagen or proteoglycan synthesis [19], the treatment of many diseases before and after menopause, [20] the elderly Dementia, [21] Aplastic anemia. [22] |
| Lycii fructose, Discourage rhizome, Cucuta semen | Yam spleen and kidney, meddler, dodder kidney essence. Adjudication of the monarch has a "Treating Yin for Yang," the meaning of the various drugs can be used to warm kidney yang, fill the best. For sexual dysfunction, chronic nephritis and diabetes are kidney deficiency, life-fiery who. [23] |
| Rehmanniae radix praeparata, Ginseng radix et rhizome, Cuscutae semen, Cyathulae radix | With antlers glue medicine, nourishing yin and reinforcing kidney. Clinically used for male and female reproductive system diseases, chronic nephritis, plastic anemia, neurasthenia and lumbar muscle strain such as kidney yin deficiency, yin deficiency or heat syndrome. [24] |
| Rehmanniae radix praeparata, Cinnamomi cortex, Ephedrae herbed, Baijiezi, Glycyrrhizae radix et rhizome | Antagonizing drugs and antlers, warming yang and nourishing kidney, cold pass stagnation. In clinical application of bone metastases, orthopedic inflammation, peritoneal tuberculosis, lobular hyperplasia of women, dysmenorrheal and other evidence of Yang deficiency and cold coagulation. [25] |

## 5. Outlook

Compatibility of traditional Chinese medicine is based on the drug's drug with other drugs according to the seven emotions and together, according to the principle of monarchy with purposefully according to the needs of the disease selectively two flavor and more than two drugs used in conjunction. Rational drug compatibility can get unattainable attenuated, synergies and increase the effect of indications.

Antler as a traditional precious Chinese herbal medicine has a long history of application, as flesh and blood products, warming the liver and kidney, blood and blood effects, and no toxicity reported. Deer heaven and earth to get the most complete yang, Shangyang the top grade, used for yang weak patients. Good Yang, who must seek yin in the yin, Yang yin help and biochemical infinite. Antler Jiaogang Yuanyang often associated with Yin Yin, in order to achieve the effect of yin and yang. Tortoise the world's thickest, compatibility with turtle shell glue, in order to reconcile the yin and yang and Ren Du two veins, and Rehmannia compatibility combined with yin and yang double supplement, kidney and liver Rongrong. And then Ejiao use, yin and yang, nourishing the essence of spermatogenesis, Changbu, Bunao, anti-epilepsy increased ability to increase indications. Antler glue and angelica, Angelica blood and blood, antler blood and blood, Wenshen help Yang, blood can increase the effectiveness of two drugs Qi. Yam spleen and stomach, Sheng Jin lung, Bushen astringent, medlar qi and blood, nourishing yin yang, often as an adjuvant and antler glue and tortoiseshell glue compatibility can be used to kidney Kidney essence, help yang yang. Longevity fluid antlers, tortoises shell glue and other combinations can reduce the inhibitory effect of tripterygium on the male reproductive system.

With the ongoing scientific research, many of the antler glue medicinal functions are also being gradually found and applied to clinical, antlers plastic kidney yang, the effect of qi and blood prompted us, whether can antlers and Ganoderma lucidum, red bean shirt, toad Such anti-cancer drugs combined to assist the recovery of cancer patients. Hoping to make reasonable additions and subtractions on the basis of the ancient prescription, so that the antler can maximize the medicinal value in prescription compatibility, bringing greater social and economic benefits.

## 6. References

[1] Li Na, Qu Xiaobo, Jiang Zongwen, etc .. Animal medicine finishing research - antler [J]. Jilin Chinese medicine, 2014, 34 (1): 74-76.
[2] State Pharmacopoeia Commission. Pharmacopoeia of the People's Republic of China, a [M]. Beijing: China Medical Science and Technology Press, 2015: 322.
[3] Shi Yan, Fan Xiaolei, XIAO Xinyue, et al. Determination of four major amino acids in antler glue [J] .Chinese Journal of Pharmaceutical Analysis, 2012 (5): 783-787.
[4] Zhou Fangyan, Li Ting, Liu Li, et al.A HPLC Fingerprint of Amino Acids in Antler Glues [J]. Acta Metallurgica Sinica, 2014, 20 (9): 47-51.
[5] Huang Bisheng. Ejiao glue and antler gel in the determination of trace elements [J]. Lizhen Guo Yi Guo Yao, 2001, 12 (6): 487.
[6] Shi Yan, Xiao Xinyue, Wei Feng, et al.Determination of harmful elements in antler glue based on microwave digestion and ICP-AES [J]. Acta Metallurgica Sinica, 2013, 19 (16): 151-153.
[7] Zhang Jiuzhuo, Lin Zhe, Guangguang Fu, et al.Anti-inflammatory and analgesic effect of antler [J]. Jilin Chinese Medicine, 2014, 34 (10): 975-977.
[8] Li Jing, Li Na, Guangguang, etc. Antlers gum cyclophosphamide-induced blood deficiency model mice [J]. Jilin Chinese medicine, 2014, 34 (10): 973-975.
[9] Lin He,Lv Guangfu, Tian Wenting, et al. Antagonizing effect of antler on breast hyperplasia [J]. Jilin Chinese medicine, 2013, 33 (2): 169-171.
[10] Li Hui, Chen Junyu, Yang Qing, et al. Antagonistic effect of antlers on cerebral ischemia in rats and its mechanism [J]. Chinese Journal of Gerontology, 2016, 16 (36): 3895-3897.
[11] Meng Haiyan, Qu Xiaobo, Li Na, etc. antler and antlers on ovariectomized rats with osteoporosis [J] Chinese herbal medicine, 2009,32 (2): 179-182.
[12] Niu Xiaohui, Li Pengfei, Jia Xueting, et al. Progress of Pharmacological Actions of Deerk Animal Origin in China for Ten Years [J]. Jilin Chinese Medicine, 2012, 32 (7): 702-705.

[13] State Pharmacopoeia Commission. Pharmacopoeia of the People's Republic of China, one [M]. Beijing: China Medical Science and Technology Press, 2015: 8-322.
[14] Cao Yali. Buqi nourishing kidney method for the treatment of virtual hypertension [J]. Journal of Changchun University of Traditional Chinese Medicine, 2014, 30 (05): 879-881.
[15] Mu Guanqiao, Sun Bo. Angelica Buxuetang combined with Western medicine treatment of lumbar spinal stenosis postoperative anemia / hypoproteinemia (blood deficiency) [J] .2017, 31 (4): 2-73.
[16] Wang Faling. Integrative Medicine of the treatment of aplastic anemia efficacy [J]. Shaanxi Traditional Chinese Medicine, 2012, 33 (8): 1037-1038.
[17] Gao Feng, Niu Liwen antler glue Di Huang Wan combined with nifedipine sustained-release tablets in the treatment of hypertension with hyperlipidemia in 18 cases * [J] Journal of Traditional Chinese Medicine, 2016, 29 (1): 17-19.
[18] Wang Zhenjie, Zhang Jinduo, Li Yuehua, Li Yuehua experience of treatment of osteoporosis [J]. Jilin Chinese medicine, 2012, 32 (4): 341-343.
[19] Li Nan, Yin Yubiao, Ji Huan Health, et al. Guilu Erxian and its decomposed formulas of guinea pig chondrocytes collagen type II and proteoglycan synthesis [J] Journal of Traditional Chinese Medicine, 2011,26 (11): 2524-2528.
[20] Zhao Qingchun. The use of Guilu Erxian decoction before and after treatment of Zhu Zheng experience [J]. Chinese folk medicine, 2009, 18 (3): 163.
[21] Kong Derong. Guilu Erxian glue for the treatment of 60 cases of senile dementia [J] Journal of Traditional Chinese Medicine, 2007, 20 (10): 33-34.
[22] Xiangyang, Chang Xiaohui, Sun Feng, ET al.Traditional Chinese medicine treatment of aplastic anemia drug law [J]. Liaoning Journal of Traditional Chinese Medicine, 2015, 42 (10): 1828-1831.
[23] Hu Weidong, Chen Dening, Zou Damin, et al.Development of traditional Chinese medicine for diabetic erectile dysfunction [J]. Journal of Liaoning University of Traditional Chinese Medicine, 2013, 15 (2): 14-16.
[24] Xu Haiping.Zuogui pill combined with western medicine in the treatment of aplastic anemia randomized parallel controlled study [J] Journal of Practical Traditional Chinese Internal Medicine, 2013,27 (7): 117-118.
[25] Luo Yan, Lou Lihua, Wo Xingde, etc. Yang and Tang on the efficacy of patients with acute mastitis and T lymphocyte subsets [J] Journal of Traditional Chinese Medicine, 2013, 31 (5): 1069 -1070.

**Institute of Physics**
Dirac House, Temple Back
Bristol BS1 6BE UK

ISSN: 1755-1307
ISBN 978-1-5108-5947-0

9 781510 859470

CURRAN ASSOCIATES INC.
proceedings.com